Forest-based Orchids

Azamal Husen • Manu Pant
Editors

Forest-based Orchids

Sustainable Use and Cultivation Practices

Springer

Editors
Azamal Husen
Department of Biotechnology
Graphic Era (Deemed to be University)
Dehradun, Uttarakhand, India

Wolaita Sodo University
Wolaita Sodo, South Ethiopia
Regional State, Ethiopia

Manu Pant
Department of Biosciences
Graphic Era (Deemed to be University)
Dehradun, Uttarakhand, India

ISBN 978-981-92-0095-5 ISBN 978-981-92-0096-2 (eBook)
https://doi.org/10.1007/978-981-92-0096-2

This Springer imprint is published by the registered company Springer Nature Singapore Pte Ltd.
The registered company address is: 152 Beach Road, #21-01/04 Gateway East, Singapore 189721, Singapore

Mahhika Badoni and Mishhika Badoni
(Twin daughters of Manu Pant)

Preface

Orchids, often described as the *"royal family of plants,"* represent one of the largest and most diverse groups in the plant kingdom. Their extraordinary range of floral forms, intricate shapes, dazzling hues, and remarkable longevity make them among the most captivating of all flowering plants. The beauty and mystique of orchids have fascinated botanists, horticulturists, and plant enthusiasts for centuries. Beyond their ornamental appeal, orchids are a fascinating subject of scientific study because of their highly specialized floral morphology and intricate pollination mechanisms. Their remarkable *mimetic behavior*, which often results in *deceptive pollination*, demonstrates the extraordinary coevolution between orchids and their pollinators and has played a crucial role in their diversification and global distribution. In addition to their aesthetic and ecological significance, orchids are a valuable source of numerous phytochemicals such as alkaloids, glycosides, and other secondary metabolites that possess significant pharmacological properties. Many species are used in traditional and modern medicinal systems for their therapeutic and restorative effects. Some orchids are aromatic and widely employed in perfumery and cosmetics, while others are edible and used in food or as nutraceutical ingredients. These diverse uses have made orchids indispensable in several industries, including floriculture, herbal medicine, cosmetics, and pharmaceuticals. The growing demand for orchid-based products, coupled with their visual and cultural appeal, has elevated them to the status of both a cherished hobbyist's favorite and a profitable commercial venture. The global orchid trade contributes substantially to the economies of several countries and continues to expand with technological and biotechnological advancements in propagation and cultivation. However, despite their immense ecological and economic importance, orchids are among the most threatened plant taxa worldwide. Their survival is jeopardized by rapid agricultural expansion, urbanization, deforestation, climate change, and habitat fragmentation. The challenges are further intensified by overexploitation, unregulated collection from the wild, and illegal trade, often driven by their rarity and ornamental value. Orchids exhibit slow growth, prolonged life cycles, and a high degree of ecological specialization, which makes them particularly vulnerable. Their reproductive biology poses additional difficulties: the seeds are minute, lack endosperm, and are entirely dependent on specific fungal symbionts (mycorrhizae) for germination and seedling establishment. These biological and ecological constraints have made orchid conservation an urgent global concern requiring scientific intervention and sustainable management strategies.

Forest-based orchids, in particular, are of profound ecological and commercial importance. They occur in diverse habitats ranging from tropical rainforests to temperate woodlands and contribute significantly to forest biodiversity and ecosystem stability. Many of these orchids are used as ornamental plants, medicinal resources, or raw materials for industrial applications. Their conservation and sustainable utilization, therefore, demand an in-depth understanding of their taxonomy, ecology, physiology, and reproductive biology. Propagation of orchids through seeds under natural conditions remains a major challenge due to their physiological dependence on compatible mycorrhizal fungi. To address this, researchers have developed alternative propagation approaches, such as tissue culture and micropropagation techniques, which have enabled large-scale multiplication and ex situ conservation of rare and economically important species.

The integration of biotechnological tools into orchid cultivation has revolutionized their production systems. Tissue culture methods, including asymbiotic seed germination, protocorm culture, and meristem culture, have made it possible to propagate orchids efficiently and sustainably. These techniques are now widely adopted in commercial nurseries, research institutions, and conservation programs. In situ conservation strategies, focusing on habitat protection and restoration, complement these ex situ efforts to maintain genetic diversity and ensure the long-term survival of natural populations. The successful implementation of both approaches has been demonstrated in several countries, particularly in the Netherlands, Thailand, Singapore, and Malaysia, where the orchid industry plays a pivotal role in the national economy. The economic success of these nations has inspired developing countries with favorable climatic conditions to explore orchid cultivation as a sustainable livelihood option. In such regions, forest-based orchids have the potential to contribute simultaneously to biodiversity conservation, rural development, and economic growth. This book is an effort to consolidate current scientific understanding, innovative approaches, and practical strategies related to the sustainable use, conservation, and propagation of forest-based orchids. It provides a multidisciplinary perspective encompassing orchid ecology, taxonomy, physiology, biotechnology, and economic potential. It also emphasizes sustainable utilization and management practices that balance conservation priorities with commercial and societal interests. The book opens with an overview of sustainable uses, cultivation techniques, and production strategies for forest-based orchids. It then presents a detailed account of orchid diversity in the Indian subcontinent and highlights the urgent need for effective conservation measures. The status, threats, and conservation strategies for orchids in Nepal are discussed, followed by an examination of the taxonomy, ecology, and floral biology of orchids in the Nainital region of Uttarakhand, India. The chapters on pollination ecology provide insight into the intricate plant–pollinator interactions in the Western Himalayan landscape and their implications for conservation. Further chapters explore in situ and ex situ conservation techniques, emphasizing approaches suitable for threatened and endemic orchid species. Tissue culture-based propagation techniques are elaborated upon, particularly for floriculturally important forest-based orchids. The role of fungal mycorrhizae in orchid germination and growth is discussed in detail, highlighting their ecological significance and potential for large-scale cultivation. DNA barcoding is presented as a modern molecular tool for accurate species identification, assisting both taxonomy and conservation biology. The book also explores recent trends in the global orchid industry, focusing on export-oriented production systems and their relevance to developing nations. Detailed discussions on *Cymbidium* cut-flower production, orchid cultivation as pot plants, and value addition of orchid-based products illustrate the economic potential of orchid farming. The final chapters cater to hobbyists and beginners, providing practical guidance on orchid cultivation and maintenance. By compiling research insights from across disciplines, this book aims to bridge scientific knowledge and field-level application. It will serve as a valuable reference for botanists, ecologists, conservationists, horticulturists, and entrepreneurs working in the field of orchid biology and cultivation. The interdisciplinary content is designed to inspire sustainable practices, encourage responsible trade, and strengthen global efforts toward orchid conservation.

We express our deepest gratitude to all the contributing authors whose scholarly and comprehensive chapters have made this volume possible. Their expertise and commitment have ensured the scientific rigor and depth of this work. We also extend our sincere appreciation to the publishers for their unwavering support in bringing forth this publication on such a vital and timely topic. We hope that the book in hand will serve as a significant resource for advancing research, promoting conservation awareness, and guiding sustainable orchid cultivation globally. It is our earnest belief that this book will not only enrich scientific understanding but also inspire future innovations in preserving the remarkable diversity, beauty, and ecological importance of orchids.

Dehradun, Uttarakhand, India Azamal Husen
Dehradun, Uttarakhand, India Manu Pant

Contents

About the Editors

Azamal Husen has served as Full Professor and Head of the Department of Biology at the University of Gondar, Ethiopia. He is currently affiliated as a Foreign Delegate with Wolaita Sodo University, Ethiopia, and holds an Adjunct Professorship at the Department of Biotechnology, Graphic Era (Deemed to be University) in Dehradun, Uttarakhand, India. His earlier academic roles include Full Professor at Sankalchand Patel University in Visnagar, India, and Visiting Faculty at the Forest Research Institute and Doon College of Agriculture and Forest in Dehradun, India. With more than 25 years of experience in academia and research, he is highly regarded for his expertise in nanobiotechnology, plant physiology, environmental botany, medicinal plants, and plant propagation. His research has been supported by prominent national and international organizations, including the World Bank, the National Agricultural Technology Project, the Indian Council of Agricultural Research, the Indian Council of Forestry Research and Education, and the Japan Bank for International Cooperation. He is a prolific scholar, having published more than 350 academic works, including research articles, books, and book chapters. His editorial contributions are equally noteworthy; he serves on the editorial boards and peer review panels of numerous prestigious journals published by Elsevier, Frontiers Media, Springer Nature, Taylor & Francis, RSC, Oxford University Press, the Royal Society, CSIRO, MDPI, PLOS, Sciendo, John Wiley & Sons, and UPM Journal. He is also a member of the advisory board of Cambridge Scholars Publishing, UK. He is a fellow of the Plantae group under the American Society of Plant Biologists and holds membership in several scientific organizations, including the International Society of Root Research, the Asian Council of Science—Editors, and the International Natural Product Sciences Taskforce. Throughout his career, he has delivered invited lectures, keynote addresses, and technical presentations at numerous international conferences, symposia, and workshops. In the field of academic development, he has made significant contributions through the design and delivery of orientation programs for university faculty, the development of educational modules, and the formulation of curricula at the undergraduate, postgraduate, and doctoral levels. In recognition of his contributions to teaching, research, and community service, he was honored with a special award from the University of Gondar. He currently serves as the Editor-in-Chief of the *American Journal of Plant Physiology* and is the Series Editor for several major academic collections, including *Exploring Medicinal Plants* (Taylor & Francis, USA), *Plant Biology, Sustainability and Climate Change* (Elsevier, USA), and *Smart Nanomaterials Technology* (Springer Nature, Singapore). He also leads the *Handbook Series on Nanotechnology and Its Applications* (Springer Nature, Singapore). His outstanding contributions to science have earned him international acclaim, including the prestigious distinction of being named among the "World's Top 2% Scientists" for three consecutive years (2021 onwards) by Stanford University, a recognition published in the Elsevier Data Repository. In addition to his regular academic responsibilities, he has served as Guest Editor for several high-impact special issues, including *Advances in Plant Metabolomics: Insights, Applications, and Integration* (Plant Physiology and Biochemistry), *Exploring Nanotechnology in Medicinal Plants* (Journal of Herbal Medicine), and *Integrating Nanotechnology with Plant Compounds: Enhancing Abiotic Stress Resilience and Crop Improvement via Signaling Molecules* (Frontiers in Plant Science), among others. His work continues to shape contemporary research at the intersection of plant science, biotechnology, and sustainable development. His academic leadership, research excellence, and international collaborations have positioned him as a prominent figure in global scientific circles.

Manu Pant is serving as a Full Professor and Head of the Department of Biosciences at Graphic Era (Deemed to be University), Dehradun, Uttarakhand, India. She completed her master's degree from Banasthali Vidyapith in Rajasthan, India, and her doctorate from HNB Garhwal University (A Central University) in Srinagar, India. She has extensive experience in the field of plant sciences, having worked as a research fellow at the Forest Research Institute and the Indian Council of Forestry Research and Education on nationally funded projects. Her research work on different plant species has been funded by the state governments, earning her recognition and awards at the national level. She has several peer-reviewed international papers to her credit, published patents, and delivered numerous oral and poster presentations, keynote addresses, and invited talks at national and international scientific events. She has also edited four books in the series *Exploring Medicinal Plants* (Taylor & Francis Group, USA), and some more important book projects are near to completion, which will be published by a reputed publishing house. Prof. Pant has conducted many national and international conferences, seminars, and trainings related to the thrust areas in plant sciences. She has been supervising PhD, B.Tech, MSc, and BSc students in their research work on different aspects of orchids, medicinal plants, and indigenous crop species. For her work in the field of plant sciences, Dr. Pant has also received appreciation from state- and national-level bodies.

Sustainable Uses of Forest-Based Orchids, Cultivation Practices and Production

1

Aarushi Gautam, Nidhi Nautiyal, Pranjall Sharma, Amrit Sahni, Manu Pant, Swarnika Chaddha, and Azamal Husen

1.1 Introduction

Orchids belong to the largest and most widely distributed families of beautifully diverse flowering plants (Givnish et al. 2015) (Fig. 1.1). Several thousand of these species are prized as ornamental plants, both potted plants and cut flowers, and cultural or religious uses. Orchids also include edible species used to make products like *chikanda* cake, *salep* beverage, and ice creams (Hossain et al. 2010). Many orchids like *Dactylorhiza hatagirea*, *Dendrobium* spp., *Cymbidium iridioides*, and others have exceptional medicinal properties that are highly valued in the traditional systems of medicine like Ayurveda, Chinese, Siddha, Unani medicine (Husen and Rahman 2003; Husen and Faisal 2005; Wani Panda and Mandal 2013; Husen and Pant 2024; Lal et al. 2023, 2024; Negi et al. 2024).

Due to their high demand in both domestic and international markets, orchids are among the most overharvested plants. This has led to a significant decline in the orchid populations, including extinction of some of the important orchids (Hinsley et al. 2018; Liu et al. 2015). Further, there are challenges of species-level

A. Gautam
Department of Biotechnology, Graphic Era (Deemed to be University), Dehradun, Uttarakhand, India

N. Nautiyal · P. Sharma · A. Sahni · M. Pant (✉)
Department of Biosciences, Graphic Era (Deemed to be University), Dehradun, Uttarakhand, India
e-mail: manupant@geu.ac.in

S. Chaddha
Martin Luther Universität, Halle-Wittenberg, Germany

A. Husen
Department of Biotechnology, Graphic Era (Deemed to be University), Dehradun, Uttarakhand, India

Wolaita Sodo University, Wolaita Sodo, South Ethiopia Regional State, Ethiopia

A. Husen, M. Pant (eds.), *Forest-based Orchids*,
https://doi.org/10.1007/978-981-92-0096-2_1

Fig. 1.1 Blooms of different species of orchids

identification in orchids. These factors have resulted in orchid species most heavily regulated under the Convention on International Trade of Endangered Species of Fauna and Flora (CITES), which bans commercial trade of these species. In many countries, the harvest of orchids from the wild is strictly restricted, like in the United States of America, Mexico, Thailand, Myanmar. However, illegal collection and trade continue to be prevalent in many parts of the world (Hinsley et al. 2018; Jiménez-López et al. 2019).

The global orchid market is robust and expanding, driven by ornamental preferences, floriculture industry growth, cultural demand, and a thriving international trade. In the year 2024, the global orchid market size is estimated between USD 302.5 million and USD 1.05 billion, with projections reaching up to USD 1.76 billion by 2033; the market shows compound annual growth rates (CAGR) ranging from 4% to 6.67% over the next decade. Major market regions include North America, Europe, and Asia-Pacific, with the latter leading growth due to diverse species, advanced cultivation techniques, and strong breeding programs, especially in Thailand, Taiwan, Singapore, and Vietnam. International trade statistics place Thailand as the world's largest orchid exporter, responsible for nearly 37.6–47% of global shipments, followed by the Netherlands, Vietnam, and Malaysia. The leading importers include Japan, the United States, Vietnam, Italy, and China. Orchids are a major segment of international floriculture—used for cut flowers, potted plants, and decorative arrangements—with trade regulated through CITES permits to ensure sustainability and prevent illegal collection from the wild. India plays an important role in the global orchid trade, both as an exporter and importer, though its export volumes remain modest compared to major producers like Thailand and the Netherlands. Recent data indicate that India exports orchids primarily to neighbouring Asian countries and some European markets, focusing on cut flowers and potted plants of popular genera such as *Cymbidium*, *Dendrobium*, and *Mokara*. The Indian orchid export market benefits from rich native biodiversity, coupled with increasing adoption of cultivation and micropropagation techniques to meet quality standards for international trade. For import, India sources orchids mainly from Thailand, Taiwan, and the Netherlands to satisfy growing domestic demand driven by horticultural enthusiasts and commercial floriculture sectors (Tiwari et al. 2024).

Imported orchids largely target metropolitan flower markets and ornamental gardening, reinforcing India's dynamic role as both a supplier and consumer within the international orchid network. Propagation techniques play an important role in meeting market demand. Conventional methods—division, stem and leaf cuttings, seed sowing are praised for their simplicity and reliability but are slow and yield limited number of plants. To meet commercial requirements, in vitro propagation (micropropagation, asymbiotic seed germination, temporary immersion systems) is now widely adopted because it enables rapid, large-scale multiplication of genetically uniform, disease-free plants. These advanced tissue culture methods are especially important for exporting healthy plants and for scaling up production with minimal seasonal limitations. Micropropagated orchids are more easily transported, have fewer quarantine concerns, and can be tailored to market needs in terms of variety and quality (Cardoso et al. 2020).

1.2 Orchids Botany and Botanical Features

Orchids are one of the most diverse and extensive groups of flowering plants. They are found almost everywhere on the Earth except for Antarctica. They belong to the Orchidaceae family and are further subdivided into five unique subfamilies (*Apostasioideae, Cypripedioideae, Vanilloideae, Orchidoideae, and Epidendroideae*), making up around 8% of all the flowering plant species. Until now, 29,000 species of orchids have been officially recognized (De 2020). Orchids are mostly found in the tropical regions, especially in Asia, South and Central America as shown in Fig. 1.2. The distribution of orchids varies across the different continents:

- Eurasia has about 40–60 orchid genera.
- North America has about 20–30 orchid genera.
- Tropical America is home to about 300–350 orchid genera.
- Tropical Africa has about 125–150 orchid genera.
- Tropical Asia has about 250–300 orchid genera.
- Oceania has about 50–70 orchid genera (De 2023).

Orchids are monocots and, like most monocots, they usually have simple leaves with parallel veins (parallel venation), though some, as in the *Vanilloideae* group, the leaves show net-like pattern (Reticulate venation) (De 2020). The leaves of orchids show a variety in shape and size, often growing alternatively on the stipules. The size and structure of the leaves also depend on the environmental factors, with the smaller leaves better suited for the sunny and exposed areas. Orchids may shed off their leaves seasonally or adjust the number based on their growth habits (Ackerman et al. 2023).

Orchids typically bloom from flower clusters known as inflorescences, which can be spikes, racemes, or panicles, though some species, like the *Paphiopedilum* bear just a single flower. The position of the stalk generally varies, the position of the stalk

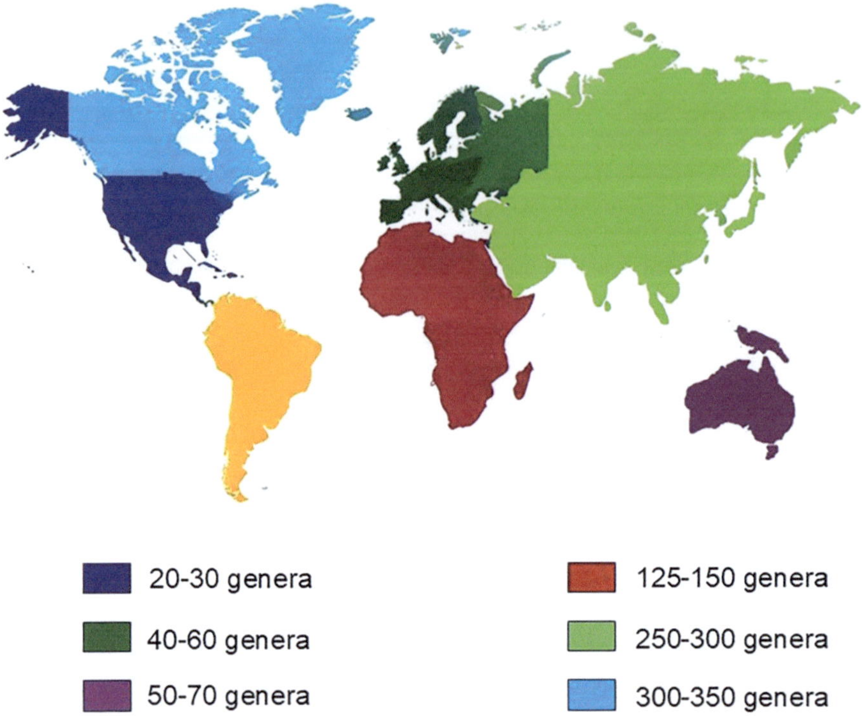

Fig. 1.2 Orchid diversity around the world

being terminal in the terrestrial species and laterally positioned in many of the epiphytic species. The flowers in the orchids are unique made up of three sepals, three petals, and a centrally located structure called the column or the gynostemium which is fusion of the male and female part of the flower (Ackerman et al. 2023). Flowers of orchids rotate to an angle of 180° during the development, a process known as the resupination so that the labellum (lip) ends up at the bottom, thereby creating a landing platform for the pollinators. In *Paphiopedilum*, flower shape that is the lip staminode, the flower petal and the width of the petal are mostly inherited, but the colour of the flowers varies across species. The seeds of orchids are tiny and numerous; the epiphytic seed types have more air space to help them spread further due to buoyancy. Since the seeds of orchids lack the endosperm, orchids rely on the mycorrhizal fungi for the supply of nutrients during growth and germination, making them mycoheterotrophic early in life (Zhang et al. 2018).

1.3 Habit and Habitat Distribution

Orchids grow in a variety of ways; they can grow by clinging to trees, called epiphytic, when growing on rocks, the orchids are called lithophytic, when they live in soil, they are called terrestrial and when the orchids derive their nutrition from the decaying organic matter, they are known as saprophytic. Most of the orchids, about 70% prefer life off the ground (i.e., epiphytic), while 25% grow in soil, and a small 5% thrive in mixed environments (De 2020).

Based on the growth, orchids can also be classified either as sympodial orchids exhibiting the lateral, that is, the sideways growth with multiple stems, as in *Cattleya* and *Oncidium* or the monopodial exhibiting the vertical growth upwards form a single central stem, adding new leaves and flowers alternatively, as in *Vanda* and *Phalaenopsis*.

Epiphytic orchids often have thick fleshy leaves with tough cell walls and sunken stomata to reduce the loss of water and are capable of CAM photosynthesis. Interestingly, the placement of the leaf stomata varies with different species—*Paphiopedilum* has them sunken, while *Cypripedium* has them sticking out (Sailo et al. 2014). Many orchids, like *Cymbidium* and *Dendrobium,* also develop swollen stem parts called the pseudobulbs that play a role in the retention of water and nutrients, helping them survive during the dry spells. In many sympodial epiphytic orchids, the stem base thickens into pseudobulbs, which store water and nutrients to help plants survive the dry periods. These vary in shape and size across genera like *Cattleya, Cymbidium, Dendrobium*, and *Oncidium*. The pseudobulbs can either be composed of a single internode called heteroblastic pseudobulb, for example: *Cattleya* and *Miltonia*, or maybe composed of two or more internodes of the same or different lengths, example: *Eria* and *Dendrobium*. As they age and lose leaves, they become back bulbs, which can still be used for the growth of the new plants (Ng and Hew 2000).

Since the majority of the orchids are epiphytes, they grow on trees, not as parasites, deriving their nutrition from trees, but simply for support. These orchids thrive in filtered light, good air flow, and absorb nutrients from rainwater and decaying organic matter around them using the bark of the tree only as support (de Faria et al. 2018). The roots in the epiphytic orchids have specialised spongy tissue called the velamen; it is a thick layer made of dead cells that helps in the absorption of moisture when it is present in the surrounding environment and prevents water loss during dry conditions. The velamen also prevents the nutrients from washing away during the rainfall and protects the root from the harmful UV rays from the sun (Gao et al. 2025). Substrates are the medium that helps in growth of the plants outside the soil and can also be used for growth purposes. Epiphytic orchids can be grown in a variety of substrates, including plant-based materials like sphagnum moss, charcoal, coffee husk, carbonised rice husk, coconut fibre, and bark; minerals like gravel, vermiculite and expanded clay, and even synthetics like phenolic foam or polystyrene (de Faria et al. 2018).

Terrestrial orchids grow directly in the soil, absorbing water and nutrients from it. The terrestrial orchids usually have thinner leaves with C3 photosynthesis

Fig. 1.3 Types of orchids based on mode of nutrition

(Ackerman et al. 2023). The terrestrial orchids have roots built in a way that allows for easy access to moisture, nutrients and are more stable. Their roots generally have a simple epidermal cell layer, and the inner tissues, like the endodermis and the exodermis, are incorporated with the Casparian strips, regulating the movement of water and ions. The terrestrial orchids generally have thicker cortical tissue, root hair and an elaborated root network for better exploration of the soil (Gao et al. 2025). Some terrestrial orchids like *Orchis* and *Ophrys* have underground tubers or corms and are rhizomatous that serve a similar storage and regenerative function as the pseudobulbs in the epiphytes (De 2023). The terrestrial orchids are much fewer in number as compared to the epiphytic orchids. Some popular types include *Cymbidium, Paphiopedilum*, and *Arundina*.

Lithophytic orchids are those that naturally grow on rocks or cliffs. They anchor their roots in the rock cracks or on the irregular rock surfaces where water and nutrients essential for growth are assembled. Examples include *Epidendrum, Hoffmannseggella,* and *Bifrenaria* (De 2023).

The saprophytic orchids lack chlorophyll, and they are solely dependent on the dead decaying organic matter, like the dead remains of the plants, for their nutritional requirements, for example, *Rhizanthella gardneri* (De 2023) as shown in Fig. 1.3.

1.4 Medicinal Uses of Orchids

Orchids have been used medicinally since ancient times, as mentioned in Table 1.1, with records dating back to 2800 B.C. in China and the Vedic period in India (2000–600 B.C.). In Ayurveda, species such as *Flickingeria macraei* and *Vanda tessellata* have been used to treat ailments like asthma, bronchitis, fever, and digestive problems (Gogoi and Sen 2023). The Siddha system in Tamil Nadu (India) makes less use of orchids, but tribal communities there and across India make wide use of them. For example, the Dongaria Kondha tribes in Odisha use 16 orchid species to treat 33 types of illnesses, while in Nagaland, 15 species are used for conditions such as rheumatism, cholera, and tuberculosis (Choudhary et al. 2023).

In China, around 300 orchid species are used in Traditional Chinese Medicine (TCM), including *Gastrodia elata* for memory, *Dendrobium* spp. for blood circulation and throat issues, and *Bulbophyllum* spp. for lung problems and injuries (Shengji and Zhiwei 2018).

Mediterranean and Middle Eastern cultures used orchid tubers to make salep, a flour for drinks and tonics traditionally consumed for coughs, diarrhoea, and low energy. Orchids, both wild harvested and cultivated, are used globally for their healing potential, treating conditions like tuberculosis, arthritis, and snakebites (Bazzicalupo et al. 2023).

Table 1.1 Medicinal properties of orchids

Orchid species	Plant part used	Ethnopharmacological application	Key references
Acampe papillosa	Root	Treats rheumatism, uterine diseases, neuralgia, arthritis etc.	Wani Panda and Mandal (2013)
Aerides multiflora	Leaves, pseudobulb, flower, seed	Antibacterial properties, Treatment of cuts and wounds	Gautam et al. (2025)
Arpophyllum spicatum	Pseudobulbs and stem	Gastrointestinal disorders	Teoh (2019)
Arundina graminflora	Root, stem, pseudobulb	Treat diabetes, tumours, hepatitis, and antibacterial activity	Gautam et al. (2025)
Bulbophyllum odoratissimum	Stem, pseudobulb, whole plant	Treat fractures, tuberculosis and chronic inflammation	Gautam et al. (2025)
Calanthe calanthoides	Corms and flowers	Cicatrize wounds and burns	Teoh (2019)
Cymbidium aloifolium	Root	Treat cancer, anti-emetic, treat diarrhoea and paralysis	Gautam et al. (2025)
Cyrtopodium macrobulbon (Lex.)	Pseudobulbs	Cicatrized wounds, burns, urinary ailments, and inflamed bowels	Morales-Sánchez et al. (2014)
Dendrobium macraei	Whole plant	Treat snake bite, asthma, bronchitis, fever, and used as an aphrodisiac	Gautam et al. (2025)

(continued)

Table 1.1 (continued)

Orchid species	Plant part used	Ethnopharmacological application	Key references
Epidendrum rigidum	Stems	Hydration	Hernández-Romero et al. (2005)
Habenaria floribunda	Leaves	Vaginal haemorrhage	Asseleih et al. (2015)
Isochilus major	Leaves	Cure inflamed blows	Teoh (2019)
Laelia autumnalis	Pseudobulbs, leaves and flowers	Hypertension, respiratory disorders, gastrointestinal disorders	Vergara-Galicia et al. (2010)
Luisia trichorrhiza	Leaves, whole plant	Treat muscular pain and jaundice	Gautam et al. (2025)
Onicidium graminifolium	Pseudobulbs	Treat renal disorders	Castillo-Pérez et al. (2024)
Rhyncostylis retusa	Root, whole plant	Treat cramps, asthma, kidney stones, menstrual disorders, rheumatic disease and wounds	Gautam et al. (2025)
Spathoglottis plicata	Leaves	Treat rheumatic disease	Gautam et al. (2025)
Stanhopea tigrina	Pseudobulbs, leaves and roots	Treat sunstroke, renal disorders, and mental disorders	Castillo-Pérez et al. (2021)
Trichocentrum brachyphyllum	Leaves	Treat artery hypertension, as a hallucinogen, treat abdominal colic	Teoh (2019)
Vanda roxburghii	Leaf, roots	Antipyretic; roots used to treat rheumatism, sciatica	Gautam et al. (2025)
Vanilla planifolia	Fruits and flowers	Fatigue, lower fever, gastrointestinal disorders, treat skin tumours, hysteria, impotence and rheumatism	Hernández and Ochoterena (1942), Chang et al. (2024)

1.5 Pollination in Orchids

Orchids represent one of the most diverse plant families known for their unique strategies to attract pollinators, which contribute to their vast morphological and taxonomic variety. These strategies include offering rewards like nectar, lipids, fragrances, pollen, resins, and sleep sites, as well as employing deceptive tactics that manipulate pollinators' sensory perceptions through chemical, visual, or tactile signals (Ackerman et al. 2023). Orchids exhibit remarkable diversity in pollination strategies, relying on both rewards and deception. About one-third of the estimated 30,000 orchid species use deception for pollination. Deception in orchids primarily involves three methods:

1. **Food Deception:** Flowers imitate nectar- or pollen-producing plants (Batesian mimicry) or exploit innate foraging instincts of insects, attracting visitors without offering actual rewards. Food-deceptive orchids tend to have low pollinator visitation, making field studies challenging (Carlomagno et al. 2024).
2. **Sexual Deception:** To attract male pollinators, orchids mimic texture, fragrance or the scent of female insects, luring them into pseudocopulation, during which pollination occurs. This strategy is common in Australia and Europe but may be more widespread in regions such as South/Central America, Asia, and the Pacific (Gaskett 2011). Sexual deception often targets specific male insects, particularly solitary, polygynous species (e.g. *Andrena nigroaenea*—pollinator of *Ophrys sphegodes, Colletes cunicularius*—pollinator of *Ophrys arachnitiformis, Zaspilothynnus nigripes*—pollinators of Drakaea).
3. **Brood-Site Deception:** Less frequent among orchids, this involves imitating egg laying or larval food sites (e.g., fungi or carrion) to draw in female flies.

While chemical mimicry of insect pheromones has been studied extensively, the role of visual and tactile traits, like colour or surface texture, remains less understood (Gaskett 2011).

Bees are the most common pollinators (40–55% of species), followed by flies, with other insects and hummingbirds also playing significant roles—particularly in the Oncidiinae subtribe, where about 84.7% of recorded species are bee-pollinated, while smaller proportions attract wasps, hummingbirds, butterflies, or flies (Castro and Singer 2019). Bees are the primary pollinators, using their hairy bodies to carry pollinia; some species like *Coryanthes* attract euglossine bees with oils, while others such as *Ophrys* mimic female bees to induce pseudocopulation (Ackerman et al. 2023; Johnson et al. 2020). Wasps are similarly exploited, with genera like *Drakaea* mimicking female thynnid wasps to attract males (Kamaladhasan et al. 2020). Moths are key nocturnal pollinators, attracted to pale, fragrant flowers with long nectar spurs, exemplified by Darwin's orchid (*Angraecum sesquipedale*) and its hawkmoth pollinator (Micheneau et al. 2009). Butterflies favour brightly coloured flowers with nectar tubes, where orchids like *Bonatea* attach pollinia to their proboscis (Johnson 1996). Flies play dual roles: some visit for nectar, while others are deceived by carrion- or fungus-mimicking orchids such as *Bulbophyllum*. Beetles, among the earliest orchid pollinators, are attracted to fleshy, strong-scented flowers, often transferring pollinia while feeding (Givnish et al. 2015). Finally, birds, especially hummingbirds, pollinate tubular, brightly colored orchids with dilute nectar adapted to avian physiology (Martén-Rodríguez et al. 2015). Collectively, these interactions underscore Hymenoptera as dominant pollinators while highlighting the evolutionary adaptability of orchids to diverse ecological niches (Table 1.2).

Although most orchids are self-compatible, some populations develop autogamy (self-pollination) in environments with limited pollinator availability, such as high-latitude regions or islands, supporting the idea of reproductive assurance.

Orchid pollination strategies, driven by both rewards and deception, are key to their evolutionary success and diversity, influencing reproductive isolation and speciation. The distribution of pollination strategies among orchid subfamilies

Table 1.2 Roles of different pollinator groups in orchid pollination

Pollinator group	Mechanism/strategy	Examples	Approx. contribution in orchids	Key references
Bees (Hymenoptera)	Reward (nectar, oils) and deception (sexual mimicry); highly efficient due to hairy bodies	*Coryanthes* (bucket orchids), *Ophrys apifera*	~60–70% of orchid pollination events globally	Johnson et al. (2020)
Wasps	Sexual deception via visual and chemical mimicry (pseudocopulation)	*Drakaea* (hammer orchids), *Ophrys insectifera*	~10–15% in specialized lineages	Peakall and Whitehead (2014)
Moths (Lepidoptera – nocturnal)	Attracted to pale, scented flowers with long spurs	*Angraecum sesquipedale*	~10–12% (especially in tropical orchids)	Micheneau et al. (2009)
Butterflies (Lepidoptera – diurnal)	Bright flowers with landing platforms, nectar tubes; pollinia attach to proboscis	*Bonatea cassidea*	~5–7% (more common in open habitats)	Johnson (1996)
Flies (Diptera)	Sapromyophily (deception via carrion/ fungi odours); nectar visits; sometimes trapping	*Bulbophyllum*, *Pterostylis*	~10–15% overall; higher in some genera (up to 30%)	Micheneau et al. (2009)
Beetles (Coleoptera)	Attracted to fleshy, strong-scented flowers; feed on tissues/pollen	*Selenipedium*, some *Bulbophyllum*	~5–10% (among earliest pollinators)	Givnish et al. (2015)
Birds (Aves)	Bright tubular flowers with dilute nectar; adapted to hummingbird metabolism	Some *Epidendrum*, *Elleanthus*	<3% of orchids (rare, mostly Neotropics)	Martén-Rodríguez et al. (2015)

demonstrates clear patterns of specialization. Across orchid subfamilies, Hymenoptera emerge as the dominant pollinators, accounting for the majority of interactions in all groups. Epidendroideae and Orchidoideae also involve Diptera and Lepidoptera, with Orchidoideae showing the highest Lepidoptera association. Cypripedioideae rely mainly on Hymenoptera with secondary Diptera involvement, while Vanilloideae are characterized by Hymenoptera dominance alongside notable autogamy and occasional bird pollination. Apostasioideae display a simpler pattern, relying largely on Hymenoptera and autogamy. These patterns reflect both the central role of Hymenoptera and the diversification of pollination strategies within specific orchid lineages (Table 1.3).

Table 1.3 Distribution of pollination strategies among orchid subfamilies (values = proportion of observations %)

Subfamily	Autogamy	Hymenoptera	Coleoptera	Lepidoptera	Diptera	Aves
Epidendroideae	9.54	53.85	8.45	6.02	20.48	1.67
Orchidoideae	4.51	49.25	8.76	26.52	10.37	0.55
Cypripedioideae	2.89	74.91	–	0.96	21.22	–
Vanilloideae	14.82	81.48	–	–	–	3.7
Apostasioideae	22.22	77.78	–	–	–	–

1.6 Commercial Propagation of Orchids

Orchids, unremarkably investigated as angel flowers, are considered nature's admirable creations due to their attractiveness and long-lasting blooms. Orchids approximately constitute 9% of India's flora and are considered to be the largest botanical family, with almost 1300 species and 140 genera. With over 25,000–35,000 species and more than 100,000 hybrids, the flower showcases remarkable diversity in shape, size, and colour. *Cymbidium* orchids, which ranked among the top ten cut flowers, are mainly grown in the north-eastern region of India. *Dendrobium* orchids are usually grown as tropical orchids (Singh and Pal 2024).

The worldwide orchid fabrication has been active in tropical, subtropical countries, many with automated and efficiently managed large greenhouses. *Phalaenopsis* remains the most highly exported orchid as potted plant, bolstered by exceptional breeding, micropropagation efforts that promise year-round production and flowering under controlled conditions. Countries like Thailand, Netherlands, Taiwan, Singapore, and Italy have made a significant contribution in this area. The orchid cultivation is now gaining momentum globally and the orchid industry is exploring new possibilities, with orchids being sold around the world both as beautiful cut flowers and as plants grown through artificial propagation (Fig. 1.4).

According to cognitive market research, the market size of orchid globally was USD 302.5 million in 2024, which is predicted to rise to USD 398.06 million by 2031. In the global market (2024), North America held a market share of more than 40% of the worldwide revenue followed by 30% revenue by European nations (Fig. 1.5).

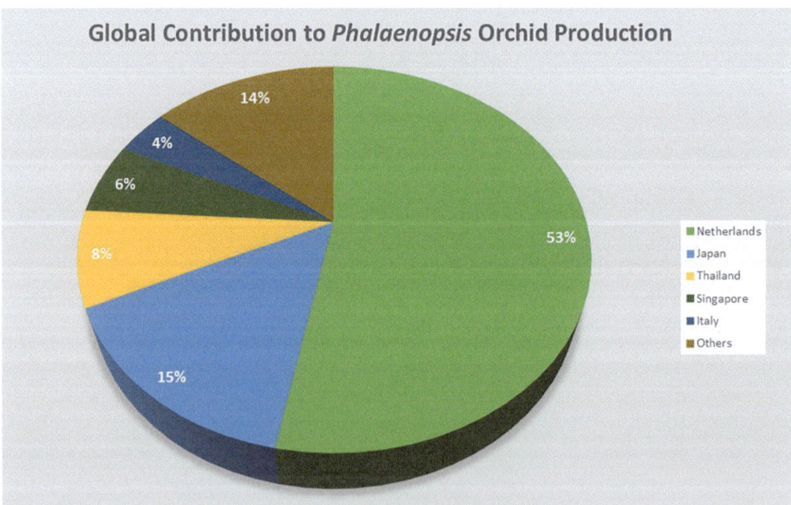

Fig. 1.4 Phalaenopsis orchid production worldwide (global contribution)
Source: https://www.cognitivemarketresearch.com/orchid-market-report

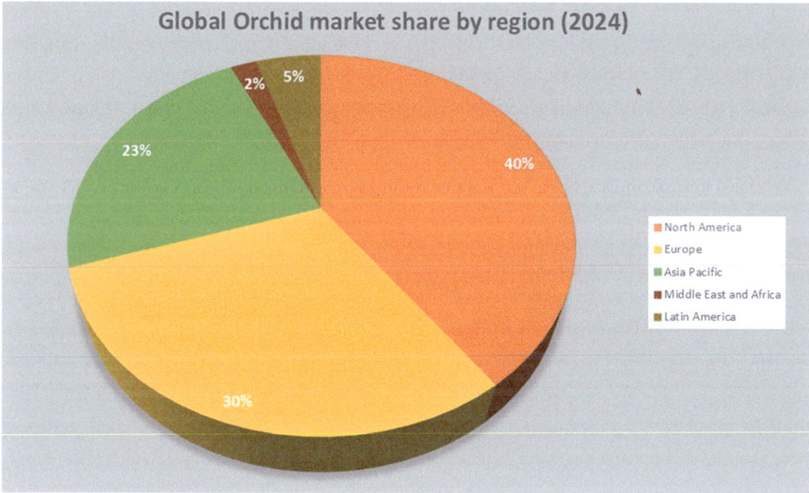

Fig. 1.5 Global orchid market share
Source: https://www.cognitivemarketresearch.com/orchid-market-report

1.7 Propagation Methods of Orchids

1.7.1 Conventional Methods

Conventional methods of orchid propagation rely heavily on seed propagation, vegetative division, offshoots (keikis), back bulbs and stem cuttings. Each method is applied based on the growth habit of orchids (sympodial or monopodial). The methods are explained as follows:

Division: The simplest technique, division is used to propagate sympodial orchids by excising out rhizomes between pseudobulbs, preferable during the early spring season. The plantations are such that each piece has a minimum of three pseudobulbs, and one dormant bud. This method ensures vigorous growth of good quality shoots. Some of the orchids propagated by this method include *Brassavola*, *Calanthe*, *Cattleya*, *Cymbidium*, *Dendrobium*, *Laelia*, *Miltonia*, *Oncidium*, *Odontoglossum*, *Paphiopedilum*, etc.

Back bulbs: In this method, the previously flowered back bulbs with roots are those with rhizomes are cut right above. These are put in pots filled with growing material (substratum) carefully such that the cut end of the bulb is near the edge of the pot. Back bulbs are known to develop new shoots within 2–3 months.

Keikis: A keiki is a small plantlet that arises from the nodes due to accumulation of plant growth hormones during the growing stage. These are called regular keikis, and may be a natural growth or through artificial application of cytokinin hormones paste at the node. The basal keikis are the baby plants that grow from the base of the mother plant, and may remain attached to it. These basal keikis are separated and used for further propagation of orchids. Dendrobium, Phalaenopsis, Epidendrum are some of the orchids that are normally propagated through keikis.

1.7.2 In Vitro Propagation

Conventional propagation techniques of orchids are slow and tedious and fail to meet the ever-growing market demand of these commercially important flowering species. To overcome these challenges and meet the demand of the floriculture industry, in vitro propagation techniques are heavily relied upon. The technique ensures rapid mass propagation of orchids, overcoming the slow growth rate through seed germination and division that might take upto 3–5 years to produce viable plants (Chugh et al. 2009). Tissue culture-based propagation produces uniform, disease-free and genetically stable plantlets through asymbiotic seed germination and direct organogenesis from explants like shoot tips, pseudobulbs, protocorm-like bodies (PLBs), leaves, thin-layer cells (tLCs) (Kiaheirati et al. 2024). Moreover, micropropagation techniques also ensure the conservation of rare, endangered and medicinal orchid species, enabling year-round mass production of elite genotypes.

Micropropagation of orchids has been practised for decades and has shown promising results for the mass propagation of orchids in a small space devoid of seasonal dependency (Chugh et al. 2009; Nongdam et al. 2023). Various explants like seeds, shoots, shoot and root tips, rhizome, nodal segments, inflorescences, and pseudobulbs have been employed for in vitro propagation of several orchids (Nirmala et al. 2006; Bhattacharjee and Islam 2014; Lal and Singh 2020; Sonia et al. 2022).

Orchid seeds, though present in large amounts, have either a slow or a poor germination rate; moreover, they require an asymbiotic association with mycorrhizal fungi (Arditti and Ghani 2000). The first report of successful in vitro asymbiotic seed germination was recorded without the help of fungus in Knudson C (KC) medium in the case of *Cattleya* and *Laelia* orchids (Knudson 1922). Since then, orchid micropropagation through seeds has been tried within species. Nongdam and Chongtham (2012) reported optimum seed germination on Mitra (M) medium consisting of Naphthalene acetic acid (NAA) (0.7 mg/L) of *Cymbidium dayanum*. Maximum seed germination (93%) of *Renanthera imschootiana* when cultured on ¼ MS (Murashige and Skoog) medium augmented with NAA (0.5 mg/L), coconut water (20%), peptone (1 g/L), sucrose (10 mg/L) (Wu et al. 2014). In a study conducted on *Aerides ringens*, it was reported that the best in vitro seed germination was observed on KC medium fortified with 6-Benzylaminopurine (BAP- 1 mg/L) and peptone (500 mg/L) (Srivastava et al. 2015). According to Rodrigues et al. (2015), 100% in vitro seed germination was observed in *Cyrtopodium saintlegerianum* when cultured on KC medium supplemented with activated charcoal (AC 3 g/L). Additionally, Mercado and Jaimes (2022) reported that seeds of *Cattleya gaskelliana* and *Cattleya warscewiczii* showed optimum seed germination of 90.7% and 92.1% respectively, when inoculated on MS medium supplemented with pineapple juice (200 ml/L).

In monopodial orchids, in vitro leaf culture is preferred as the entire mother plant is not sacrificed and the explant is available throughout the year (Hardjo and Savitri 2016). In *Rhyncostylis gigantea*, the maximum in vitro raised plantlet regeneration potential of the whole leaf segments was observed in M medium consisting of 5 mg L^{-1} Kinetin (Kn) (Pathak et al. 2017). In a study conducted by Chookoh et al. (2019), they observed that *Tolumnia* Snow Fairy leaf explant (inner expanding), when inoculated on MS medium fortified with 4 mg L^{-1} Benzyladenine (BA- 4 mg/L) and NAA (0.5 mg/L) produced protocorm-like bodies (PLBs), while the outer leaves failed to show in vitro growth. Similarly, the best PLB formation from leaf explants in *Aeridis multiflora* was observed in MS medium augmented with 1 mg L^{-1} Indole acetic acid (IAA- 1 mg/L) and BAP (2 mg/L) (Bhowmik and Rahman 2020). Manokari et al. (2022) reported that upon culturing leaf explants of *Vanilla planifolia* on MS medium augmented with BAP (1 mg/L), Kn (1 mg/L) and silver nitrate (2.5 mg/L) resulted in producing the maximum number of somatic embryos (30.0 per explant).

The first report on employing the apical meristem as an explant was published by Morel (1960) in *Cymbidium* orchids. After that, many reports have been published on using meristem explants for in vitro propagation of orchids. In a study conducted by Priya et al. (2013) in *Dendrobium* orchids, it was reported that early bud break from shoot tip explants was observed in ½ strength MS medium fortified with BA (4 mg/L) and the optimum number of shoots (4.66) was produced when inoculated on medium containing Kn (2 mg/L) and NAA (0.1 mg/L). In a study conducted on *Dendrobium* Red Bull, newly developed shoot buds (keikis) were inoculated on MS medium containing various concentrations of BAP and NAA. Maximum shoots (7.66) were obtained in MS medium augmented with BAP (3 mg/L) and NAA (1.5 mg/L) (Mamun et al. 2018). Farrokhzad et al. (2022) reported that shoot tip explants were inoculated on ½ MS medium supplemented with 3% (w/v) sucrose, myoinositol (0.1 mg/L), thidiazuron (TDZ- 2 mg/L) and BAP (1 mg/L) to initiate in vitro shoot production in *Phalaenopsis amabilis*. In *Vanda brunnea*, shoot tip explants, when cultured on Orchimax culture medium fortified with BA (0.5 mg/L) resulted in the highest plant regeneration rate (100%) (Nowakowska et al. 2022).

Besides these explants, several other explants have been used for in vitro propagation of orchids. For instance, Kerbauy (1984) first used root tip explants of *Catasetum* orchids to directly form PLBs on MS medium without the formation of callus. In another study conducted by Sharma (2019), root tip explants of *Rhyncostylis retusa* were inoculated on M medium supplemented with Kn (3 mg/L) and 1 mg/L, resulted in the highest plant regeneration rate (31%), producing 28 plantlets in 15 weeks. Poobathy et al. (2019) cultured nodal explants of *Ludisia discolor* on ½ strength MS medium containing TDZ (0.1 mg/L), NAA (1 mg/L), AC (0.2%), sucrose (3%) and banana cultivar homogenate (8%), resulting in maximum survival (42%) and plant growth rate (19.6%). In *Malaxis acuminata*, transverse thin cell layer (tCLs) of about 1–4 mm (excised from pseudobulb) was inoculated on MS medium augmented with meta-Topolin (mT- 1.5 mg/L) and 5 mg L^{-1} chitosan (5 mg/L) for maximum shoot proliferation (21 micro shoots per explant) (Bhattacharyya et al. 2022).

1.8 Conclusion

The orchid market continues to flourish worldwide, fuelled by rising consumer preference for ornamental and therapeutic uses, supported by technological advancements in propagation methods. Conventional propagation remains vital for maintaining genetic diversity and traditional cultivars, while in vitro techniques enable rapid, large-scale production meeting commercial quality and regulatory standards. Despite rapid market growth, sustainability considerations regarding wild orchid harvesting and habitat destruction underscore the need for responsible cultivation practices. Integrating sustainable forestry management, advanced propagation, and market-driven production can ensure the longevity of orchid biodiversity, bolster rural livelihoods, and meet growing global demand. Increasing contributions of developing countries in orchid export and import reflect emerging opportunities within this dynamic sector. Continued research and innovation play a vital role in balancing market success with conservation imperatives in orchids' multifaceted uses worldwide.

References

Ackerman JD, Phillips RD, Tremblay RL, Karremans A, Reiter N, Peter CI, Bogarín D, Pérez-Escobar OA, Liu H (2023) Beyond the various contrivances by which orchids are pollinated: global patterns in orchid pollination biology. Bot J Linn Soc 202(3):295–324

Arditti J, Ghani AKA (2000) Numerical and physical properties of orchid seeds and their biological implications. New Phytol 146(3):569–569

Asseleih LMC, García RAM, Cruz JYSR (2015) Ethnobotany, pharmacology and chemistry of medicinal orchids from Veracruz. J Agric Sci Technol A 5:745–754

Bazzicalupo M, Calevo J, Smeriglio A, Cornara L (2023) Traditional, therapeutic uses and phytochemistry of terrestrial European orchids and implications for conservation. Plants (Basel) 12(2):257

Bhattacharjee B, Islam SS (2014) Effects of plant growth regulators on multiple shoot induction in Vanda tessellata (Roxb.) Hook. Ex G. Don an endangered medicinal orchid. Int J Sci Nat 5(4): 707–712

Bhattacharyya P, Lalthafamkimi L, Sharma T, Kumar D, Kumar A, Kumar D, Kumar S (2022) Metabolic and tissue-specific expression profiling in micropropagated plants of Malaxis acuminata: an endangered medicinal orchid. Plant Cell Tissue Organ Cult 151(3):535–549

Bhowmik TK, Rahman MM (2020) In vitro study of medicinally important orchid Aerides multiflora Roxb. from nodal and leaf explants. J Pharmacogn Phytochem 9(4):179–184

Cardoso JC, Zanello CA, Chen JT (2020) An overview of orchid protocorm-like bodies: Mass propagation, biotechnology, molecular aspects, and breeding. Int J Mol Sci 21:985

Carlomagno F, Lanzino M, Mendicino F, Bonacci T, Pellegrino G (2024) Pollinator diversity of the food-deceptive orchids in southern Italy. Plant Biol (Stuttg) 26(7):1144–1115

Castillo-Pérez LJ, Martínez-Soto D, Fortanelli-Martínez J, Carranza-Álvarez C (2021) Asymbiotic seed germination, in vitro seedling development, and symbiotic acclimatization of the Mexican threatened orchid Stanhopea tigrina. Plant Cell Tissue Organ Cult 146(2):249–257
Castillo-Pérez LJ, Ponce-Hernández A, Alonso-Ac AJ, Solano R, Fortanelli-Martínez J, Lagunez-Rivera L, Carranza-Álvarez C (2024) Medicinal orchids of Mexico: a review. Pharmaceuticals 17(7):907
Castro JB, Singer RB (2019) A literature review of the pollination strategies and breeding systems in Oncidiinae orchids. Acta Bot Brasil 33(4):618–643
Chang HH, Chang AY, Tsai BC, Chen YJ, Wu SG, Chen LJ, Lin YX, Hsueh YS (2024) Ethanol extract of Vanilla planifolia stems reduces PAK6 expression and induces cell death in glioblastoma cells. J Cell Mol Med 28(17):e70065
Chookoh N, Chiu YT, Chang C, Hu WH, Dai TE (2019) Micropropagation of Tolumnia orchids through induction of protocorm-like bodies from leaf segments. HortScience 54(7):1230–1236
Choudhary D, Mashkey VK, Goutam E, Shrivastava M, Rawat M, Kumari A, Tripathi V (2023) Medicinal orchids: traditional uses and recent advances. Ann Phytomed 12(1):1–9
Chugh S, Guha S, Rao IU (2009) Micropropagation of orchids: a review on the potential of different explants. Sci Hortic 122(4):507–520
De LC (2020) Morphological diversity in orchids. Int J Bot Stud 5(5):229–238
De LC (2023) Taxonomy and morphology of Orchids. Sabujeema 3(8):50–56
de Faria RT, Stegani V, Bertoncelli D, Alves GAC, de Assis AM (2018) Substrates for the cultivation of epiphytic orchids. Semin Ciênc Agrár 39(6):2851–2866
Farrokhzad Y, Babaei A, Yadollahi A, Kashkooli AB, Mokhtassi-Bidgoli A, Hessami S (2022) Informative title: development of lighting intensity approach for shoot proliferation in Phalaenopsis amabilis through combination with silver nanoparticles. Sci Hortic 292:110582
Gao TY, Zhang W, Zhang SB (2025) Differentiation in water adaptation strategy between epiphytic and terrestrial species of Cymbidium, Orchidaceae. AoB Plants 17(4):plaf030
Gaskett AC (2011) Orchid pollination by sexual deception: pollinator perspectives. Biol Rev Camb Philos Soc 86(1):33–75
Gautam S, Sahu R, Sahu TL, Harmukh N, Khunte N (2025) Ethnomedicinal uses of orchids and the conservation concerns of their threatened species: a review. Int J Adv Biochem Res 9(7): 1421–1426
Givnish TJ, Spalink D, Ames M, Lyon SP, Hunter SJ, Zuluaga A, Iles WJD, Clements MA, Arroyo MTK, Leebens-Mack J, Endara L, Kriebel R, Neubig KM, Whitte WM, Williams NH, Cameron KM (2015) Orchid phylogenomics and multiple drivers of their extraordinary diversification. Proc R Soc B Biol Sci 282(1814):20151553
Gogoi P, Sen S (2023) Orchids of Dibru-Saikhowa: a systematic review on their traditional use, pharmacological activity and phytochemistry. Curr Top Med Chem 23(24):2277–2299
Hardjo PH, Savitri WD (2016) Somatic embryo from basal leaf segments of Vanda tricolor Lindl. var. pallida. KnE Life Sci 2017:173–179
Hernández F, Ochoterena I (1942) Historia de las plantas de Nueva España. Imprenta Univ 3:695
Hernández-Romero Y, Acevedo L, de Los Ángeles Sánchez M, Shier WT, Abbas HK, Mata R (2005) Phytotoxic activity of bibenzyl derivatives from the orchid Epidendrum rigidum. J Agric Food Chem 53(16):6276–6280
Hinsley A, De Boer HJ, Fay MF, Gale SW, Gardiner LM, Gunasekara RS, Kumar P, Masters S, Metusala D, Roberts DL, Veldman S, Wong S, Phelps J (2018) A review of the trade in orchids and its implications for conservation. Bot J Linn Soc 186(4):435–455
Hossain MM, Sharma M, Teixeira da Silva JA, Pathak P (2010) Seed germination and tissue culture of Cymbidium giganteum Wall ex Lindl. Sci Hortic 123:479–487
Husen A, Faisal M (2005) Orchid diversity and its prospects of cultivation in Uttaranchal, India. Ann For 12:275–280
Husen H, Pant M (2024) Exploring medicinal orchids. Taylor & Francis Group, LLC, Broken Sound Pkwy NW. https://doi.org/10.1201/b23375

Husen A, Rahman MF (2003) Orchids an important group of plants for traditional system of medicine in India. Indian Forester 129:651–653

Jiménez-López DA, Pérez-García EA, Martínez-Melendez N, Solano R (2019) Wild orchids traded in a traditional market in Chiapas, Mexico. Bot Sci 97(4):691–700

Johnson S (1996) Bird pollination in South African species of Satyrium (Orchidaceae). Pl Syst Evol 203(1):91–98

Johnson SD, Sivechurran J, Doarsamy S, Shuttleworth A (2020) Dung mimicry: the function of volatile emissions and corolla patterning in fly-pollinated Wurmbea flowers. New Phytol 228(5):1662–1673

Kamaladhasan N, Mohan Raj R, Krishnankutty N, Indhar Saidanyan R, Soundararajan N, Saravanan S (2020) Evolution of organismal female Wasp mimics in sexually deceptive orchid genus Chiloglottis (Orchidaceae). In: Khasim S, Hegde S, González-Arnao M, Thammasiri K (eds) Orchid biology: recent trends & challenges. Springer Singapore

Kerbauy GB (1984) Regeneration of protocorm-like bodies through in vitro culture of root tips of Catasetum (Orchidaceae). Z Pflanzenphysiol 113(4):287–291

Kiaheirati H, Hashemabadi D, Kaviani B (2024) In vitro propagation of the orchid Phalaenopsis circus via organogenesis and somatic embryogenesis using protocorm and thin cell layer explants. Ital Botanist 18:29–50

Knudson L (1922) Nonsymbiotic germination of orchid seeds. Bot Gaz 73(1):1–25

Lal N, Singh M (2020) Prospects of plant tissue culture in orchid propagation: a review. Int J Biosci 7(2):103–110

Lal A, Pant M, Kumar A, Palni LMS, Singh A, Wabaidur SM, Husen A (2023) Clonal propagation and hardening of epiphytic orchids (*Aerides multiflora* and *Rhyncostylis retusa*) for sustainable conservation and utilization. S Afr J Bot 160(2023):402–411. https://doi.org/10.1016/j.sajb.2023.09.029

Lal A, Pant M, Kumar A, Palni LMS, Singh A, Siddiqui MR, Husen A (2024) DNA barcoding of epiphytic orchids for species level identification. J Hortic Sci Biotechnol:1–11. https://doi.org/10.1080/14620316.2024.2371591

Liu Q, Chen J, Corlett RT, Fan X, Yu D, Yang H, Gao J (2015) Orchid conservation in the biodiversity hotspot of southwestern China. Conserv Biol 29(6):1563–1157

Mamun AA, Islam MM, Ahmed M, Ghosh GC (2018) In vitro mass propagation from shoot tip of Dendrobium Red Bull-an endangered epiphytic orchid. Plant Tissue Cult Biotechnol 28(2):161–169

Manokari M, Priyadharshini S, Cokulraj M, Dey A, Faisal M, Alatar AA, Alok A, Shekhawat MS (2022) Exogenous implications of silver nitrate on direct and indirect somatic embryogenesis and germination of cold stored synseeds of Vanilla planifolia Jacks. ex Andrews. S Afr J Bot 150:129–138

Martén-Rodríguez S, Quesada M, Castro AA, Lopezaraiza-Mikel M, Fenster CB (2015) A comparison of reproductive strategies between island and mainland Caribbean Gesneriaceae. J Ecol 103:1190–1204

Mercado SAS, Jaimes YMO (2022) Inclusion of organic components in culture medium to improve the in vitro propagation of Cattleya warscewiczii and Cattleya gaskelliana. S Afr J Bot 148:352–359

Micheneau C, Johnson SD, Fay MF (2009) Orchid pollination: from Darwin to the present day. Bot J Linn Soc 161(1):1–19

Morales-Sánchez V, Rivero-Cruz I, Laguna-Hernández G, Salazar-Chávez G, Mata R (2014) Chemical composition, potential toxicity, and quality control procedures of the crude drug of Cyrtopodium macrobulbon. J Ethnopharmacol 154(3):790–797

Morel GM (1960) Producing virus-free Cymbidium. AOS. Bulletin 29:495–497

Negi A, Chauhan K, Adhikari H, Husen A, Pant M (2024) Medicinal orchids potential, conservation, and healing traditions. In: Husen A (ed) Exploring medicinal orchids. Taylor & Francis Group, LLC, Broken Sound Pkwy NW. https://doi.org/10.1201/9781003436638-1

Ng CKY, Hew CS (2000) Orchid pseudobulbs–false bulbs with a genuine importance in orchid growth and survival. Sci Hortic 83(3–4):165–172
Nirmala C, Nongdam P, Tewari R (2006) Biotechnological and molecular approaches for improvement of orchids. Plant Cell Biotechnol Mol Biol 7(1&2):1–10
Nongdam P, Chongtham N (2012) In vitro seed germination and mass propagation of Cymbidium dayanum Reichb: an important ornamental orchid of North-East India. Trends Hortic Res 2(2): 28–37
Nongdam P, Beleski DG, Tikendra L, Dey A, Varte V, El Merzougui S, Pereira VM, Barros PR, Vendrame WA (2023) Orchid micropropagation using conventional semi-solid and temporary immersion systems: a review. Plants 12:1136
Nowakowska K, Marciniak P, Pacholczak A (2022) A protocol for efficient micropropagation of rare orchid Vanda brunnea Rchb f. S Afr J Bot 150:233–239
Pathak P, Verma S, Prakash A, Mahant KC (2017) Regeneration competence of an ornamentally important epiphytic orchid, Rhynchostylis gigantea (Lindl.) Ridl. through leaf segments: a study in vitro. J Orchid Soc India 31:97–101
Poobathy R, Zakaria R, Murugaiyah V, Subramaniam S (2019) Surface sterilization and micropropagation of Ludisia discolor. Biocatal Agric Biotechnol 22:101380
Priya KI, Sabina GT, Rajmohan K (2013) Influence of plant growth regulators on in vitro clonal propagation of Dendrobium Sonia 'Earsakul'. J Bio Innov 2(2):51–58
Rodrigues LA, Paiva Neto VBD, Boaretto AG, Oliveira JFD, Torrezan MDA, Lima SFD, Otoni WC (2015) In vitro propagation of Cyrtopodium saintlegerianum rchb. f.(orchidaceae), a native orchid of the Brazilian savannah. Crop Breed Appl Biotechnol 15:10–17
Sailo N, Rai D, De LC (2014) Physiology of temperate and tropical orchids overview. Int J Sci Res 3(12):3–8
Sharma V (2019) In vitro conversion of root meristem into protocorm like bodies in Rhynchostylis retusa-an endangered fox tail orchid. Ecol Environ Conserv 25(4):1812–1816
Shengji P, Zhiwei Y (2018) Orchids and its uses in Chinese medicine and health care products. Med Res Innov 2(1):1–3
Singh DR, Pal R (2024) Smart solutions for harnessing orchid potential. Ind Hort 69(5):21–25
Sonia P, Apana N, Tikendra L, Dey A, Jamir I, Nongdam P (2022) Molecular clonal fidelity assessment of micropropagated orchids using DNA markers. In: Biotechnology and crop improvement. CRC Press, pp 143–179
Srivastava D, Gayatri MC, Sarangi SK (2015) In vitro seed germination and plant regeneration of an epiphytic orchid Aerides ringens (Lindl.) Fischer. Indian J Biotechnol 14(4):574–580
Teoh ES (2019) An ancient fantasy: Salep as an aphrodisiac. In: Orchids as Aphrodisiac, medicine or food. Springer, Singapore, pp 13–54
Tiwari P, Sharma A, Bose SK, Park KI (2024) Progress in orchid biology: Biotechnological achievements, translational success, and business outcomes. Horticulturae 10(2):152
Vergara-Galicia J, Aguirre-Crespo F, Castillo-Espana P, Arroyo-Mora A, López-Escamilla AL, Villalobos-Molina R, Estrada-Soto S (2010) Micropropagation and vasorelaxant activity of Laelia autumnalis (Orchidaceae). Nat Prod Res 24(2):106–114
Wani Panda AK, Mandal D (2013) The folklore medicinal orchids of Sikkim. Ancient Sci Life 33(2):92–96
Whitehead MR, Peakall R (2014) Pollinator specificity drives strong pre-pollination reproductive isolation in sympatric sexually deceptive orchids. Evolution 68(6):1561–1575
Wu K, Zeng S, Lin D, Teixeira da Silva JA, Bu Z, Zhang J, Duan J (2014) In vitro propagation and reintroduction of the endangered Renanthera imschootiana Rolfe. PLoS One 9(10):e110033
Zhang S, Yang Y, Li J, Qin J, Zhang W, Huang W, Hu H (2018) Physiological diversity of orchids. Plant Divers 40(4):196–208

Orchid Diversity in the Indian Subcontinent and the Need for Conservation

2

Gopalakrishnan Aazhivaendhan, Noyaljohnson Tinoammini, Periyasamy Shakthidhasan, Govindarajan Siva, Thiruppathi Senthil Kumar, Prem Kumar Ganesan, Ponnusamy Baskaran, and Mariappan Muthukumar

2.1 Introduction

The Indian subcontinent is one of the distinct geographical regions of the world with unique floral and faunal diversity. This remarkable geographical region was formed by the tectonic collision between Indian and Eurasian plate approximately between 70–50 million years ago (Besse et al. 1984; Jaeger et al. 1989; Yin and Harrison 2000). The collision resulted in the formation of the Himalayas and separated the Indian subcontinent from the rest of Eurasia. The isolated geographic positioning of the Indian subcontinent, combined with its diverse topography and climatic heterogeneity, has played a crucial role in shaping its rich species diversity and high levels of endemism. The barriers created by oceans, mountain ranges, and varied ecological zones have limited gene flow and promoted speciation, making the region a global biodiversity hotspot. The isolated geographic positioning of Indian subcontinent has played a crucial role in shaping its biodiversity and high levels of endemism.

G. Aazhivaendhan · N. Tinoammini · T. S. Kumar
Department of Botany, Bharathidasan University, Tiruchirappalli, Tamil Nadu, India

P. Shakthidhasan
Noorul Islam Centre for Higher Education, Kumaracoil, Kanyakumari, Tamil Nadu, India

G. Siva · M. Muthukumar (✉)
University Science Instrumentation Centre, Bharathidasan University, Tiruchirappalli, Tamil Nadu, India

P. K. Ganesan
Department of Plant Science and Landscape Architecture, University of Maryland, College Park, MD, USA

P. Baskaran
Oil Palm Research Center, Univanich Palm Oil Public Company Ltd., Plai Phraya, Krabi District, Thailand

A. Husen, M. Pant (eds.), *Forest-based Orchids*,
https://doi.org/10.1007/978-981-92-0096-2_2

Politically, this region comprises the modern nations of India, part of Pakistan, Nepal, Bhutan, Bangladesh, and Sri Lanka, and in broader definitions sometimes include the Maldives and part of Myanmar (Bose and Jalal 2022). Unlike many other regions of the world, the Indian subcontinent harbours an exceptionally diverse range of forest types and ecosystems, extending from the hot arid deserts of Rajasthan to the alpine meadows and coniferous forests of the Himalayas. This ecological spectrum also includes tropical rainforests of the Western Ghats and Northeast India, low-lying plains flourished by perennial rivers in central India, mangroves of the Sundarbans and coastal as well as island ecosystems. The Indian subcontinent hosts one of the world's most complex and diverse monsoon systems, shaped by its distinctive geography, ocean–land interactions, and major climatic drivers (Thomson et al. 2021). Such heterogeneity in the habitats, vegetations and microclimatic conditions provides a wide array of ecological niches, thereby supporting remarkable levels of Orchid diversity and endemism (Perez et al. 2024; Ackerman et al. 2025). Orchids are a unique group of plants known for their aesthetic floral architecture and unique medicinal properties. Among the most advanced plant families, Orchidaceae is exhibiting remarkable evolutionary adaptations within the plant kingdom and represents 10% of the entire flowering plants (Roberts and Dixon 2008), comprising 31,577 species under 730 genera (POWO 2025). Orchids exhibit diversity in multiple biological aspects, such as morphology, pollination mechanism, mycorrhizal symbiosis, ecological adaptations, etc. Their specific adaptation mechanisms have drawn the attention of researchers for centuries and continue to remain a mystery in resolving many evolutionary puzzles.

Orchids exhibit remarkable ecological adaptability, which enables them to thrive in diverse habitats. They grow as epiphytes, growing on the tree trunks and branches without deriving nutrients from the hosts; as lithophytes, colonizing rock surfaces and cliffs where soil is scarce; as terrestrials in the forest floors, grasslands and alpine meadows; and as mycoheterotropics, lacking photosynthetic pigments and depend on symbiotic association with mycorrhizal fungi for their nutrition (Fig. 2.1). This

Fig. 2.1 Diverse habitats of orchids: (**a**) Epiphytic, (**b**) Lithophytic, (**c**) Terrestrial, (**d**) Mycoheterotropic

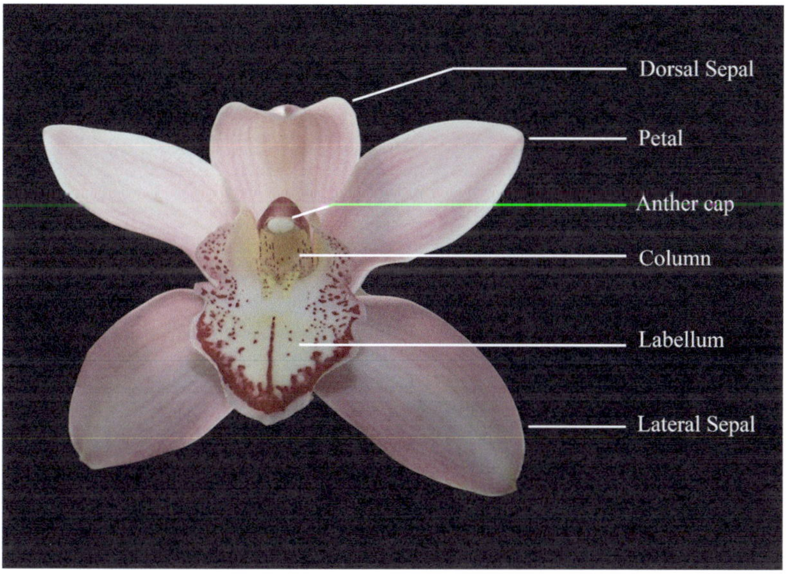

Fig. 2.2 Orchid floral morphology

habitat adaptability not only reflects evolutionary success but also contributes to their extraordinary diversity and distribution across tropical, subtropical and temperate regions of the world (Pridgeon et al. 1999).

In addition, orchid flowers manifest extraordinary diversity in their floral traits, showcasing unmatched variations in floral architecture, colour, and size (Lal et al. 2020). Their unique ability for inter-generic hybridization makes them one of the most important groups in floriculture. Wild orchids serve as valuable reservoirs of genetic material that can be retrieved and modified for developing novel hybrids. Furthermore, their floral structures are highly modified to ensure pollination mechanisms. Thereby facilitating successful reproduction and enhancing their evolutionary success (Fig. 2.2).

Beyond their medicinal and horticultural values, this group of plants plays a significant role in maintaining ecological balance. Due to their habitat specificity and sensitivity to environmental changes, orchids act as indicators of ecosystem health (Martin-Forés et al. 2022). Globally, this peculiar plant family is facing severe threats from habitat loss, overexploitation, and climate change (Lal et al. 2023; Tinoammini et al. 2024). Therefore, comprehensive documentation and analysis of orchid diversity, along with factors threatening their survival, are essential, not only for enriching the concepts of plant biology but also for making effective global conservation strategies and sustainable utilization of natural resources.

2.2 Geographic Position

The Indian subcontinent is located in the Southern region of Asia with the latitude from 0° 70′ S (Gen in Maldives) up to 37 04′ N (Hindukush range) and longitude from 67° 16′ E (Kirthar range in Pakistan) up to 97° 41′ E (Kibitu in Arunachal Pradesh). Geographically, it is defined as the landmass lying south to the great Himalayas which was surrounded by oceans on its three other sides. Earlier, the term Indian subcontinent was used in a broader sense that encompassed regions like entire Pakistan, Myanmar, Malay peninsula, whereas territories like Lakshadweep and the Maldives were not considered part of it (Clarke 1898; Hooker 1907). In the present work, the boundaries of Indian subcontinent were redefined based on the geological evidences resulted by Indo-Asia collision. The hill ranges formed due to tectonic collisions were considered as the Northern, Northwestern, and Northeastern boundaries. Although, the Maldives, Lakshadweep, and Andaman and Nicobar Islands are geographically separated from the mainland, they are still considered part of Indian subcontinent as they are part of Indian tectonic plate (Fig. 2.3).

Nearly, the entire northern boundary of Indian subcontinent was formed by the Himalayas along with their associated hill ranges—the Sulaiman and Kirthar on the northwest merging into the Hindu Kush, while the northeast boundary was defined by the Purvanchal range (which includes Patkai hills, Naga hills, Manipur hills, and Mizo hills) which extended southwest and merging into Arakan Yoma range in Myanmar. The southern boundary is delimited by the Indian Ocean, while the western and eastern boundaries are bordered by the Arabian Sea and the Bay of Bengal, respectively, giving the mainland of the subcontinent a distinct peninsular form with long stretches of coastal and marine boundaries.

With the landcover of c. 41,27,771 sq. km. the Indian subcontinent is contributing the Indo-Malayan realm which is one of the 8 biogeographic realms of the world (Udvardy 1975). Although it covers only 2.7% of global land area, Indian subcontinent is contributing 2.17% total forest area of the world (FAO-FRA 2020). Due to its species richness and high endemism, it is recognized one of the 17 mega diverse regions of the world (Williams 2001). This remarkable status is further highlighted by the presence of 4 of the 36 biodiversity hotspots namely Himalaya, Western Ghats and Sri Lanka, Indo-Burma and Sunderland Nicobar Islands) making the subcontinent a critical centre for global biodiversity and conservation significance (Chitale et al. 2020).

2.3 Phytogeographic Regions

The Indian subcontinent has been divided into many phytogeographic regions for botanical reference. Workers like Clarke (1898), Hooker (1907), Champion (1936), Chatterjee (1939) have classified Indian subcontinent into different floral regions based on various climatic factors and vegetation. Even though these classifications were widely accepted by the following researchers, some modifications have been carried out based on geography and plant diversity (Fig. 2.4). With the redefined

Fig. 2.3 Map showing geographic position of Indian subcontinent

geographical boundary of the Indian subcontinent, the phytogeographic regions were classified into the following phytogeographic regions.

1. Western Himalayas
2. Central Himalayas
3. Eastern Himalayas

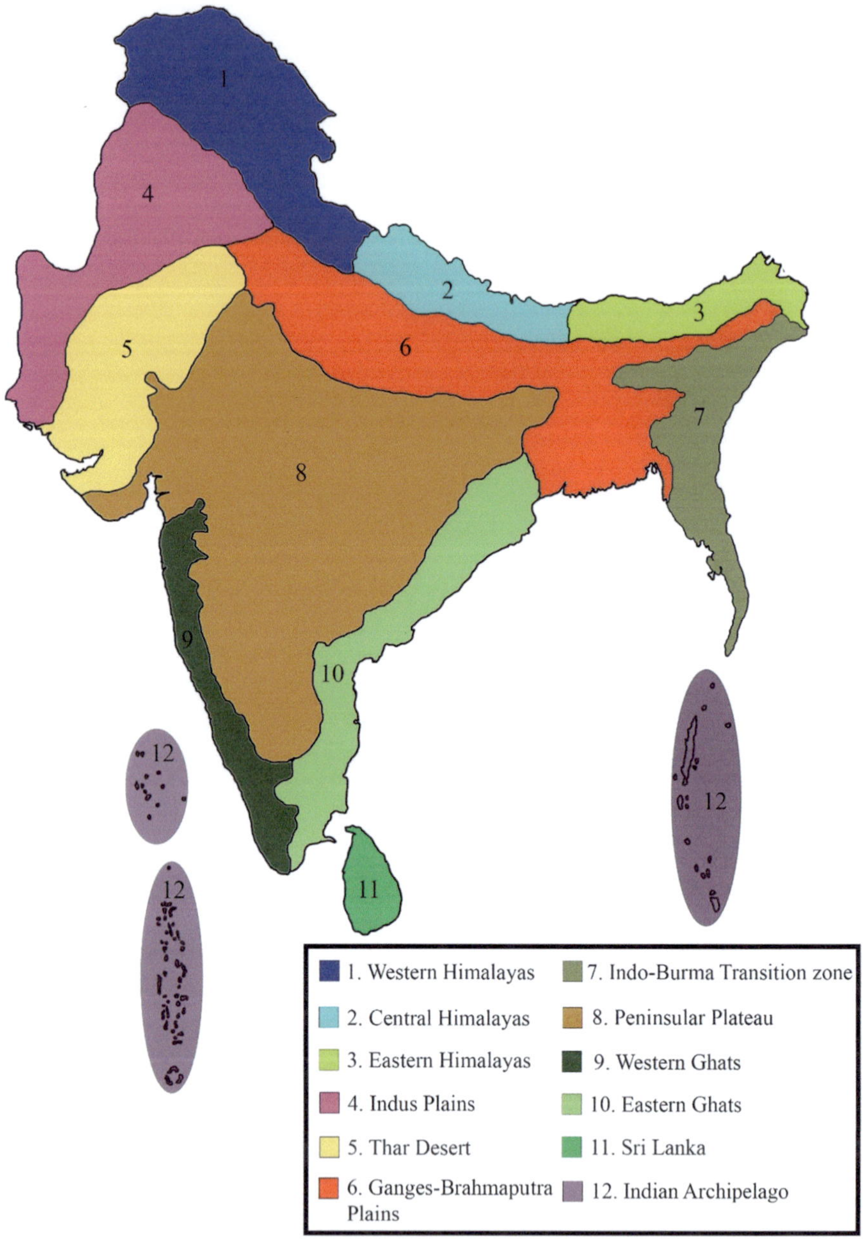

Fig. 2.4 Phytogeographic regions in Indian subcontinent

4. Indus Plains
5. Thar Desert
6. Ganges-Brahmaputra plains

7. Indo-Burma Transition zone
8. Peninsular Plateau
9. Western Ghats
10. Eastern Ghats
11. Sri Lanka
12. Indian Archipelago

2.4 Orchid Diversity in the Indian Subcontinent

Orchids inhabit almost all geographic regions of the world except extreme deserts and Polar regions. These adaptation mechanisms help them to achieve the position of the second-largest family in the entire plant kingdom. The Indian subcontinent comprises c. 1300–1400 orchid species under c. 155 genera, which is roughly 4.2% of total orchid population in the world. The mainland India contributes the majority of the orchid diversity with 1256 sp. Under 133 genera (Singh et al. 2019) followed by Bhutan 498 sp. under 100 genera (Gyeltshen et al. 2024), Nepal 437 sp. under 104 genera (Rokaya et al. 2013), Sri Lanka 194 sp. under 78 genera (Kottawa-Arachchi and Gunasekara 2020), Bangladesh 200 sp. under 73 genera and Pakistan 46 sp. under 24 genera (E-Flora of Pakistan n.d.) (Fig. 2.5).

Genera such as *Eulophia, Habenaria, Spiranthes, Zeuxine* are widely distributed across all the countries in the Indian subcontinent except the Maldives. In contrast, genera like *Bulbophyllum, Dendrobium, Eria, Oberonia, Habenaria, Liparis* represent the dominant group, contributing a high number of species to the orchid diversity of the region (Fig. 2.6).

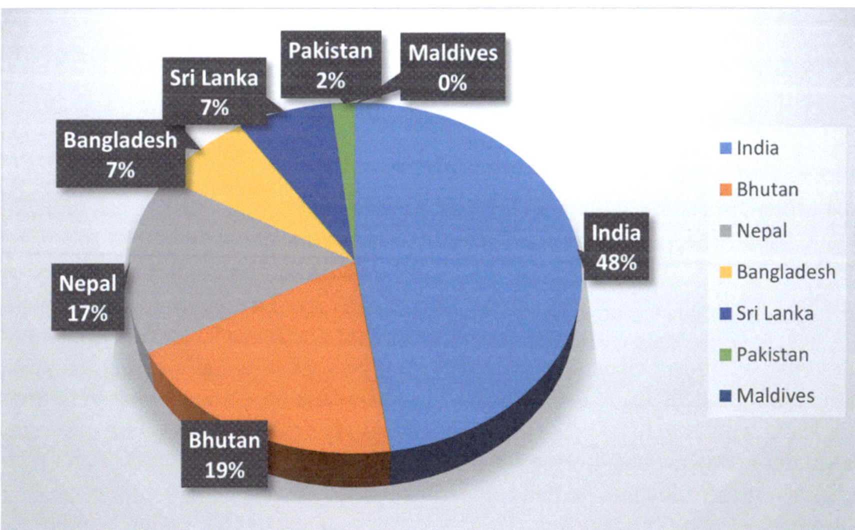

Fig. 2.5 Contribution of each country in Indian subcontinent's orchid diversity

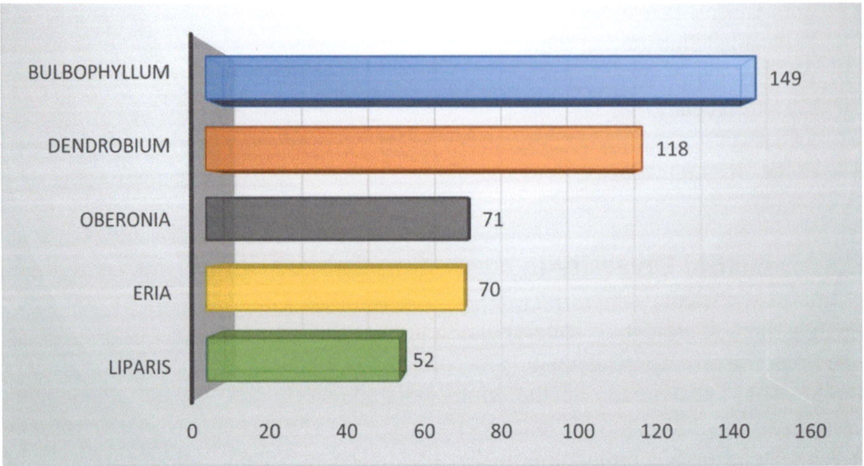

Fig. 2.6 Dominant genera of the family Orchidaceae in Indian subcontinent

2.5 Orchid Diversity in Different Phytogeographic Regions

Even though orchids are widely distributed throughout the Indian subcontinent, regions like the Himalayas, Northeast India, Western Ghats and Andaman and Nicobar Islands show high orchid diversity and richness as compared to the rest of the subcontinent. To provide a comprehensive understanding of the distributional pattern, orchid diversity has been analysed and discussed in the context of the major phytogeographic regions of the Indian subcontinent.

2.5.1 Western Himalaya

This phytogeographic region represents the Western segment of the Great Himalayan Mountain system, extending between eastern Pakistan and north-western India. It encompasses regions such as Khyber Pakhtunkhwa, Gilgit Baltistan, and Azad Kashmir in Pakistan, parts of eastern Afghanistan and the Indian territories of Jammu and Kashmir, Ladakh, Himachal Pradesh, and Uttarakhand. The Western Himalaya serves as the basin for both the Indus and the Ganga River systems, two of the most significant river networks in South Asia. This region spreads across an altitudinal gradient from 300 to 8611 m, encompassing climates that shift from tropical lowlands to nival zones (Singh Jalal and Jayanthi 2015a, b). As the western range lies in the rain-shadow region of the Indian summer monsoon (ISM) system, it receives significantly less rainfall compared to the Eastern Himalaya with the annual precipitation of 600–1800 mm. However, the climate of this region is strongly influenced by Western Disturbances (WDs), which play a crucial role in snow accumulation, glacier nourishment and sustainable water resources in the glacier river systems. Due to its sharp altitudinal and climatic gradients, the Western Himalaya supports a remarkable diversity of forest types, ranging from moist tropical forests, dry tropical forests, montane subtropical forests, montane temperate forests, sub-alpine forests and alpine forests (Champion and Seth 1968).

The Western Himalaya comprises 240 species of orchids belonging to 72 genera, of which 131 are terrestrial, 13 are mycoheterotrops and 96 are epiphytes. Orchid diversity is comparatively rich in Garhwal and Kumaon Himalayan region (232 *sp.*) rather than north-western part. Most of the regions in the Western Himalaya are open, temperate coniferous with lower canopy density and less structural complexity (Husen and Faisal 2005; Lal et al. 2024). Further, high altitude regions show extremely low temperatures with long dry winters. Epiphytic orchids are sensitive to frost and desiccation and require moist humidity, wet conditions, and high canopy cover. It requires a substrate to attach and grow further, which is scarce in high altitude regions. Therefore, most of the epiphytic and mycoheterotropic orchid species in the Western Himalaya are limited to the sub-tropical regions growing on moss-covered tree trunks of oaks, conifers, pines, as well as on rocky slopes. In contrast, the terrestrial orchids, being rooted in the soil, are better adapted to drier, seasonal moisture conditions with long hibernation mechanisms. Thus, the alpine meadows, and high-altitude grasslands of Western Himalaya are dominated with terrestrial orchid genus like *Cypripedium, Dactylorhiza*, *Dienia*, *Oreorchis,* and *Risleya*.

Even though epiphytic genus like *Dendrobium* (16 *sp.*) and *Bulbophyllum* (12 *sp.*) are represented with a greater number of species, terrestrial genus like *Habenaria* (14 sp.), *Neottia* (11 *sp.*), *Eulophia* (9 *sp.*), *Nervilia* (9 *sp.*), *Liparis* (10 *sp.*), *Calanthe* (8 *sp.*) and *Peristylis* (8 *sp.*) are the most dominant genera with high number of endemic species. Although this region comprises fewer orchid diversity than the Eastern Himalaya, it is home to several specialised endemic orchid species, namely: *Corallorhiza trifida* Châtel., *Dendrobium hesperis* (Seidenf.) Schuit. and P.B.Adams, *Eria occidentalis* Seidenf. and Arora, *Gastrochilus garhwalensis* Tsi, *Herminium kumaunensis* Deva and H.B.Naithani, *Neottia inayatii* (Duthie) Schltr., *Neottia mackinnonii* Deva and H.B. Naithani, *Neottia nandadeviensis* (Hajra) Szlach., *Nervilia gleadowii* A.N.Rao, *Nervilia pangteyana* Jalal, Kumar and G.S.Rawat, *Peristylus kumaonensis* Renz, *Ponerorchis renzii* Deva and H.B. Naithani.

2.5.2 Central Himalaya

This region is predominantly represented by the Himalayan range of Nepal, stretching between 26°12′ to 30°27′ N and 80°04′ E to 88°12′ E (Bhuju et al. 2007). With the land cover of 1,47,181 km^2 this region forms the core section of the Himalayan Mountain arc. It covers vast mountain terrain that includes the world's highest peak, Mount Everest. Nepal encompasses roughly one-third of the entire Himalayan range, which is separated from the Western Himalaya by Kali

River and from the Eastern Himalaya by Kanchenjunga- Singalila range and the Mechi River on the east. Its unique placement between India and China enhances its strategic importance in the South Asian region. The Central Himalaya serve as the basin for three major rivers that contribute to the Gangetic plains, namely Koshi in the east, Gandak in the centre and Karnali in the west. Altitude within this region ranges from 60 m msl to 8848 msl which supports the forests ranging from tropical rainforests in the southern region to alpine scrub forest and nival zones in north (Chaudhary 1998).

Orchidaceae is the largest plant family in Central Himalaya, comprising 437 species under 104 genera (Rokaya et al. 2013). Among the recorded orchids, 251 species are epiphytic, 63 lithophytic, 211 terrestrial and 21 saprophytic, with several species exhibiting mixed habits across these categories. They are distributed in wide altitudinal range from 60 to 5200 msl (Rokaya et al. 2013). The unique geography of the Nepal Himalaya supports 17 endemic orchid species representing a notable share of the 115 orchid species endemics to the entire Himalayan region. The orchid flora of the Central Himalaya is dominated by a few major genera, notably *Bulbophyllum* (38 *sp.*) and *Dendrobium* (31 *sp.*), along with considerable representation from *Habenaria* (20 *sp.*), *Oberonia* (19 *sp.*) and *Peristylis* (19 *sp.*). Species such as: *Eria annapurnensis* L. R. Shakya and M. R. Shrestha, *Eria baniae* Bajrach., L. R. Shakya and M. K. Chettri, *Eria nepalensis* Bajrach. and K. K. Shrestha, *Eria pokharensis* Bajrach., Subedi and Shrestha, *Gastrochilus calceolaris var. biflora* L. R. Shakya and M. R. Shrestha, *Herminium hongdeyuanii* B. B. Raskoti, *Malaxis dolpensis* M. R. Shrestha, L. R. Shakya et S. K. Ghimire, *Malaxis monophyllos var. obtusa* H. Tsukaya and H. Ohba, *Malaxis tamurensis* Tuyama, *Neottia chandrae* B. B. Raskoti, J. J. Wood and Rita Ale, *Neottia nepalensis* (N. P. Balakr.) Szlach., *Oberonia nepalensis* L. R. Shakya and R. P. Chaudhary, *Oreorchis porphyranthes* Tuyama, *Panisea panchaseensis* Subedi, *Pleione coronaria* P. J. Cribb and C. Z. Tang, *Stelis biflora* Sm., *Sunipia nepalensis* B. B. Raskoti and R. Ale represent some of the endemic orchid species reported from Central Himalaya.

2.5.3 Eastern Himalaya

Among the Himalayan divisions, the Eastern Himalaya stands out as the most significant geographic region, with remarkable ecological and biogeographic significance. It extends from Sikkim through North Bengal, Bhutan, and Arunachal Pradesh to the northern region of Myanmar and includes southern and south-eastern Tibet. Eastern Himalaya gives rise to major river systems like Brahmaputra, Tista, Subansiri which play a crucial role in maintaining the ecological integrity of Gangetic-Brahmaputra plains.

This region exhibits a remarkable altitudinal gradient, with the elevations ranging from c. 300 msl. in the foot hills regions to c. 8500 msl at peaks. From the misty lowlands to snow-capped summits, shifting altitudes and climates give rise to a tapestry of diverse forest types, including Tropical wet and semi-evergreen forests, sub-tropical forests, mixed-deciduous forests, temperate forests, sub-alpine and alpine meadows, all packed within relatively short distances. The monsoon system of the Eastern Himalaya is a key factor for its climate, ecology and biodiversity. This region is strongly influenced by the Indian summer monsoon, which results in comparatively higher annual rainfall than the Western Himalaya, ranging from c. 2000–5000 mm in some areas. Further, the Eastern Himalaya is a zone of overlap between Indo-Burma, Sino-Himalayan, and Indo-Malay floras. This mixed geographic position, along with altitudinal and climatic variation, diverse forests, and rugged topography, renders the Eastern Himalaya a global hotspot for orchid diversity, both in species richness and endemism (Rawat et al. 2025).

The Eastern Himalaya harbours approximately 920 orchid species distributed across c. 128 genera (Hegde 2003; De and Medhi 2014; Dalström 2017; Singh et al. 2019), making it the richest orchid reservoir in the Indian subcontinent. About 87 species of this flora are regional endemics, many of them are restricted to a narrow belt and specialised habitats. Among the Eastern Himalayan regions, Arunachal Pradesh exhibits the highest orchid diversity, as it encompasses within a single state, the full range of elevations, climatic regimes, and forest types represented across the other three regions. Comparatively, epiphytic orchids are dominant, constituting two-thirds of the total orchid diversity in Eastern Himalaya. Mid-elevation forests with high rainfall, persistent cloud cover, and structurally complex canopies create ideal conditions for epiphytic orchid growth and development. Terrestrial orchids, though comparatively fewer in diversity, are widely distributed across forest floors, while a smaller assemblage of mycoheterotropic orchids flourishes in shaded, undisturbed habitats. This life-form composition reflects the combined influence of topographic, diverse climatic regimes, and ecological interactions, which result in exceptional orchid richness and endemism of the Eastern Himalaya. Orchid diversity in Eastern Himalaya is dominated by the epiphytic genera like *Bulbophyllum, Dendrobium, Cymbidium, Coelogyne, Eria*, *Aerides*, *Phalaenopsis,* and *Vanda*. Furthermore, terrestrial genera such as Calanthe, Goodyera, Habenaria, Liparis, Crepidium, Spathoglottis, and *Neottia* are also abundant in this region. Several genera, including *Amitostigma, Bhutanthera, Cleisocentron, Corybas, Diglyphosa,* and *Biermannia* are strictly confined to this region. Moreover, genera such as *Apostasia, Paphiopedilum, Gastrodia,* and *Rhomboda* having most of the species endemic to Eastern Himalaya. *Amitostigma puberulum* (King and Pantl.) Schltr., *Bhutanthera himalayica* is (D.J. Middleton and R.R. Smith) Raskoti., *Biermannia arunachalensis* A.N. Rao, *Calanthe keshabii* Lucksom, *Cheirostylis sherriffii* N. Pearce and P.J. Cribb, *Coelogyne pantlingii* Lucksom, *Crepidium aphyllum* (King and Pantl.) A.N. Rao, *Cymbidium whiteae* King and Pantl., *Eulophia stenopetala* Lindl.**,** *Goodyera alveolata* Pradhan, *Hetaeria pelota* N. Pearce and P.J. Cribb, *Liparis chungthangensis* Lucksom, *Myrmechis bakhimensis* D. Maity, N. Pradhan and G. Maiti, *Neottia pantlingii* (W.W. Sm.) Tang and Wang, *Oberonia lobulata* King and Pantl., *Pomatocalpa bhutanicum* N.P.Balakr., *Taeniophyllum arunachalense* A.N. Rao and J. Lal are some of the orchid species endemic to Eastern Himalaya.

2.5.4 Indus Plains and Thar Desert

The Indus plains are among the oldest riverine tracts of South Asia, running across Pakistan and northwest India. This region is enriched by the alluvial deposits of the Indus and its major tributaries, including the Jhelum, Chenab, Ravi, Beas, and Sutlej, from the Western Himalayas. In terms of spatial extent, the plains stretch from the Western Himalayan foothills to the Arabian Sea, covering much of the Punjab and Sindh provinces of Pakistan, as well as the Indian states of Punjab, western Haryana, and northern Rajasthan. Most of the Indus plains fall into the semi-arid category, with annual rainfall ranging from 500 to 700 mm. This region depends largely on the southwest monsoon, which contributes 60–80% of total annual rainfall, while the remainder comes from Western disturbances. With the combination of freshly deposited soils, water availability and climatic variations, Indus plains support Riparian forests, Grasslands and Pastures, Scrub Jungles, and Wetlands.

The Great Indian Desert is the prominent arid zone of the Indian subcontinent, located in north-western India. While the major portion lies in Rajasthan, it also extends into northern Gujarat in the south and parts of Haryana in India, and further into the eastern provinces of Sindh and Punjab in Pakistan. With the land cover over 2,00,000 km^2, it forms an ecological transition zone between the arid southwest Asia and humid south Asia. It holds particular importance as a floristic transition zone, where elements of the Indo-Malayan flora intersect with those of the Irano-Turanian and Saharo-Sindian regions, resulting in a unique assemblage of flora with ecological adaptations. Its extreme aridity, low rainfall and dry climatic conditions result in vegetation dominated by xerophytes, succulents, thorny shrubs, grasses and halophytes.

Orchids are habitat-sensitive plants that require specific ecological conditions for survival. They depend on consistent moisture to support germination and a stable microclimate with moderate temperature and humidity for their development. Further, they depend on suitable habitats such as humus-rich soil for terrestrial and tree trunks for epiphytes. In contrast, deserts and semi-arid plains experience extreme climatic fluctuations with very high summer temperatures (above 50 °C) and extreme cold winters near freezing. The soil in these regions has low organic matter and poor water retention. The sparse vegetation in these regions has limited canopy cover, and the trees have no mossy bark to support epiphytic orchid species. The low rainfall, sandy plains, shifting dunes, saline tracts, excessive evapotranspiration, and prolonged dry seasons with extreme temperatures render the Indus plains and Thar Desert uninhabitable for orchid species.

2.5.5 Ganges-Brahmaputra Plains

This region represents the most extensive and fertile landforms of the Indian subcontinent, situated just below the great Himalayas, connecting Northern and Eastern India. This region covers approximately 560,000 km^2 of landmass and spans the Indian states of Uttar Pradesh, Bihar, West Bengal, Assam, as well as southern Nepal and Bangladesh. This vast fertile expanse is enriched by the centuries of alluvial deposits from the Himalayas by the Ganga, Brahmaputra and its tributaries, which form the world's largest delta. The geographical boundaries of this region are defined by the Himalayas in the north, the Peninsular Plateau in the south, the Indus plains and the Thar Desert in the west, and the north-eastern states of Arunachal Pradesh, Nagaland, and Meghalaya in the east. This region serves as an ecological transect, linking the Terai grasslands of the Himalayas with the moist evergreen and dry deciduous forests of the middle plains, as well as swamp, riparian forest, and grasslands in the east. The elevation is generally very low and uniform, ranging between 30 and 130 msl. The elevation of this region is a little higher towards the Himalayan foothills, while it gradually slopes towards the south and east. This gradient significantly shaped the drainage pattern, sediment deposition and species richness. This region experiences a southwest monsoonal climate with rainfall variation from west to east. The western plains are relatively dry, receiving 600–1000 mm of annual rainfall, while the eastern plains receive more than 2000 mm of annual rainfall.

This region harbours 269 species of orchids under 75 genera (Rahman 2017). The low elevation of the plains restricts diversity compared to adjoining areas. Still, the high rainfall in the eastern part and the Brahmaputra valley create favourable niches that sustain a relatively rich orchid flora. Moreover, orchid diversity in this region is strongly influenced by adjoining phytogeographic zones, resulting in the absence of strict endemic species. Orchid diversity in the Ganges-Brahmaputra plains increases eastward with higher rainfall and also rises along the elevation gradient. The flat terrain and warm conditions largely favour terrestrial orchid species over epiphytic ones. *Eulophia, Habenaria, Zeuxine, Calanthe, Bulbophyllum* and *Dendrobium* are some of the dominant genera that contribute a greater number of species to this region. *Aerides odorata* Lour., *Bulbophyllum roxburghii* (Lindl.) Rchb.f. *Calanthe sylvatica* (Thouars) Lindl., *Eulophia mackinnonii* Duthie, *Eulophia dabia* (D.Don) Hochr., *Geodorum densiflorum* (Lam.) Schltr., *Habenaria commelinifolia* (Roxb.) Wall. ex Lindl., *Habenaria plantaginea* Lindl., *Luisia zeylanica* Lindl., *Nervilia plicata* (Andrews) Schltr., *Rhynchostylis retusa* (L.) Blume, *Vanda tessellata* (Roxb.) Hook. ex G.Don, *Zeuxine strateumatica* (L.) Schltr. are some of the common species found in this region.

2.5.6 Indo-Burma Transition Zone

This region forms a transition zone between the Peninsular shield of India and Southeast Asia, stretching from Meghalaya through southern Assam, Nagaland, Manipur, Mizoram, Tripura and the adjoining parts of Bangladesh and Myanmar. The young mountains formed by the Indo-Asia collision create a complex landscape that resulted in a rich blend of ecosystems. It exhibits drastic geographical variation, with dissected hills, plateaus, and narrow valleys within a small land cover, which facilitates ecological diversity and species richness. Geographically, this region is situated between the Brahmaputra Valley in the north and the Purvanchal,

Arakan-Yoma hills of Myanmar in the east and southeast, and the Tripura plains that open into the Bengal lowlands in the southwest. The region exhibits striking altitudinal variation, from the Assam plains at 30 msl. and the Meghalaya Plateau at 300–1950 msl. to the Arakan-Yoma ranges, which exceed 3500 msl. The climate of this region is influenced by the Southwest Monsoon, which delivers an average annual rainfall of 2000–4000 mm. Due to the funnelling effect of the Himalayas and the Meghalaya plateau regions, like Mawsynram and Cherrapunji, receive the world's highest annual rainfall of 10,000–12,000 mm. This region exhibits wide variety of forest ecosystems ranging from tropical evergreen and semi-evergreen forest in Nagaland, Mizoram and Tripura to subtropical broadleaved forests in Meghalaya and Manipur, sub-tropical Pine forests in Meghalaya and grasslands in Assam plains (Dikshit and Dikshit 2013).

Home to 692 species across 124 genera, the orchid flora of this region highlights its significance within Indo-Burma biodiversity hotspot and its remarkable species richness. Among the phytogeographic regions of the Indian subcontinent, this region ranks second in endemic orchid species richness, hosting 67 species. Complex topography, high rainfall, and ecological gradients across borders create diverse habitats that support both epiphytic and terrestrial orchids, making it one of the orchid-rich zones in South Asia. The orchid diversity of this region is dominated by epiphytic orchid genera, including *Bulbophyllum, Dendrobium, Eria, Coelogyne, Luisia, Oberonia, Phalaenopsis,* and *Pholidota*. The forest floors and open grasslands are occupied by the tyerrestrial genera like *Anoectochilus, Arundina, Calanthe, Crepidium, Cymbidium, Goodyera, Habenaria, Liparis, Paphiopedilum, Phaius, Pleione, Zeuxine*. Further, the humus rich forest floor of tropical evergreen forests is also occupied by mycoheterotropic orchid genera such as *Didymoplexis, Epipogium, Galeola, Gastrodia.*

Agrostophyllum flavidum Phukan, *Biermannia quinquecallosa* King and Pantl., *Bulbophyllum kipgenii* (Kishor, Chowlu and Vij) J.J. Verm., *Coelogyne ghatakii* T.K. Paul, S.K. Basu and M.C. Biswas, *Corybas purpureus* J. Joseph and Yogan., *Crepidium meghalayensis* M. Murugesan, *Dendrobium tuensangense* Odyuo and C. Deori, *Eria crassicaulis* Hook. f., *Eulophia santapaui* Panigrahi and Kataki, *Flickingeria abhaycharanii* Phukan and A.A. Mao, *Gastrochilus carnosus* Z.H. Tsi, Nervilia khasiana (King and Pantl.) Schltr., *Odontochilus tetrapterus* (Hook.f.) Av. Bhattacharjee and H.J. Chowdhery, are some of the strict endemics species belongs to this phytogeographic region.

2.5.7 Peninsular Plateau

The peninsular plateau forms a triangular landmass at the centre of peninsular India, characterised by its ancient geological origin. Occupying central and southern India, this region played a significant role in shaping the Indian subcontinent's drainage system and biodiversity. It consists of igneous and metamorphic rocks that constitute the Indian Shield's stable core. The territories of this region are defined by the Aravalli range in the northwest, the Ganges plains in the north, the Rahr plains in the east and the Western Ghats and Eastern Ghats in the southwest and southeast, respectively. It is the largest phytogeographic region in the Indian subcontinent, covering the Indian states of Gujarat, south-eastern Rajasthan, Madhya Pradesh, Chhattisgarh, Jharkhand, Odisha, Telangana, Maharashtra, Andhra Pradesh, Karnataka, and Tamil Nadu. Geographically, it is broadly divided into the Central Highlands and the Deccan Plateau, which are separated from each other by the Vindhya and Satpura ranges.

The central highlands comprise several small dissected plateaus, including the Marwar plateau, the Malwa plateau, the Bundelkhand Upland, the Bhagelkhand plateau, and the Chotanagpur Plateau, while the Deccan plateau include the Maharashtra plateau, the Karnataka plateau and the Telangana plateau. Each of these regions has unique geological substrates that influence the soil type, pH, and nutrient profile, ultimately resulting in a varied plant diversity. This region exhibits a wide range of geological variations, including discontinuous plateaus, ancient hill ranges (Aravalli, Vindhya, Satpura), mounds, valleys and riverine landscapes, which make this region a unique combination of natural landforms. The elevation of this region ranges from 300–900 msl., increasing gradually from east to west. As this region lies in the rain-shadow zone of the southwest monsoon, it receives only about 500–900 mm of annual rainfall. Nevertheless, its riverine landscapes are sustained and enriched by major peninsular rivers such as Godavari, Krishna, Cauvery, Mahanadi, and Narmada. The forest types of this region categorized into tropical dry deciduous, moist deciduous, scrub, thorn and semi-evergreen forests.

This phytogeographic region comprises approximately 196 orchid species, distributed across 58 genera, including 30 endemic species. The orchid diversity of this region reflects a mixture of elements from both the Western and Eastern Ghats; therefore, only one species, Eulophia campbellii Prain, is strictly endemic to this region. Genera such as *Habenaria, Eulophia, Dendrobium, Acampe, Aerides, Peristyles* and *Zeuxine* are dominant, contributing the majority of species recorded here. *Aerides maculosa* Lindl., *Eulophia pratensis* Lindl., *Habenaria gibsonii* Hook. f. and *Habenaria grandifloriformis* Blatt. and McCann are some of the endemic species having wide distribution range. *Aerides multiflora* Roxb., *Cymbidium aloifolium* (L.) Sw., *Dendrobium herbaceum* Lindl., *Dendrobium crepidatum* Lindl. and Paxton., *Geodorum densiflorum* (Lam.) Schltr., *Habenaria digitata* Lindl., *Pelatantheria insectifera* (Reichb. f.) Ridl., *Peristylus constrictus* (Lindl.) Lindl., *Rhynchostylis retusa fo. albiflora* (I. Barua and Bora) Christenson, *Zeuxine strateumatica* (L.) Schltr. Are some of the common orchid species found in this region.

2.5.8 Western Ghats

The Western Ghats are one of the ancient mountain ranges in the Indian subcontinent formed due to tectonic uplift and faulting associated with the break-up of Gondwana. Extending nearly 1600 km parallel to the western coast of Peninsular India, they cover the states of Maharashtra, Karnataka, Kerala, and Tamil Nadu. The elevation of this region ranges between 200 and 2600 msl. The central portion of this range near Goa is relatively lower in elevation, whereas the northern and southern segments are characterized by higher average elevation of 900–1500 msl. The southern part is particularly crowned with many of the highest peaks of this range, including Anamudi (2695 m) in Kerala, the tallest peak in peninsular India and Doddabetta (2637) in Tamil Nadu.

Functioning as the principal watershed of the Indian subcontinent, the Western Ghats give rise to several perennial rivers such as the Godavari, Krishna, and Kaveri, which sustain vast portions of peninsular India. Positioned along the western coast, the Ghats exert a profound influence on the regional climate. Acting as an orographic barrier, they intercept and block the southwest monsoon winds, causing heavy rainfall on the western slopes ranging from 2000 mm to 9000 mm annually, while creating a rain-shadow effect on the eastern side where rainfall may drop to <600 mm in some areas. The topographic and climatic variations of the region support a remarkably diverse range of ecosystems, including tropical evergreen forests at the high elevations, moist deciduous forests in the lower elevations, dry deciduous forests in the rain shadow eastern slopes, scrub forests at the foothills, swamps and montane shola grasslands in the valleys. The distinctive geographical features and wide climatic gradients of this region create a mosaic of habitats, promoting high levels of orchid diversity. Its exceptional biodiversity and endemism classify the region as one of the eight hottest biodiversity hotspots in the world (Chinnasamy and Honap 2023).

Documented orchid diversity of this region is composed of 402 species across 73 genera, including 134 endemic species and 4 endemic genera. The southern Western Ghats harbour higher orchid diversity than the northern region. Higher elevations, complex topography and dense evergreen and montane shola forests in the south of regions provide diverse microhabitats, resulting in higher species diversity and endemism. The orchid diversity of this region is primarily contributed to by the genera Bulbophyllum, Dendrobium, Habenaria, Eria, Oberonia, and *Liparis*. Genera such as *Aenhenrya, Diplocentrum, Ipsea, Simthsonia, Xenikophyton* are endemic to this region.

Orchid species, such as *Acampe congesta* (Lindl.) Lindl., *Aenhenrya rotundifolia* (Blatt.) C.S. Kumar and F.N. Rasm., *Brachycorythis wightii* Summerh., *Bulbophyllum aureum* (Hook. f.) J.J. Sm., *Cheirostylis seidenfadeniana* C.S. Kumar and F.N. Rasm., *Crepidium crenulatum* (Ridl.) Sushil K. Singh, Agrawala and Jalal, *Dendrobium anamalayanum* Chandrab., V. Chandras and N.C. Nair, *Disperis monophylla* Blatt. ex C.E.C. Fisch., *Eulophia emilianae* Saldanha, *Habenaria flabelliformis* Summerh. ex C.E.C. Fisch., *Liparis beddomei* Ridl., *Paphiopedilum druryi* (Bedd.) Stein, *Robiquetia josephiana* Manilal and C.S. Kumar, *Schoenorchis manilaliana* M. Kumar and Sequiera, *Trias bonaccordensis* C.S. Kumar, *Xenikophyton seidenfadenianum* M. Kumar, are endemic to Southern Western Ghats. In addition, this region is home to several mycoheterotropic species, such as *Aphyllorchis montana* Rchb.f., *Didymoplexis pallens* Griff., *Epipogium roseum* (D.Don) Lindl., *Gastrodia exilis* Hook.f., among which *Didymoplexis seidenfadenii* C.S. Kumar and Ormerod, *Gastrodia silentvalleyana* C.S. Kumar et al. are endemic to this phytogeographic region.

2.5.9 Eastern Ghats

The Eastern Ghats are fragmented hill ranges extending along the eastern margin of peninsular India. Starting from the southeast Chotanagpur plateau, the Eastern Ghats runs for about 1750 km southwards and merges with the Nilgiris, covering Odisha, Andhra Pradesh, Tamil Nadu and parts of Karnataka and Telangana. The major hill ranges of this region include Simlipal, Kandhamal, Niyamgiri, Mahendragiri, Deomali in Odisha; Palakonda, Gurthedu, Kondapalli, Kondaveedu, Nallamala, Seshachalam, Velikonda in Andra Pradesh; Biligiriranga range in Karnataka; Javadu, Kalvarayan, Shevaroy, Kolli hills, Pachamalai, and Sirumalai in Tamil Nadu.

The elevation of this region ranges between 200 and 1600 msl. The climate of this region is influenced by both southwest and northeast monsoon. The southern region of Eastern Ghats receives major portion of its rainfall from the northeast monsoon, while the northern part is largely dependent on southwest monsoon. The annual rainfall of this region ranges between 600 and 1500 mm. Together with the Western Ghats, this region plays a significant role regulating the rainfall patterns of the Peninsular Plateau. This region harbours wide range of forest ecosystems ranging from tropical evergreen forests, tropical semi-evergreen forests, tropical moist-deciduous forests, tropical dry-deciduous forests, savannah forests and scrub forests. The diverse forest types accommodate many endemic floral and faunal species.

The Eastern Ghats comprises 197 orchid species under 66 genera, among them 22 species are endemic. The northern part of this region harbours a higher number of orchid species compared to the southern portion. More than 50% of the orchid species of this region are contributed by the following genera *Dendrobium, Bulbophyllum, Habenaria, Oberonia, Zeuxine, Liparis, Eulophia, Nervilia* and *Lusia.* Among the endemic orchids, species such as *Bulbophyllum panigrahianum* S. Misra, *Eria meghasaniensis* (S. Misra) S. Misra, *Odisha cleistantha* S. Misra, *Habenaria panigrahiana* S. Misra, *Habenaria panigrahiana var. parviloba* S. Misra, *Habenaria ramayyana* Ram. Chary and J.J. Wood, *Liparis vestita* Rchb. f., *Zeuxine mooneyi* S. Misra, have distribution restricted solely to the Eastern Ghats.

2.5.10 Sri Lanka

Sri Lanka is a tropical island nation in the Indian Ocean, separated from mainland India by the Palk Strait and Gulf of Mannar. Known as 'emerald jewel of Indian Ocean', it harbours unique biodiversity found nowhere else in the world. With the landcover of 65,610 $km^{2,}$ it unfolds like a miniature world, where complex topography, varied climates and evolutionary process have given rise to remarkable biodiversity. In combination with Western Ghats, it forms one of the unique biodiversity hotspots of the world with many endemic flora and fauna (Pullaiah 2024).

The geography of this region is broadly categorized into three regions: (1) Central highlands (2) The northern plains and (3) Coastal zone. The elevation ranges from seal level to 900 msl. in north, northeast and southeast plains and reaches up to 2200 msl. in the central highlands. The climate of this region is strongly influenced by the southwest and northeast monsoons which delivers the annual rainfall of 900–5000 mm. The southeast and central highlands receive maximum rainfall (2500–5000 mm) comparing the northern dry region. The Central Highlands form a primary watershed of the country, giving rise to more than 100 rivers, while the rivers in the southwestern lowlands are predominantly perennial. Together, these two regions harbour exceptional species diversity and are collectively recognized as a 'super biodiversity hotspot' (UNESCO 2010). The changing landscapes and shifting climates create a rich mosaic of ecosystems, including tropical rainforests, montane cloud forests, dry mixed evergreen forests, scrublands, wetlands and coastal mangroves. Its diverse geographic and climatic conditions make this region a global biodiversity hotspot with an exceptional level of species richness and endemism.

Orchidaceae is the largest plant family in this phytogeographic region, distributed in all the vegetation types across the nation. This region represents 194 orchid species under 78 genera, of which 55 species and 1 genus (*Adrorhizon* Hookf.) are endemic. The orchid flora of this region demonstrates significant phytogeographic affinities with the Western Ghats and Malayan Archipelago. Genera such as *Stichorkis, Ipsea* and *Taprobanea* are confined to the Western Ghats-Sri Lanka complex, whereas *Bromheadia* and *Octarrhena* exhibit distribution restricted to Sri Lanka and Southeast Asia. A large proportion of the endemic orchid flora of this region is contributed by genera such as Bulbophyllum, Dendrobium, Eria, Habenaria, Liparis, Oberonia, and *Peristyles.* Among them, *Adrorhizon purpurascens* (Thwaites) Hook. f., *Agrostophyllum zeylanicum* Hook. f., *Anoectochilus regalis* Blume, *Arundina minor* Lindl., *Bromheadia srilankensis* Kruizinga and de Vogel, *Eria articulata* Lindl., *Erythrodes latiloba* Ormerod, *Gastrodia zeylanica* Schltr., *Goodyera stelidifera* Ormerod, *Ipsea speciosa* Lindl., *Liparis thwaitesii* Hook. f., *Malaxis thwaitesii* Bennet, *Oberonia claviloba* Jayaweera, *Podochilus saxatile* Lindl., *Phreatia jayaweerae* Ormerod, *Pteroceras viridiflorum* (Thwaites) Holttum, *Schoenorchis tortifolia* (Jayaweera) Garay, *Thrixspermum pugionifolium* (Hook. f.) Schlechter, *Vanilla moonii* Thwaites, *Zeuxine regia* (Lindl.) Trimen are some of the orchid species strictly restricted to this region.

2.5.11 Indian Archipelago

The Indian Archipelago comprises three major island groups located in the Indian Ocean. The Andaman and Nicobar Islands in the Bay of Bengal, the Lakshadweep Islands and the Maldives in the Arabian Sea.

2.5.11.1 Andaman and Nicobar Islands

Andaman and Nicobar Islands are volcanic arc-island chain made up of 572 small islands, islets and rocks positioned at the juncture of Bay of Bengal and Andaman Sea. These islands are the continuation of Arakan–Yoma Mountain range of Myanmar, formed due to the collision of Indian tectonic plate with Eurasia. The topography consists of steep hills, ridges and narrow valleys with elevation ranging from sea level to 700 msl. The climate of this region is highly influenced by Southwest and Northeast monsoon which delivers the annual rainfall of 2500–3000 mm. Compared to Andaman Islands, the Nicobar Islands are relatively lower in elevation and receives lesser rainfall. The unique topography and tropical maritime climate with warm temperature, high humidity and heavy rainfall have given rise to varied forest types including tropical wet evergreen forests, tropical semi evergreen forests, tropical moist deciduous forests, hilltop shola forests, littoral and swamp forests.

The flora of Andaman and Nicobar Islands is enriched with 157 orchid species belonging to 65 genera, including 25 endemic species. The orchid diversity of this region shows more phytogeographic affinity with the Southeast Asia than with Indian subcontinent. In this region, epiphytic orchids are more abundant than terrestrial species. The high rainfall and humidity of this region, together with tropical forests having complex canopy cover, favour epiphytic orchids growth. In contrast, the saline and porous nature of the soil does not facilitate the proliferation of terrestrial species. In terms of orchid diversity, the Andaman and Nicobar Islands are considered separate groups, as they differ in geography, elevation, climate and forest types. Further, the Ten Degree Channel which separates the Andaman and Nicobar groups, acts as a natural barrier, limiting gene flow between orchid populations in these two regions. Additionally, their isolated geography and unique phytogeographic affinities resulted in distinct endemic orchid species in each group. Among the endemic orchids, 15 species are restricted only to Andaman Islands and 7 species are restricted only to Nicobar Islands and only one species has distribution in both the groups. Even though, orchid diversity of this region is dominated by genera such as *Bulbophyllum, Dendrobium, Eulophia* and *Lusia,* majority of the endemics are contributed by the genera *Habenaria, Dendrobium* and *Anoectochilus*. Among them, *Aerides emericii* H.G. Reichenbach, *Anoectochilus nicobaricus* N.P. Balakrishnan and P. Chakraborty, *Appendicula nicobarica* J. Jayanthi, R. Sumathi and K. Karthigeyan, *Bulbophyllum serratotruncatum* G. Seidenfaden, *Dendrobium gunnarii* P.S.N. Rao, *Eria andamanica* J.D. Hooker, *Habenaria andamanica* J.D. Hooker, *Luisia balakrishnanii* S. Misra, *Malleola andamanica* N.P. Balakrishnan and N. Bhargava, *Taeniophyllum andamanicum* N.P. Balakrishnan and N. Bhargava, *Trichoglottis orchidea* (J. Koenig) L.A. Garay, *Vanilla andamanica* R.A. Rolfe, *Zeuxine andamanica* G. King and R. Pantlingare are some of the orchid species endemic to this region.

2.5.11.2 Lakshadweep and Maldives Islands

Lakshadweep and the Maldives are coral atolls formed on submerged volcanic bases, situated in the Arabian Sea, southwest of India. Geographically Lakshadweep comprises 36 islets and lies between 8° N to 12° N latitude and 71° E to 74° E longitude, while the Maldives comprises 26 islets and extended between 1° S to 8° N latitude and 72° E to 74° E longitude. In comparison with the Andaman and Nicobar Islands, these island groups are geologically much younger. The soil in these regions, composed primarily of coral sand and limestone, is nutrient-poor and alkaline in nature. This region receives annual rainfall of 1500–2500 mm with southwest monsoon. Due to porous nature of soil and lack of topography to store water the rain water drains quickly. The elevation of these regions is only 2–5 m above sea level. Further, these regions lack mountains, valleys and river systems which resulted a flat terrain without geographical variations.

There is no wild orchid species have been reported from Lakshadweep and the Maldives due to their isolated coralline geology, poor soil type, lack of fresh water, absence of forest types and microclimatic variations. Only hybrid orchids are cultivated in these regions mainly for horticultural and tourism purposes.

2.6 Distribution and Endemism

Among the 12 phytogeographic regions in Indian subcontinent, orchids diversity is concentrated in Eastern Himalaya, Western Himalaya, Central Himalaya, and Western Ghats regions (Fig. 2.7). The richness is primarily shaped by the unique geography of these regions, which provide diverse habitats, steep altitudinal variations, abundant rainfall, favourable climatic conditions and high forest and vegetation diversity. The ecological stability of these regions maintains suitable microhabitats, promoting the survival, proliferation and diversification of a wide spectrum of orchid species.

The Indian subcontinent is a unique geographic region where the Afrotropical, Palearctic and Indo-Malayan realms converge, creating a zone of high species richness. This strategic position contributes substantially to the orchid diversity.

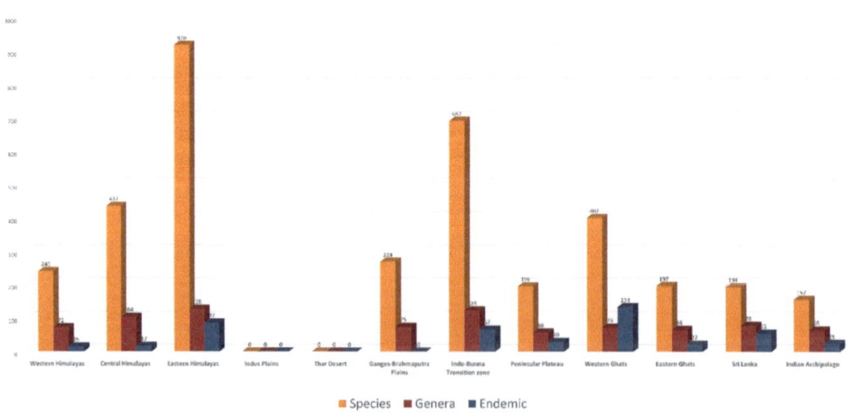

Fig. 2.7 Orchid diversity in different phytogeographic regions of Indian subcontinent

Table 2.1 Endemic orchid genera in Indian subcontinent

S. No.	Genus	No. of species
1.	*Bhutanthera* Renz	6
2.	*Bulleyia* Schltr.	1
3.	*Cottonia* Wight	1
4.	*Diplocentrum* Lindl.	2
5.	*Aenhenrya* Gopalan	1
6.	*Jejosephia* A.N. Rao and Manilal	1
7.	*Odisha* S. Misra	2
8.	*Plectoglossa* (Hook.f.) K. Prasad and Venu	1
9.	*Seidenfadeniella* C.S. Kumar in C.S. Kumar and Manilal	4
10.	*Sirhookera* Kuntze	2
11.	*Smithsonia* Saldanha	3
12.	*Xenikophyton* Garay	2

The orchid flora of Northeast region has the affinity with Indo-Malayan diversity, while orchids species present in.

Indian subcontinent harbours 388 endemic orchid species furthermore, 12 orchid genera are endemic to this landmass (Table 2.1). Further, genera such as *Amitostigma* Schltr., *Biermannia* King and Pantl., *Cleisocentron* Bruh., *Cryptochilus* Wall., *Diplomeris* D.Don, *Ipsea* Lindl.**,** *Uncifera* Lindl. are endemic to the Indo-Malayan realm with many endemic species to Indian subcontinent.

The level of endemism in Indian subcontinent is relatively lower compared to Southeast Asia, which represents high orchid diversity. Within this region, the Peninsular India exhibits a notable degree of endemism. The distinct tropical climate, prolonged ecological stability and geographical isolation of this region facilitate the evolution and persistence of several narrow-endemic orchid taxa.

2.7 Conservation Priorities

In the twenty-first century, nearly all the wild plants and animal species are facing threat due to human-made changes in the earth and orchids are no exceptions. Among the world populations, 39.4% of plants are now threatened with extinction, which is doubled the amount of 2016's estimation (Antonelli et al. 2023). The population explosion of the human race places immense pressure on natural resources, particularly in the regions with rich biological diversity (Kennedy 2006). Scientists warn that, plant species may disappear from the plant faster than the science can find and study as the extinction rate is 1000–10,000 times faster than the possibilities of new speciation (Hilton-Taylor and Brackett 2000).

Orchids are one among the highly evolved plant species which are well adaptive to the changing environment. However, they continue to struggle against the drastic changes occurring in the Anthropocene era. Among all plant families, Orchidaceae has highest number of genera with high proportion of threatened species (Swarts and

Dixon 2009). Among the 2123 orchid species evaluated by International Union for Conservation of Nature (IUCN 2025), nearly 1062 species fall under the threatened categories. Of these, 274 species are Critically endangered, 516 are Endangered, 272 are vulnerable and 48 species are possibly extinct.

From the Indian subcontinent, only one species, *Eulophia stenopetala* Lindl., which was endemic to Bhutan, has been declared as extinct by IUCN. Furthermore, species such as *Aerides ringens* (Lindl.) C.E.C. Fischer, *Habenaria viridiflora* (Rottler ex Sw.) Lindl., *Pinalia mysorensis* (Lindl.) Kuntze, *Pinalia polystachya* (A.Rich.) Kuntze, *Smithsonia maculata* (Dalzell) C.J.Saldanha, *Trias stocksii* Benth. Ex. Hook.f., all of which are endemic to peninsular India, are expected to be extinct (Jalal and Singh 2015). In addition, *Diplomeris josephii* A.N.Rao and Swamin., *Spathoglottis arunachalensis* Tsering and K.Prasad, endemic to Eastern Himalaya, are categorized as possibly extinct (IUCN 2025-1). Comparing Epiphytic species, Terrestrials and mycoheterotrophs are more prone to Extinction. They are comparatively sensitive to the changes in the environment. According to the IUCN (2025-1), among the assessed orchid species, six have been declared extinct globally, while four species from Indo-Malayan realm are listed as possibly extinct. Among these, 80% of orchids are terrestrials.

Threats leading to extinction of orchid species are more strongly driven by direct anthropogenic activities then by indirect factors such as climate change. According to IUCN (2025) nearly 85% of threats for orchid species caused due to direct human interventions. The major threats are: (1) land clearing for residential, commercial and agricultural purposes, (2) Energy production including hydropower projects and mining, (3) Transportation and service corridors, (4) Collection for biological resources, (5) Human intrusions and disturbances, (6) Natural system modification, (7) Pollution. Apart from the direct threats, climate change also significantly contributes to drive the orchid species towards extinction. Rising global temperatures, melting ice, sea level rise, fluctuations in weathers, shifting rainfall patterns, prolonged droughts, intensified storms are largely influenced by indirect anthropogenic activities like burning fossil fuels. The non-anthropogenic threats can be categorised into two types: (1) factors that affect the growth of the species such as natural calamities, lack of fungal associators and interspecific competitions; (2) factors affect the reproduction such as lack of pollinators. These two factors affect the orchid diversity independently or in combination (Kumar et al. 2024; Kaladharan et al. 2024).

Orchid-rich regions in the Indian subcontinent is under serious threats due to various developmental activities. The Eastern Himalaya and Western Himalaya are highly disturbed due to road widening, logging, and tourism-based infrastructure developments. Forests in the Himalayan foothills are further impacted by hydro power projects, dam constructions and deforestation for agriculture purposes. In addition, species belonging to genera like, *Dendrobium, Paphiopedilum, Cypripedium, Vanda, Renanthera, Cymbidium,* and *Coelogyne* are vigorously collected from the wild for domestication and horticultural purposes. Terrestrial orchids from the genera such as *Crepidium, Habenaria, Eulophia, Geodorum* are highly collected and sold in local markets for their medicinal value. Apart from that species like

Arundina graminifolia (D.Don) Hochr., *Geodorum densiflorum* (Lam.) Schltr., *Spathoglottis plicata* Blume, *Eulophia ochreata* Lindl. are traditionally used as cattle fodder in rural communities (Tandon and Kumaria 2012).

Although developmental activities are essential for a country's economic growth, it is equally important to prioritize the conservation of natural resources. Ecological specialization which contributes the Orchids diversity, also increases their vulnerability to extinction (Cribb et al. 2003). Protecting these sensitive organisms within their natural habitats offers the most effective assurance of their survival. Since each population of a single species may experience different threats, conservation efforts should move beyond single-species approaches and instead focus on enhancing habitat quality and protecting entire natural ecosystems. Such holistic strategies will ensure not only the survival of orchids but also the preservation of overall biodiversity.

2.8 Conclusion

The Indian subcontinent represents rich of orchid diversity in the world. The unique geography of this region hosts remarkable array of species with wide range of morphological variations. This region exhibits a wide range of variations in forest types, soil composition, rainfall patterns, altitudinal gradients and climatic conditions, making it a unique centre for speciation diversification, and conservation. Conservation and production of orchid diversity is more essential for sustainable utilisation of their medicinal and economic values and for ecosystem management. For the same, numerous protected areas have been designated in Indian subcontinent for the protection of natural habitats. Furthermore, the establishment of the two National Orchidarium in India and the preservation of sacred groves by tribal communities of the northeast India have significantly helped in in-situ conservation. Significantly, several orchid species once thought to be extinct have been rediscovered in recent years, highlighting the need for intensive field surveys and rigorous taxonomic studies to resolve confusions and update species records. In addition, international treaties such as CITES and conservation frameworks under IUCN have recognized orchids high priority to prevent illegal trade and ensure their survival.

For effective conservation along with policy frame works, community participation is also equally important. A lack of awareness among local peoples often results in habitat degradation and unsustainable utilization. Educating and involving local communities in conservation practices not only benefits orchids but also protects the entire web of biodiversity. Conservation of orchids is not merely about preserving ornamental or medicinal plants—it is about protecting ecosystems and maintaining ecological balance on which life depends.

References

Ackerman JD, Tremblay RL, Arias T, Zotz G, Sharma J, Salazar GA, Kaur J (2025) Persistent habitat instability and patchiness, sexual attraction, founder events, drift and selection: a recipe for rapid diversification of orchids. Plants 14(8):1193

Antonelli A, Fry C, Smith RJ, Eden J, Govaerts RH, Kersey P, Nic Lughadha E, Onstein RE, Simmonds MS, Zizka A, Ackerman JD (2023) State of the world's plants and fungi: tacking the nature emergency: evidence gaps and priorities. Royal Botanic Gardens, Kew Report

Besse J, Courtillot V, Pozzi JP, Westphal M, Zhou YX (1984) Palaeomagnetic estimates of crustal shortening in the Himalayan thrusts and Zangbo suture. Nature 311(5987):621–626. https://doi.org/10.1038/311621a0

Bhuju UR, Shakya PR, Basnet TB, Shrestha S (2007) Nepal biodiversity resource book: protected areas, Ramsar sites, and World Heritage sites. International Centre for Integrated Mountain Development (ICIMOD) and Ministry of Environment, Science and Technology (MOEST), Government of Nepal, in Cooperation with UNEP, Kathmandu

Bose S, Jalal A (2022) Modern South Asia: History, culture, political economy. Routledge

Champion HG (1936) A preliminary survey of the forest types of India and Burma. Indian For Rec 1:1–286

Champion HG, Seth SK (1968) A revised survey of the forest types of India. Manager of Publications

Chatterjee D (1939) Studies on the endemic flora of India and Burma. J Roy Asiatic Soc Bengal Sci 5(1):19–67

Chaudhary RP (1998) Biodiversity in Nepal: status and conservation. S. Devi, Saharanpur, India and Tecpress Books, Bangkok

Chinnasamy P, Honap VU (2023) Spatiotemporal variations in soil loss across the biodiversity hotspots of Western Ghats Region, India. J Earth Syst Sci 132(2):90

Chitale VS, Behera MD, Roy PS (2020) Congruence of endemism among four global biodiversity hotspots in India. Curr Sci 118(1):25–32. https://www.researchgate.net/publication/338547597

Clarke CB (1898) On the Subsubareas of British India, illustrated by the detailed distribution of the Cyperaceœ in that Empire. Bot J Linn Soc 34(235):1–146

Cribb PJ, Kell SP, Dixon KW, Barrett RL (2003) Orchid conservation: a global perspective. In: Orchid conservation. Royal Botanic Gardens, Kew, pp 1–24

Dalström S (2017) Orchid checklist of Bhutan. National Herbarium, National Biodiversity Centre, Ministry of Agriculture, Serbithang, Thimphu, Bhutan. http://www.bhutanbiodiversity.net/, accessed 2025-08-29

De LC, Medhi RP (2014) Diversity and conservation of rare and endemic orchids of North East India-a review. Indian J Hill Farming 27(1):138–153

Dikshit KR, Dikshit JK (2013) Natural vegetation: forests and grasslands of North-East India. In: North-East India: land, people and economy. Springer Netherlands, Dordrecht, pp 213–255

E-flora of Pakistan (n.d.) (http://www.efloras.org/flora)

FAO (2020) Global forest resources assessment 2020—key findings. FAO, Rome. https://doi.org/10.4060/ca8753en

Gyeltshen P, Rabgay K, Tobgay K, Namgyel P (2024) Addiction of five orchid species to the flora of Bhutan. Lankesteriana 24(2):205–216. https://doi.org/10.15517/lank.v24i2.61573

Hegde SN (ed) (2003) Arunachal Pradesh State biodiversity strategy and action plan, National Biodiversity Strategy and Action, vol 2. State Forest Research Institute, Department of Environment and Forests

Hilton-Taylor C, Brackett D (2000) 2000 IUCN red list of threatened species

Hooker JD (1907) Flora of British India, vol I-VII. The Indian Empire, The Imperial Gazetteer of India, London

Husen A, Faisal M (2005) Orchid diversity and its prospects of cultivation in Uttaranchal, India. Ann Forest 12:275–280

IUCN (2025) The IUCN red list of threatened species. Version 2025–1. https://www.iucnredlist.org. Accessed August 2025

Jaeger JJ, Courtillot V, Tapponnier P (1989) Paleontological view of the ages of the Deccan Traps, the Cretaceous/Tertiary boundary, and the India-Asia collision. Geology 17(4):316–319. https://doi.org/10.1130/0091-7613(1989)017<0316:PVOTAO>2.3.CO;2

Jalal JS, Singh P (2015) Threatened orchids of Maharashtra: a preliminary assessment based on IUCN regional guidelines and conservation prioritisation. J Orchid Soc India 29:1–14

Kaladharan S, Rengasamy A, Chinnaiyan R, Mariappan M, Thiruppathi SK (2024) In vitro asymbiotic seed germination and micropropagation of Dendrobium heyneanum Lindl.–an endemic orchid of Western Ghats, India. Plant Cell Tissue Organ Culture (PCTOC) 157(2):31

Kennedy D (ed) (2006) Life on a human dominated planet in Science magazine's state of the planet 2006–2007. Island Press

Kottawa-Arachchi JD, Gunasekara RS (2020) Research priorities and future directions in conservation of wild orchids in Sri Lanka: a review. Nat Conserv Res (Заповедная наука) 5(Suppl. 1): 34–45. https://doi.org/10.24189/ncr.2020.029

Kumar J, Katoch D, Thakur A, Pathania A, Anand A, Choudhary K (2024) A comprehensive review on threats and conservation status of orchids. J Appl Biol Biotechnol 12(2):43–47

Lal A, Pant M, Palni LMS, Kumar A (2020) Development of rapid micropropagation protocol for germplasm conservation of two orchid species- *Aerides multiflora* Roxb. And *Rhyncostylis retusa* (L.) Blume. Asian. J Conserv Biol 9(2):341–347

Lal A, Pant M, Pant G, Palni LMS, Datta A, Kumar G (2023) ISSR marker assisted genetic diversity assessment in natural populations of two endemic orchids Aerides multiflora and Rhynchostylis retusa from Uttarakhand, India. South Afr J Bot 157:151–158. https://doi.org/10.1016/j.sajb.2023.03.031

Lal A, Pant M, Kumar A, Palni LMS, Singh A, Siddiqui MR, Husen A (2024) DNA barcoding of epiphytic orchids for species level identification. J Horticult Sci Biotechnol:1–11. https://doi.org/10.1080/14620316.2024.2371591

Martín-Forés I, Bywaters SL, Sparrow B, Guerin GR (2022) Simultaneous effect of habitat remnancy, exotic species, and anthropogenic disturbance on orchid diversity in South Australia. Conserv Sci Pract 4(4):12652. https://doi.org/10.1111/csp2.12652

Pérez-Escobar OA, Bogarín D, Przelomska NA, Ackerman JD, Balbuena JA, Bellot S, Antonelli A (2024) The origin and speciation of orchids. New Phytol 242(2):700–716

POWO (2025) Plants of the world online. Facilitated by the Royal Botanic Gardens, Kew. Published on the Internet; https://powo.science.kew.org/ Retrieved on September 2025

Pridgeon AM, Cribb PJ, Chase MW, Rasmussen FN (1999) Genera Orchidacearum: Volume 1: General introduction, Apostasioideae, cypripedioideae, vol 1. Oxford University Press

Pullaiah T (ed) (2024) Biodiversity hotspot of the Western Ghats and Sri Lanka. CRC Press

Rahman MA, Huda MK, Rashid ME (2017) Orchid species diversity in Bangladesh and their revised nomenclatural updates. Biodivers Bullet Bangladesh 10:1–70

Rawat R, Joshi C, Arya AL, Tiwari D (2025) Diversity, ethnobotany, and conservation status of orchids in the Himalayan Botanical Garden, Nainital, India. J Biodivers Conserv 9(3):80–88

Roberts DL, Dixon KW (2008) Orchids. Curr Biol 18(8):325–329

Rokaya MB, Raskoti BB, Timsina B, Münzbergová Z (2013) An annotated checklist of the orchids of Nepal. Nord J Bot 31(5):511–550. https://doi.org/10.1111/j.1756-1051.2013.01230.x

Singh Jalal J, Jayanthi J (2015a) An annotated checklist of the orchids of western Himalaya, India. Lankesteriana 15(1):7–50

Singh Jalal J, Jayanthi J (2015b) An annotated checklist of the orchids of western Himalaya, India. Lankesteriana 15(1):07–50

Singh SK, Agrawala DK, Jalal JS, Dash SS, Mao AA, Singh P (2019) Orchids of India—a pictorial guide. Botanical Survey India, Kolkata

Swarts ND, Dixon KW (2009) Terrestrial orchid conservation in the age of extinction. Ann Bot 104(3):543–556

Tandon P, Kumaria S (2012) Orchid resources of the North East India and their sustainable utilzation. In: Biotechnology for sustainable development: achievements and challenges. Tata McGraw Hill Education Private Limited, New Delhi, India, pp 183–119

Thomson JR, Holden PB, Anand P, Edwards NR, Porchier CA, Harris NB (2021) Tectonic and climatic drivers of Asian monsoon evolution. Nat Commun 12(1):4022. https://doi.org/10.1038/s41467-021-24244-z

Tinoammini N, Aazhivaendhan G, Kumar TS (2024) Invitro germination and optimization of basal media for protocorm-like bodies proliferation in *Dienia ophrydis* (J. Koenig) Seidenf. Rhizosphere 29(11):100854

Udvardy MD (1975) A classification of the biogeographical provinces of the world. https://portals.iucn.org/library/node/6582

UNESCO (2010) Central Highlands of Sri Lanka. In: Report of the decisions adopted by the World Heritage Committee at its 34th session. WHC-10/34.COM/20

Williams J (2001) Biodiversity theme report—Australia State of the environment report (theme report). Published by CSIRO on Behalf of the Department of the Environment and Heritage. http://www.environment.gov.au/node/21579 (accessed on September 2025)

Yin A, Harrison TM (2000) Geologic evolution of the Himalayan-Tibetan orogen. Annu Rev Earth Planet Sci 28(1):211–280. https://doi.org/10.1146/annurev.earth.28.1.211

Orchids in Nepal: Status, Threats and Conservation Strategies

3

Keshab Raj Rajbhandari, Mukti Ram Paudel, Tirtha Raj Pandey, Mahendra Thapa, and Bijaya Pant

3.1 Introduction

Orchids, belonging to the family *Orchidaceae*, represent one of the largest groups of flowering plants. Globally, *Orchidaceae* is the largest family of vascular plants, comprising approximately 31,480 species across 758 genera (Elliott et al. 2025). Nepal, with its diverse terrain, climatic variability, and remarkable elevation gradient, harbors a rich diversity of orchids found throughout the country—from lowland plains to high mountain regions. Among these, the midhills serve as the most orchid-rich habitats. Within Nepal's flowering plant diversity, *Orchidaceae* is the largest family, encompassing 93 genera, 451 species, and 6 varieties, followed by *Asteraceae* with 129 genera and 442 species (Rajbhandari et al. 2024a, b), and *Poaceae* with 126 genera and 410 species (Rajbhandari and Rai 2017). Most orchid species in Nepal are epiphytic, although terrestrial species also constitute a significant portion. In recent years, numerous new species and records of orchids have been documented. However, a comprehensive systematic investigation is still lacking. The *Orchidaceae* of Nepal is currently being compiled as Volume 11 of the *Flora of Nepal*, with some genera already published online. This paper aims to present the

K. R. Rajbhandari (✉)
Kathmandu, Nepal

M. R. Paudel · B. Pant (✉)
Central Department of Botany, Tribhuvan University, Kirtipur, Kathmandu, Nepal
e-mail: bijaya.pant@cdb.tu.edu.np

T. R. Pandey
National Herbarium and Plant Laboratories, Godawari, Lalitpur, Nepal

M. Thapa
Annapurna Research Center, Maitighar, Kathmandu, Nepal

A. Husen, M. Pant (eds.), *Forest-based Orchids*,
https://doi.org/10.1007/978-981-92-0096-2_3

current status of Nepal's orchid flora and highlight the threats it faces, along with potential conservation strategies.

3.2 Status of Orchid Flora

Nepalese orchids were among the first plant groups to be scientifically explored in Nepal. In 1802, Buchanan-Hamilton collected numerous orchid specimens from the Kathmandu Valley, including areas such as Swoyambhu, Narayanhiti, and Shivapuri. Later, in 1820, Nathaniel Wallich expanded the collections across central Nepal. These early specimens were incorporated into David Don's landmark publication *Prodromus Florae Nepalensis* (Don 1825), which listed 52 orchid species from Nepal, many of them were new to science. Subsequent documentation continued with *An Enumeration of the Flowering Plants of Nepal Volume 1* by Hara et al. (1978), which recorded 315 species under 90 genera. Since then, several notable publications have enriched the understanding of Nepalese orchids (Banerji 1978; Banerji and Pradhan 1984; White and Sharma 2000; Rajbhandari and Bhattarai 2001; de Milleville and Shrestha 2004; Raskoti 2009; Bajracharya 2010; Rajbhandari 2015; Samuel 2022).

Regional orchid inventories have also emerged, including Tribhuvan Rajpath and Chitwan (Makwanpur) by White and Sharma (2000), Raja Rani (Morang) by Bajracharya (2011), Shivapuri National Park (Kathmandu) by Shakya and Bajracharya (2013), Hanspokhari (Ilam) by Bhardowaj et al. (2015), Central Nepal (Karunabhumi Community Forest, Daman and Chandragiri Community Forest, Chitlang of Makwanpur) by Pant et al. (2016), Panchase (Kaski) by Subedi et al. (2007) and Bhandari et al. (2020b), Godawari (Lalitpur) by Samuel (2022), Shiureni (Syangja) by Kafle and Pathak (2023), and Annapurna Conservation Area by Tiruwa et al. (2024a,b). Comprehensive checklists have been compiled over time, including those by Rajbhandari and Dahal (2004) (100 genera and 377 species), Rokaya et al. (2013) (104 genera and 437 species), Rajbhandari (2014) (107 genera and 450 species), and Rajbhandari and Rai (2017) (96 genera and 382 species). Taxonomic research continues actively, with recent additions reported by Kurzweil et al. (2017), Subedi et al. (2018), Bhandari et al. (2019a, b, 2020b), Raskoti and Ale (2019a, b, 2023), Paudel et al. (2020), Samuel (2023), Kafle et al. (2024, 2025), and Gupta et al. (2025).

Based on these sources and updated nomenclature, the present paper reports 93 genera, 451 species, and 6 varieties from Nepal (Appendix 1). Of these, 225 species are epiphytic, 214 terrestrial, and 12 mycotrophic—though some epiphytes exhibit terrestrial tendencies.

Distribution studies by Rajbhandari et al. (2000) revealed that epiphytic orchids are most abundant in the lower temperate zone (1101–2000 m), terrestrial orchids peak in the upper temperate zone (2001–3000 m), and subalpine and alpine zones show low orchid diversity. Five genera in Nepal have more than 20 species; *Bulbophyllum* (47 species), *Dendrobium* (30 species), *Herminium* (25 species),

Coelogyne and *Liparis* (24 species each). Phytogeographically, central Nepal hosts the highest orchid diversity, followed by eastern Nepal, with western Nepal having the fewest species (Rajbhandari et al. 2000; Acharya et al. 2011).

3.3 Ornamental Orchids

More than 100 species of Nepalese orchids possess ornamental value, belonging to genera such as *Acampe*, *Aerides*, *Arundina*, *Bulbophyllum*, *Calanthe*, *Coelogyne*, *Cymbidium*, *Cypripedium*, *Dendrobium*, *Habenaria*, *Luisia*, *Panisea*, *Papilionanthe*, *Pelatantheria*, *Phalaenopsis*, *Pinalia*, *Pleione*, *Rhynchostylis*, *Satyrium*, *Spathoglottis*, *Thunia*, and *Vanda*. Several species are cultivated for ornamental purposes, including *Aerides multiflora*, *Bulbophyllum leopardinum*, *Calanthe masuca*, *Calanthe plantaginea*, *Calanthe tankervilleae*, *Calanthe tricarinata*, *Coelogyne cristata*, *Cymbidium elegans*, *Cymbidium iridioides*, *Dendrobium densiflorum*, *Dendrobium moschatum*, *Dendrobium nobile*, *Pleione praecox*, *Rhynchostylis retusa*, *Vanda ampullacea*, *Vanda cristata*, and *Vanda tessellata*. Some of these are collected from forests around the Kathmandu Valley, including Phulchoki, and sold in local markets.

A comprehensive list of Nepalese orchids with ornamental value is provided in Appendix 1. Among them, *Dendrobium* and *Cymbidium* species are particularly sought after in the floral market, commanding high prices. One notable species, *Dendrobium aphyllum*—known locally as Salhesh Phool—holds both ornamental and cultural significance in Nepal. Every year, during the Nepalese New Year (mid-April), thousands of visitors from Nepal and neighbouring regions of India gather to witness its blooming at the Phulbari Mela in Salhesh Phulbari (Salhesh Garden), located in Lahan Municipality, Siraha District (central Nepal) at an elevation of 120 m.

Cultural Significance: According to legend, King Salhesh and his consort Malini once strolled through this garden. Each year, Malini—adorned in full makeup and garlands of *Dendrobium aphyllum*—awaits the return of her beloved king. The orchid blooms on the Haram tree (*Bischofia javanica*), beginning in the last week of Chaitra (second week of April) and reaching full bloom on New Year's Day (1st Baisakh). The flowering of *Dendrobium aphyllum* symbolises Malini's enduring love and is the centrepiece of the celebration, drawing crowds to honour this romantic folklore (Rajbhandari and Pandey 2024b).

3.4 Medicinal Orchids

In addition to their ornamental appeal, Nepalese orchids hold significant medicinal value. Pant and Raskoti (2013) documented medicinal uses for 90 species, while Teoh (2019) listed 82 medicinal orchid species from Nepal. Indigenous communities have long used these orchids to treat a wide range of ailments, including general debility, stomachache, bone fractures, colds, wound healing, and other conditions.

Some *Dendrobium* species have shown promising anticancer properties (Paudel et al. 2020; Pant et al. 2021a, b; Joshi et al. 2024). Orchids are traditionally used either alone or in combination with other medicinal plants (Acharya and Rokaya 2010; Pun 2014).

Many species are valued as tonics, with local names in parenthesis reflecting their cultural relevance. Examples include: *Brachycorythis obcordata* (Gaamdol), *Calanthe plantaginea*, *Calanthe tankervilleae*, *Cephalanthera longifolia*, *Coelogyne articulata*, *Cypripedium cordigerum* (Jibre), *Dactylorhiza hatagirea* (Paanch Aunle), *Dendrobium densiflorum* (Sungaabhaa), *Dendrobium eriiflorum*, *Dendrobium fugax* (Jivanti), *Dendrobium nobile*, *Dienia cylindrostachya*, *Eulophia dabia*, *Galearis roborowskii*, *Hemipilia calcicola*, *Herminium monorchis*, *Platanthera sikkimensis*, *Satyrium nepalense* (Gaamdol), *Smitinandia micrantha* (Thurjo), and *Spiranthes australis*. For cuts and wounds, commonly used species include: *Aerides multiflora* (Sirami Jhaar), *Aerides odorata*, *Coelogyne corymbosa*, *Coelogyne cristata* (Chaandigaabhaa), *Coelogyne porrecta*, *Cymbidium iridioides* (Thur), *Dendrobium heterocarpum*, *Gymnadenia orchidis* (Nakkali Hattajadi), *Habenaria marginata*, *Luisia zeylanica* (Harjona), *Pleione humilis* (Haathi-taauke), *Pleione praecox* (Lasun-paate), *Rhynchostylis retusa* (Ghogegaabhaa), and *Zeuxine strateumatica*. For bone fractures and dislocations, the following orchids are traditionally used: *Bulbophyllum odoratissimum* (Thurjo), *Coelogyne alba*, *Coelogyne fusca*, *Cymbidium aloifolium* (Banharchul), *Dendrobium crepidatum*, *Dendrobium moschatum*, *Dendrobium transparens*, *Papilionanthe teres* (Harjor), *Thunia alba* (Chhade Phool) and *Vanda cristata* (Harchur). A complete list of the Nepalese orchids with medicinal value is provided in Appendix 1.

3.5 Endemic Orchids

Recent taxonomic research has identified 16 endemic orchid species and 2 varieties exclusive to Nepal, including *Bulbophyllum nepalense*, *Calanthe himalaicum*, *Habenaria palpensis*, *Habenaria plantaginea var. nepalensis*, *Habenaria sandiegoensis*, *Herminium fimbriatum*, *Herminium hongdeyuanii*, *Liparis langtangensis*, *Malaxis dolpensis*, *Malaxis monophyllos var. obtusa*, *Malaxis tamurensis*, *Neottia nepalensis*, *Odontochilus nandae*, *Oreorchis porphyranthes*, *Pinalia annapurnensis*, *Pinalia baniae*, *Pinalia pokharensis*, and *Pleione coronaria*. These endemic orchids exhibit extremely limited distributions, with most species known from a single collection event.

3.6 Threats to Nepalese Orchids

Wild orchid populations in Nepal face ongoing and severe threats due to extensive harvesting for ornamental and medicinal purposes. Large quantities of medicinal orchids such as *Anoectochilus roxburghii*, *Dactylorhiza hatagirea*, *Dendrobium*, *Cymbidium*, *Coelogyne* and *Satyrium* species are collected from their natural habitats and exported to neighbouring countries (Pun 2013, 2014; Pant et al. 2018).

Overexploitation has placed many species, especially *Dactylorhiza hatagirea*, *Coelogyne*, *Dendrobium*, and *Pleione*, at high risk of extinction (Pant et al. 2016). Illegal trade is escalating, with a report by Bashyal et al. (2023) documenting 36 confiscated consignments of orchids totalling 28,315 kg between 2010 and 2020. Simultaneously, 49,789 kg of orchids, primarily *Dendrobium* stems, were legally exported from 2008 to 2016, with Thailand receiving 80%, followed by Switzerland (16%) and China (4%). Habitat degradation, especially in subtropical and lower temperate zones below 2000 m, further threatens orchid populations. These regions, heavily used for agriculture, are losing orchid diversity annually. Despite strict legal bans, traders have adopted deceptive practices, disguising orchids as unprotected plants, for instance, *Pleione* species are traded as 'Paani Amalaa' (*Nephrolepis cordifolia*), often semi-processed to evade identification (Rajbhandari and Pandey 2024a). The cumulative impact of trade, habitat loss, and misidentification has pushed many orchid species toward extinction. Conservation has become an urgent priority for Nepal (Rajbhandari et al. 2000; Pant 2014). The ornamental and medicinal species listed in Appendix 1 are now increasingly rare and demand immediate protection.

3.7 Conservation Strategies

Orchids are among Nepal's most treasured natural resources and precious jewels of its rich biodiversity, but their wild populations now face unprecedented threats. Despite their ecological and cultural significance, many orchid species remain understudied and vulnerable to extinction. Systematic research conducted at the National Herbarium, Department of Plant Resources, has documented approximately 270 species of Nepalese orchids (Rajbhandari and Pandey 2024a). However, numerous additional species are housed in herbaria around the world, and a complete national inventory is still incomplete. The conservation of Nepalese orchids demands urgent attention and coordinated efforts from government agencies, private sectors, research institutions, non-governmental organizations, orchid growers, and local communities (Pant and Raskoti 2013). Both *in situ* conservation, protecting orchids in their natural habitats, and *ex situ* approaches such as cultivation and seed banking are essential to safeguard these species for future generations.

3.7.1 In Situ Conservation

While wild orchids in Nepal are declining at an alarming rate, National Parks and Conservation Areas offer promising places for their preservation. For instance, Shakya and Bajracharya (2013) report that orchids in Shivapuri National Park face no major threats due to well-conserved forests and strict protection rule. Field observations support this, making Shivapuri a model for orchid and forest conservation. Similar conservation potential exists across Nepal's protected areas. A recent checklist of seed plants in Khaptad National Park (Rajbhandari et al. 2024a, b) documents 9 orchid species thriving there, including three epiphytes (*Coelogyne articulata*, *Dendrobium eriiflorum*, *Pleione humilis*), five terrestrial species

(*Dactylorhiza hatagirea*, *Habenaria commelinifolia*, *Hemipilia chusua*, *Oreorchis foliosa*, *Platanthera japonica*), and one mycotrophic species (*Neottia listeroides*). To expand conservation efforts, the establishment of Orchid Reserves, both within and beyond protected areas, is crucial. Several orchid-rich sites outside National Parks merit protection, such as Raja Rani in Morang (Bajracharya 2011), Phulchoki in Lalitpur (Samuel 2022), and Panchase in Kaski (Bhandari et al. 2020b).

Nepal's 20 protected areas, including 12 National Parks, 1 Wildlife Reserve, 1 Hunting Reserve, and 6 Conservation Areas, span diverse ecological zones and have significantly contributed to orchid conservation. Complementing these efforts are robust legal frameworks. International trade is regulated under CITES, while domestic legislation such as the Forest Act (2019), National Parks and Wildlife Conservation Act (1973), Environment Protection Act (2019), and Customs Act (2007), along with associated policies and directives, provide essential tools to protect orchids in their natural habitats and monitor their trade.

3.7.2 Ex Situ Conservation

Botanical gardens serve as vital hubs for the *ex situ* conservation of orchid species in Nepal. The Department of Plant Resources, Government of Nepal, has established 11 botanical gardens across various districts. Among them, the Mountain Botanical Garden in Daman (2320 m) conserves around 40 orchid species, primarily from genera such as *Bulbophyllum*, *Coelogyne*, *Dendrobium*, *Eria,* and *Pleione*. Other notable gardens include Mulpani Botanical Garden in Salyan (28 species), Maipokhari Botanical Garden in Ilam (30 species) (Lamichhane 2014), and Dewahariya Botanical Garden in Kailali (17 species) (Shahi et al. 2025). The National Botanical Garden in Godawari, Lalitpur, holds the largest collection with 88 orchid species (Parmar et al. 2024).

The Central Department of Botany at Tribhuvan University (CDB, TU) has initiated germplasm storage and *ex situ* conservation of medicinally and ornamentally significant orchids. Advanced *in vitro* techniques, including immature seed culture, protocorm culture, explant-based propagation, and synthetic seed production, have been applied to over 30 species such as *Aerides odorata*, *Bulbophyllum leopardinum*, *Coelogyne alba*, *Cymbidium aloifolium*, *Dendrobium crepidatum*, *Papilionanthe uniflora*, *Rhynchostylis retusa*, *Thunia alba*, and *Vanda tessellata* (Pant et al. 2019) (Figs. 3.1, 3.2, 3.3, 3.4, 3.5, 3.6, 3.7, 3.8, 3.9, 3.10, 3.11, 3.12, 3.13, 3.14, 3.15, 3.16, 3.17, 3.18, and 3.19).

In addition to these efforts, the Annapurna Research Center (ARC) has played a pivotal role in orchid conservation through tissue culture propagation and acclimatization (Pant et al. 2022). Supported by various organizations and trusts, ARC has successfully reintroduced species like *Aerides odorata*, *Cymbidium aloifolium*, and *Rhynchostylis retusa* to locations including Lumbini, agroforestry lands in Gorkha and Ilam, the Central Zoo in Lalitpur, and the premises of Tribhuvan University.

ARC also maintains a diverse germplasm collection and has achieved successful acclimatization of species such as *Cymbidium* species, *Coelogyne* species,

Fig. 3.1 *Aerides odorata*

Fig. 3.2 *Bulbophyllum guttulatum*

Fig. 3.3 *Coelogyne flaccida*

Fig. 3.4 *Cymbidium elegans*

Dendrobium species, and *Vanda* species, ensuring long-term genetic preservation (Pant et al. 2022).

Community-based initiatives are also gaining momentum. Groups like the Orchid Floriculture Group of Nepal are actively domesticating ornamental orchids including *Dendrobium*, *Cymbidium*, *Calanthe*, and *Vanda* through training programs and knowledge exchange on cultivation under varying environmental conditions. This integrated approach combining scientific propagation, institutional conservation, and community engagement could be a robust framework for sustainable ex situ orchid conservation in Nepal (Fig. 3.20).

Fig. 3.5 *Cymbidium lanceifolium*

Fig. 3.6 *Dactylorhiza hatagirea*

Fig. 3.7 *Dendrobium fimbriatum*

Fig. 3.8 *Dendrobium transparens*

Fig. 3.9 *Eria coronaria*

Fig. 3.10 *Hemipilia chusua*

Fig. 3.11 *Herminium albomarginatum*

Fig. 3.12 *Oberonia anthropophora*

Fig. 3.13 *Papilionanthe teres*

Fig. 3.14 *Pelatantheria insectifera*

3.8 Conclusion

Wild orchid populations in Nepal are facing a serious threat and are rapidly declining at an alarming rate. To mitigate this crisis, strict regulations must be enforced to prohibit the illegal collection of orchids from their natural habitats, except for legitimate research purposes. Conservation efforts should focus on introducing

Fig. 3.15 *Pleione humilis*

Fig. 3.16 *Rhyncostylis retusa*

rare, endangered, and endemic orchid species into botanical gardens and research institutions to safeguard their genetic resources. Simultaneously, the cultivation of commercially valuable wild orchids, both ornamental and medicinal, should be prioritized to reduce pressure on wild populations. However, a significant gap in knowledge remains regarding orchid cultivation technologies. Despite this, commercial orchid farming for cut flowers and potted plants is gaining momentum in Nepal

Fig. 3.17 *Satyrium nepalense*

Fig. 3.18 *Vanda ampullaceal*

Fig. 3.19 In vitro propagation and acclimatization stages of *Cymbidium aloifolium*. (**a**) Longitudinal section of a mature seed pod showing the enclosed seeds. (b) Germination of seeds into the protocorm stage under aseptic culture conditions. (**c**) Initiation of shoot development from the protocorm. (**d**) Formation of complete plantlets with well-developed shoots and roots. (**e**) Acclimatization of plantlets in a controlled environment to adapt them to ex vitro conditions. (**f**) Re-potting of hardened plantlets into polybags, prepared them for reintroduction into natural conditions. (Source: Annapurna Research Center)

and beginning to make its mark in international markets. *Cymbidium* is currently the most widely cultivated genus for commercial purposes. Recognizing the importance of sustainable orchid management, the Government of Nepal has prioritized domestication, research, and cultivation of orchid species. Continued systematic research, particularly in tissue culture techniques, is essential to support both conservation and commercial goals.

Appendix 1. List of Orchids of Nepal

(Source: Species list from Rajbhandari and Rai (2017) and other recent publications. Medicinal use from Pant and Raskoti (2013). (West—from western border to 83° E longitude; Central—from 83° E longitude to 86°30″ E longitude and East—from 86°30″ E longitude to eastern border; Orn.—ornamental; Med.—medicinal).

Fig. 3.20 Hardening of micropropagated orchid species and establishment in green house (upper two) and natural habitat (lower two). Source: Annapurna Research Center

Acampe **Lindl**

Acampe praemorsa (Roxb.) Blatt. & McCann

Epidendrum praemorsum Roxb.

Acampe papillosa (Lindl.) Lindl.

Epiphytic.

Central, East; 200–1200 m.

Distribution: India, Nepal, Bhutan, Bangladesh, Sri Lanka, SE Asia.

Use: Orn. Med.—Root in rheumatism.

Acampe praemorsa (Roxb.) Blatt. & McCann var. **longepedunculata** (Trimen) Govaerts

Acampe wightiana Lindl. var. *longepedunculata* Trimen

A. rigida (Buch.-Ham. ex Sm.) P.F.Hunt

Epiphytic.

Central, East; 200–1500 m.
Distribution: India, Nepal, Bhutan, Bangladesh, Sri Lanka, China, SE Asia.
Use: Orn.
Acanthephippium Blume
Acanthephippium striatum Lindl.
Terrestrial.
Central; 400–1500 m.
Distribution: India, Nepal, China, Japan, SE Asia.
Aerides Lour.
Aerides multiflora Roxb.
Epiphytic.
West, Central, East; 200–1100 m.
Distribution: India, Nepal, Bhutan, Bangladesh, SE Asia.
Use: Orn. Med.—Leaf paste on cuts and wounds.
Aerides odorata Lour.
Epiphytic.
West, Central, East; 200–1200 m.
Distribution: India, Nepal, Bhutan, Bangladesh, China, SE Asia.
Use: Orn. Med.—Leaf paste on cuts and wounds.
Agrostophyllum Blume
Agrostophyllum callosum Rchb.f.
Epiphytic.
Central, East; 2100–2200 m.
Distribution: Nepal, NE India, Bhutan, Bangladesh, China, SE Asia.
Agrostophyllum planicaule (Wall. ex Lindl.) Rchb. f.
Eria planicaulis Wall. ex Lindl.
Epiphytic.
East; 500–800 m.
Distribution: Nepal, NE India, China, SE Asia.
Ania Lindl.
Ania penangiana (Hook.f.) Summerh.
Tainia penangiana Hook.f.
Tainia hookeriana King & Pantl.
Terrestrial.
Central; 700–1700 m.
Distribution: India, Nepal, China, SE Asia, New Guinea.
Anoectochilus Blume
Anoectochilus roxburghii (Wall.) Lindl.
Chrysobaphus roxburghii Wall.
Terrestrial.
Central; 1000–1500 m.
Distribution: India, Nepal, Bhutan, Bangladesh, China, Myanmar, Thailand, Laos, Vietnam.
Use: Med.—Plant to cure tuberculosis.
Anthogonium Wall. ex Lindl.

Anthogonium gracile Wall. ex Lindl.
Terrestrial.
Central, East; 1500–2000 m.
Distribution: Nepal, NE India, Bhutan, Bangladesh, China, Myanmar, Thailand, Cambodia, Laos, Vietnam.
Aphyllorchis Blume
Aphyllorchis alpina King & Pantl.
Terrestrial.
West; 2000–2600 m.
Distribution: India, Nepal, China (Tibet).
Apostasia Blume
Apostasia wallichii R.Br.
Terrestrial.
Central; 600–900 m.
Distribution: Nepal, NE India, Bhutan, Bangladesh, Sri Lanka, China, Japan, SE Asia, New Guinea, Australia.
Arachnis Blume
Arachnis cathcartii (Lindl.) J.J.Sm.
Vanda cathcartii Lindl.
Arachnanthe cathcartii (Lindl.) Benth. & Hook.f.
Esmeralda cathcartii (Lindl.) Rchb.f.
Epiphytic.
600–2000 m.
Distribution: Nepal, NE India, Bhutan, Myanmar.
Use: Orn.
Arachnis clarkei (Rchb.f.) J.J.Sm.
Esmeralda clarkei Rchb.f.
Epiphytic.
East; 1500–1700 m.
Distribution: Nepal, NE India, Bhutan, China, Myanmar, Thailand, Laos, Vietnam.
Use: Orn.
Arundina Blume
Arundina graminifolia (D.Don) Hochr.
Bletia graminifolia D.Don
Terrestrial.
West, Central, East; 400–2300 m.
Distribution: India, Nepal, Bhutan, Bangladesh, Sri Lanka, China, SE Asia, New Guinea.
Use: Orn. Med.—Root to relieve bodyache.

Brachycorythis **Lindl.**
Brachycorythis obcordata (Lindl. ex Wall.) Summerh.
Platanthera obcordata Lindl. ex Wall.
Terrestrial.

West, Central, East; 1000–2600 m.
Distribution: India, Nepal, Bhutan, Bangladesh, Myanmar.
Use: Med.—Tuber as tonic.
Bulbophyllum Thouars
Bulbophyllum affine Wall. ex Lindl.
Epiphytic.
Central, East; 600–1800 m.
Distribution: India, Nepal, Bhutan, China, Japan, Myanmar, Thailand, Cambodia, Laos, Vietnam.
Use: Orn.
Bulbophyllum ambrosia (Hance) Schltr.
Eria ambrosia Hance
Bulbophyllum ambrosia (Hance) Schltr. subsp. *nepalensis* J.J.Wood
Epiphytic.
Central; 1400 m.
Distribution: Nepal, NE India, China, Laos, Vietnam.
Bulbophyllum bisetum Lindl.
Epiphytic.
East; 1500–2000 m.
Distribution: Nepal, NE India, China, Thailand.
Bulbophyllum careyanum (Hook.) Spreng.
Anisopetalon careyanum Hook.
Epiphytic.
Central, East; 600–2100 m.
Distribution: India, Nepal, Bhutan, Bangladesh, China, Myanmar, Thailand, Vietnam.
Use: Orn. Med.—Leaf and pseudobulb paste for skin burn.
Bulbophyllum cariniflorum Rchb.f.
Epiphytic.
Central; 1300–1800 m.
Distribution: India, Nepal, Bhutan, China (Tibet), Myanmar, Vietnam.
Bulbophyllum caudatum Lindl.
Cirrhopetalum caudatum (Lindl.) King & Pantl.
Epiphytic.
Central; 500–2000 m.
Distribution: Nepal, NE India, Bhutan, Bangladesh, China (Tibet), Myanmar.
Bulbophyllum cauliflorum Hook.f.
Phyllorkis cauliflora (Hook.f.) Kuntze
Epiphytic.
Central; 800–1800 m.
Distribution: Nepal, NE India, Bhutan, China (Tibet), Myanmar, Vietnam.
Bulbophyllum cirrhatum (Lindl.) Hook.f.
Ione cirrhata Lindl.
Bulbophyllum purpureofuscum J.J.Verm., Schuit. & de Vogel
Epiphytic.

Central; 1400–2200 m.

Distribution: Nepal, NE India, Bhutan, China, Myanmar.

Bulbophyllum crabro (C.S.P.Parish & Rchb.f.) J.J.Verm., Schut. & De Vogel

Monomeria crabro C.S.P.Parish & Rchb.f.

Monomeria barbata Lindl.

Epiphytic.

Central; 1400–1900 m.

Distribution: Nepal, NE India, China, SE Asia.

Bulbophyllum crassipes Hook.f.

Phyllorkis crassipes (Hook.f.) Kuntze

Epiphytic.

Central; 1000–1200 m.

Distribution: India, Nepal, Bhutan, Bangladesh, China, Myanmar, Thailand, Laos, Vietnam.

Bulbophyllum cylindraceum Wall. ex Lindl.

Epiphytic.

Central; 1600–2400 m.

Distribution: Nepal, NE India, Bhutan, China, Myanmar, Thailand.

Bulbophyllum elatum (Hook.f.) J.J.Sm.

Cirrhopetalum elatum Hook.f.

Epiphytic.

Central; 900–2300 m.

Distribution: Nepal, NE India, Bhutan, China, Myanmar, Vietnam.

Use: Orn.

Bulbophyllum emarginatum (Finet) J.J.Sm.

Cirrhopetalum emarginatum Finet

Epiphytic.

East; 1700–2600 m.

Distribution: Nepal, NE India, Bhutan, China, Myanmar, Vietnam.

Bulbophyllum eublepharum Rchb.f.

Epiphytic.

Central; 2000–2500 m.

Distribution: Nepal, NE India, Bhutan, China, Myanmar.

Bulbophyllum gamblei (Hook.f.) Hook.f.

Bulbophyllum leptanthum var. *gamblei* Hook.f.

Epiphytic.

Central; 1800–2400 m.

Distribution: Nepal, NE India, Bhutan, Bangladesh, China (Tibet), Vietnam.

Bulbophyllum griffithii (Lindl.) Rchb.f.

Sarcopodium griffithii Lindl.

Epiphytic.

Central; 240–2600 m.

Distribution: Nepal, NE India, Bhutan, China, Thailand, Vietnam.

Bulbophyllum guttulatum (Hook.f.) N.P.Balakrishnan

Cirrhopetalum guttulatum Hook.f.

Epiphytic.
Central, East; 1000–2300 m.
Distribution: Nepal, NE India, Bhutan, China, Myanmar, Vietnam.

Bulbophyllum helenae (Kuntze) J.J.Sm.
Phyllorkis helenae Kuntze
Epiphytic.
Central; 900–1200 m.
Distribution: India, Nepal, Bhutan, China, Myanmar, Thailand, Laos, Vietnam.

Bulbophyllum hirtum (Sm.) Lindl.
Stelis hirta Sm.
Epiphytic.
Central; 1000–2600 m.
Distribution: India, Nepal, Bhutan, Bangladesh, China, Myanmar, Thailand, Vietnam.

Bulbophyllum khaoyaiense Seidenf.
Tripudianthes khaoyaiensis (Seidenf.) Szlach. & Kras
Epiphytic.
Central; 1100–1500 m.
Distribution: Nepal, China, Myanmar.

Bulbophyllum kingii Hook.f.
Acrochaene punctata Lindl.
Epiphytic.
East; 1500–1700 m.
Distribution: Nepal, NE India, Myanmar, Thailand, Laos.

Bulbophyllum leopardinum (Wall.) Lindl.
Dendrobium leopardinum Wall.
Epiphytic.
Central, East; 1400–3200 m.
Distribution: India, Nepal, Bhutan, China, Myanmar, Thailand, Laos, Vietnam.
Use: Orn. Med.—Leaf and pseudobulb paste for skin burn.

Bulbophyllum moniliforme C.S.P.Parish & Rchb.f.
Epiphytic.
Central; 1400–1600 m.
Distribution: Nepal, NE India, Myanmar, Thailand, Cambodia, Laos, Vietnam.

Bulbophyllum muscicola Rchb.f.
Bulbophyllum hookeri (Duthie) J.J.Sm.
Epiphytic.
Central; 2100–2500 m.
Distribution: India, Nepal, Bhutan, Vietnam.

Bulbophyllum nepalense Raskoti & Ale
Epiphytic.
Central; 1600–2300 m.
Distribution: Endemic to Nepal.

Bulbophyllum odoratissimum (Sm.) Lindl.
Stelis odoratissima Sm.

Epiphytic.
Central; 1000–2000 m.
Distribution: India, Nepal, Bhutan, China, Myanmar, Thailand, Cambodia, Laos, Vietnam.
Use: Orn. Med.—Plant to treat tuberculosis and fracture.

Bulbophyllum polyrrhizum Lindl.
Epiphytic.
Central, East; 900–2100 m.
Distribution: India, Nepal, China, Myanmar, Thailand.

Bulbophyllum raskotii J.J.Verm., Schuit. & de Vogel
Sunipia nepalensis Raskoti & R.Ale
Epiphyte.
Central; 2400 m.
Distribution: Nepal, NE India.

Bulbophyllum refractum (Zoll.) Rchb.f.
Cirrhopetalum refractum Zoll.
Epiphytic.
Central; 1400–1700 m.
Distribution: India, Nepal, Bhutan, SE Asia.

Bulbophyllum reptans (Lindl.) Lindl.
Tribrachia reptans Lindl.
Epiphytic.
West, Central, East; 1300–2500 m.
Distribution: India, Nepal, Bhutan, Bangladesh, China, Myanmar, Thailand, Laos, Vietnam.

Bulbophyllum retusiusculum Rchb.f.
Epiphytic.
Central, East; 2100–3000 m.
Distribution: Nepal, NE India, Bhutan, China, SE Asia.

Bulbophyllum rigidum King & Pantl.
Epiphytic.
Central; 1000–2000 m.
Distribution: Nepal, NE India.

Bulbophyllum rolfei (Kuntze) Seidenf.
Phyllorchis rolfei Kuntze
Bulbophyllum dyerianum (King & Pantl.) Seidenf.
Epiphytic.
Central, East; 1800–2600 m.
Distribution: Nepal, NE India, Bhutan, China, Myanmar.

Bulbophyllum roseopictum J.J.Verm.
Ione bicolor (Lindl.) Lindl.
Sunipia bicolor Lindl.
Epiphytic.
West, Central, East; 1500–2500 m.

Distribution: India, Nepal, Bhutan, Bangladesh, China, Myanmar, Thailand, Vietnam.

Bulbophyllum roxburghii (Lindl.) Rchb.f.

Cirrhopetalum roxburghii Lindl.

Epiphytic.

East; 300–600 m.

Distribution: Nepal, NE India, Bangladesh.

Bulbophyllum sarcophyllum (King & Pantl.) J.J.Sm.

Cirrhopetalum sarcophyllum King & Pantl.

Epiphytic.

Central; 300 m.

Distribution: Nepal, NE India, Bhutan, Myanmar.

Bulbophyllum scabratum Rchb.f.

Epiphytic.

East; 2000–2200 m.

Distribution: Nepal, NE India, Bhutan, Bangladesh, China, Myanmar, Thailand, Laos, Vietnam.

Bulbophyllum secundum Hook.f.

Epiphytic.

Central; 1100–2200 m.

Distribution: India, Nepal, China, Myanmar, Thailand, Vietnam.

Bulbophyllum sterile (Lam.) Suresh

Epidendrum sterile Lam.

Epiphytic.

Central, East; 500–2100 m.

Distribution: India, Nepal, Bangladesh.

Bulbophyllum striatum (Griff.) Rchb.f.

Dendrobium striatum Griff.

Epiphytic.

Central; 1400–2300 m.

Distribution: Nepal, N.E.India, Bhutan, Bangladesh, China, Thailand, Myanmar, Vietnam.

Bulbophyllum sunipia J. J. Verm., Schuit. & de Vogel

Sunipia racemosa (Sm.) Tang & Wang

Sunipia scariosa Lindl.

Epiphytic.

Central, East; 1200–1800 m.

Distribution: Nepal, NE India, China, Myanmar, Thailand, Laos, Vietnam.

Bulbophyllum triste Rchb.f.

Epiphytic.

Central, East; 800–2300 m.

Distribution: India, Nepal, Bhutan, China, Thailand, Myanmar.

Bulbophyllum umbellatum Lindl.

Epiphytic.

Central, East; 300–2000 m.

Distribution: India, Nepal, Bhutan, China, Myanmar, Thailand, Laos, Vietnam.
Use: Orn.
Bulbophyllum viridiflorum (Hook.f.) Schltr.
Cirrhopetalum viridiflorum Hook.f.
Epiphytic.
Central, East; 1000–2600 m.
Distribution: India, Nepal.
Bulbophyllum wallichii (Lindl.) Rchb.f.
Cirrhopetalum wallichii Lindl.
Epiphytic.
Central, East; 1000–2500 m.
Distribution: India, Nepal, Bhutan, Myanmar, Thailand, Laos, Vietnam.
Bulbophyllum yoksunense J.J.Sm.
Cirrhopetalum brevipes Hook.f.
piphytic.
Central; 2000–2600 m.
Distribution: India, Nepal, Bhutan, China, Myanmar.
Bulbophyllum yunnanense Rolfe
Bulbophyllum otoglossum Tuyama
Epiphytic.
East; 1400–2800 m.
Distribution: Nepal, NE India, Bhutan, China.

Calanthe **R.Br.**
Calanthe alpina Hook.f. ex Lindl.
Terrestrial.
West, Central, East; 1900–4000 m.
Distribution: India, Nepal, Bhutan, Myanmar.
Calanthe arcuata Rolfe
Terrestrial.
Central; 2000–2100 m.
Distribution: Nepal, China, Myanmar.
Calanthe brevicornu Lindl.
Terrestrial.
Central, East; 1500–2600 m.
Distribution: India, Nepal, Bhutan, China, Myanmar.
Use: Orn.
Calanthe chloroleuca Lindl.
Terrestrial.
East; 1500–3000 m.
Distribution: Nepal, NE India, Bhutan.
Calanthe davidii Franchet
Terrestrial.
Central; 2000–2100 m.
Distribution: India, Nepal, Bhutan, China, Japan, Vietnam.

Calanthe densiflora Lindl.
Terrestrial.
Central; 1400 m.
Distribution: Nepal, NE India, Bhutan, Bangladesh, China, Japan, Myanmar, Thailand, Vietnam.

Calanthe griffithii Lindl.
Terrestrial.
East; 2200–2300 m.
Distribution: India, Nepal, Bhutan, China, Myanmar.

Calanthe himalaicum Raskoti
Terrestrial.
West; 2200 m.
Distribution: Endemic to Nepal.

Calanthe longipes Hook.f.
Terrestrial.
Central; 1500–1600 m.
Distribution: Nepal, NE India, Bhutan, China, SE Asia.

Calanthe mannii Hook.f.
Terrestrial.
Central, East; 1500–2500 m.
Distribution: Nepal, NE India, Bhutan, China, Japan, Myanmar, Vietnam.

Calanthe masuca (D.Don) Lindl.
Bletia masuca D.Don
Calanthe sylvatica (Thouars) Lindl.
Terrestrial.
West, Central; 1500–2800 m.
Distribution: Madagascar, India, Nepal, Bhutan, Sri Lanka, China, Japan, SE Asia.

Calanthe metoensis Z.H.Tsi & K.Y.Lang
Terrestrial.
East; 2000 m.
Distribution: Nepal, China.

Calanthe plantaginea Lindl.
Terrestrial.
West, Central; 1500–2500 m.
Distribution: India, Nepal, Bhutan, China.
Use: Orn. Med.—Rhizome as tonic.

Calanthe puberula Lindl.
Terrestrial.
Central, East; 1600–3000 m.
Distribution: India, Nepal, Bhutan, Bangladesh, E Asia, Myanmar, Thailand, Vietnam.
Med.—Rhizome as antipyretic.

Calanthe tankervilleae (Banks) M.W.Chase, Christenh. & Schuit.
Limodorum tankervilleae Banks

Phaius tankervilleae (Banks) Blume
Terrestrial.
Central, East; 700–1800 m.
Distribution: India, Nepal, Bhutan, Bangladesh, Sri Lanka, China, Japan, SE Asia, New Guinea, Australia.
Use: Orn. Med.—Tuber for tonic.
Calanthe tricarinata Lindl.
Terrestrial.
West, Central; 1500–3200 m.
Distribution: Pakistan, India, Nepal, Bhutan, E Asia, Myanmar, Thailand.
Use: Orn. Med.—Leaf paste applied on sores.
Calanthe trulliformis King & Pantl.
Terrestrial.
Central; 2000–2600 m.
Distribution: Nepal, NE India, Bhutan, Myanmar.
Calanthe woodfordii (Hook.) M.W.Chase, Christenh. & Schuit.
Bletia woodfordii Hook.
Phaius flavus (Blume) Lindl.
Limodorum flavum Blume
Terrestrial.
Central; 200–2000 m.
Distribution: India, Nepal, Bhutan, China, Japan, SE Asia, New Guinea.
Use: Orn.
Callostylis Blume
Callostylis rigida Blume
Eria discolor Lindl.
Epiphytic.
East; 150–1700 m.
Distribution: Nepal, NE India, Bhutan, China, SE Asia
Cephalanthera Rich.
Cephlanthera erecta (Thunb.) Blume
Serapias erecta Thunb.
Terrestrial.
West, Central; 2000–2300 m.
Distribution: Nepal, NE India, E Asia.
Cephalanthera longifolia (L.) Fritsch
Serapias helleborine L. var. *longifolia* L.
Terrestrial.
West, Central; 1200–3400 m.
Distribution: Europe, Africa, N, W & C Asia, Afghanistan, Pakistan, India, Nepal, Bhutan, China, Japan, Myanmar.
Use: Med.—Rhizome as tonic and healing wound.
Ceratostylis Blume
Ceratostylis himalaica Hook.f.
Epiphytic.

East; 600–1900 m.

Distribution: Nepal, NE India, Bhutan, China, Myanmar, Laos, Vietnam.

Cheirostylis Blume

Cheirostylis griffithii Lindl.

Terrestrial.

Cenral, East; 900–2200 m.

Distribution: Pakistan, India, Nepal, Bangladesh, China, Myanmar, Thailand, Laos.

Chiloschista Lindl.

Chiloschista parishii Seidenf.

Chiloschista lunifera (Rchb.f.) J.J.Sm. ex Holttum

Epiphytic.

Central, East; 200–2400 m.

Distribution: Nepal, NE India, Bhutan, China, Myanmar, Thailand, Laos, Vietnam.

Chiloschista usneoides (D.Don) Lindl.

Epidendrum usneoides D.Don

Epiphytic.

Central; 900–2500 m.

Distribution: India, Nepal, Bhutan, Myanmar.

Chrysoglossum Blume

Chrysoglossum ornatum Blume

Terrestrial.

Central; 1200–1500 m.

Distribution: Nepal, NE India, Bhutan, Sri Lanka, China, SE Asia, New Guinea.

Cleisostoma Blume

Cleisostoma aspersum (Rchb.f.) Garay

Sarcanthus aspersus Rchb.f.

Epiphytic.

East; 600–1800 m.

Distribution: India, Nepal, Bhutan, China, Myanmar, Thailand, Vietnam.

Cleisostoma filiforme (Lindl.) Garay

Sarcanthus filiformis Lindl.

Epiphytic.

Central, East; 150–1500 m.

Distribution: Nepal, NE India, Bangladesh, China, SE Asia.

Cleisostoma racemiferum (Lindl.) Garay

Saccolabium racemiferum Lindl.

Epiphytic.

Cenral, East; 1100–2000 m.

Distribution: Nepal, NE India, Bhutan, China, Myanmar, Thailand, Laos, Vietnam.

Cleisostoma simondii (Gagnepain) Seidenf.

Vanda simondii Gagnepain

Epiphytic.

Central, East; 400–900 m.

Distribution: Nepal, NE India, Bhutan, Bangladesh, China, Myanmar, Thailand, Laos, Vietnam.

Coelogyne Lindl.

Coelogyne alba (Lindl.) Rchb.f.

Otochilus albus Lindl.

Epiphytic.

Central, East; 1500–2400 m.

Distribution: Nepal, NE India, China, Myanmar, Thailand, Vietnam.

Use: Orn. Med.—Pseudobulb in fracture.

Coelogyne articulata (Lindl.) Rchb.f.

Pholidota articulata Lindl.

Pholidota griffithii Hook.f.

Epiphytic.

West, Central, East; 500–2300 m.

Distribution: India, Nepal, Bhutan, Bangladesh, China, SE Asia.

Use: Orn. Med.—Plant as tonic.

Coelogyne corymbosa Lindl.

Epiphytic.

Central, East; 1500–2900 m.

Distribution: Nepal, NE India, Bhutan, China, Myanmar.

Use: Orn. Med.—Pseudobulb paste applied to relieve headache.

Coelogyne cristata Lindl.

Epiphytic.

West, Central, East; 1000–2600 m.

Distribution: India, Nepal, Bhutan, Bangladesh, China.

Use: Orn. Med.—Pseudobulb juice applied in wounds.

Coelogyne demissa (D.Don) M.W.Chase & Schuit.

Dendrobium demissum D.Don

Panisea demissa (D.Don) Pfitz.

Epiphytic.

Central, East; 1500–2400 m.

Distribution: Nepal, NE India, Bhutan, China, Myanmar, Thailand, Laos, Vietnam.

Use: Orn.

Coelogyne fimbriata Lindl.

Epiphytic.

Central, East; 600–2600 m.

Distribution: Nepal, NE India, Bhutan, China, SE Asia.

Use: Orn.

Coelogyne flaccida Lindl.

Epiphytic.

Central; 900–1400 m.

Distribution: Nepal, NE India, Bhutan, China, Myanmar, Thailand, Laos, Vietnam.

Use: Orn. Med.—Pseudobulb paste applied to treat headache.

Coelogyne fusca (Lindl.) Rchb.f.

Otochilus fuscus Lindl.

Epiphytic.

Central, East; 1100–2000 m.

Distribution: Nepal, NE India, Bhutan, Bangladesh, China (Yunnan), Myanmar, Thailand, Cambodia, Vietnam.

Use: Orn. Med.—Pseudobulb in fracture.

Coelogyne fuscescens Lindl.

Epiphytic.

Central, East; 1000–1900 m.

Distribution: Nepal, NE India, Bhutan, Bangladesh, Myanmar, Thailand, Laos, Vietnam.

Use: Orn. Med.—Pseudobulb paste applied for abdominal pain.

Coelogyne gardneriana Lindl.

Neogyna gardneriana (Lindl.) Rchb.f.

Epiphytic.

Central; 1500–3000 m.

Distribution: Nepal, NE India, Bhutan, Bangladesh, China, Myanmar, Thailand, Laos, Vietnam.

Coelogyne imbricata (Hook.) Rchb.f.

Pholidota imbricata Hook.

Epiphytic.

Central, East; 600–2900 m.

Distribution: India, Nepal, Bhutan, Bangladesh, Sri Lanka, China, SE Asia, Australia.

Use: Med.—Plant as tonic.

Coelogyne lancilabia (Seidenf.) R.Rice

Otochilus lancilabius Seidenf.

Epiphytic.

Central, East; 1400–2500 m.

Distribution: Nepal, NE. India, Bhutan China, Laos, Vietnam.

Use: Orn.

Coelogyne longipes Lindl.

Epiphytic.

Central, East; 1500–2300 m.

Distribution: Nepal, NE India, Bhutan, China, Myanmar, Thailand, Laos.

Coelogyne nitida (Wall. ex D.Don) Lindl.

Cymbidium nitidum Wall. ex D.Don

Epiphytic.

Central, East; 1300–2400 m.

Distribution: Nepal, NE India, Bhutan, Bangladesh, China, Myanmar, Laos, Vietnam.

Use: Orn. Med.—Pseudobulb juice for stomachache.

Coelogyne ovalis Lindl.

Epiphytic.
West, Central; 1300–1700 m.
Distribution: India, Nepal, Bhutan, China, Myanmar, Thailand, Laos, Vietnam.
Use: Orn. Med.—Pseudobulb for aphrodisiac.

Coelogyne pallida (Lindl.) Rchb.f.
Pholidota pallida Lindl.
Epiphytic.
Central; 1000–2300 m.
Distribution: Nepal, NE India, China, Myanmar, S. E. Asia.
Use: Med.—Pseudobulb and roots applied to relieve abdominal and rheumatic pain.

Coelogyne panchaseensis (Subedi) M.W.Chase & Schuit.
Panisea panchaseensis Subedi
Epiphytic.
Central; 2200–2400 m.
Distribution: Nepal, NE India (Nagaland).

Coelogyne porrecta (Lindl.) Rchb.f.
Otochilus porrectus Lindl.
Epiphytic.
West, Central, East; 1800–2400 m.
Distribution: India, Nepal, China, Myanmar, Thailand, Laos, Vietnam.
Use: Med.—Pseudobulb as tonic.

Coelogyne prolifera Lindl.
Coelogyne flavida Wall. ex Lindl.
Epiphytic.
Central; 1000–2300 m.
Distribution: Nepal, NE India, Bhutan, China, Myamnar, Thailand, Laos, Vietnam.
Use: Med.—Pseudobulb paste for boils and bachache.

Coelogyne protracta (Hook.f.) R.Rice
Pholidota protracta Hook.f.
Epiphytic.
Central, East; 500–2200 m.
Distribution: Nepal, NE India, Bhutan, China, Myanmar.
Use: Orn.

Coelogyne punctulata Lindl.
Epiphytic.
Central; 200–2900 m.
Distribution: Nepal, NE India, Bhutan, Bangladesh, China, Myanmar, Thailand, Vietnam.

Coelogyne recurva (Lindl.) Rchb.f.
Pholidota recurva Lindl.
Epiphytic.
Central; 700–1800 m.
Distribution: Nepal, NE India, China, SE Asia.

Coelogyne stricta (D.Don) Schltr.
Coelogyne elata Lindl.
Epiphytic.
Central, East; 1200–2200 m.
Distribution: India, Nepal, Bhutan, China, Myanmar, Laos, Vietnam.
Use: Orn. Med.—Pseudobulb paste to relieve headache and fever.
Coelogyne uniflora Lindl.
Panisea uniflora (Lindl.) Lindl.
Epiphytic.
Central, East; 1000–2300 m.
Distribution: India, Nepal, Bhutan, China, SE Asia.
Use: Orn.
Corallorhiza Gagnebin
Corallorhiza trifida Chatelain
Mycotrophic, leafless.
West, Central; 3800–3900 m.
Distribution: Europe, C Asia, Russia, Pakistan, India, Nepal, E Asia, N America.
Cremastra Lindl.
Cremastra appendiculata (D.Don) Makino
Cymbidium appendiculatum D.Don
Terrestrial.
Central; 1800–2300 m.
Distribution: India, Nepal, Bhutan, China, Japan, Laos, Vietnam.
Crepidium Blume
Crepidium acuminatum (D.Don) Szlach.
Malaxis acuminata D.Don
Terrestrial.
West, Central, East; 400–3000 m.
Distribution: India, Nepal, Bhutan, Bangladesh, Sri Lanka, China, SE Asia.
Crepidium biauritum (Lindl.) Szlach.
Microstylis biaurita Lindl.
Malaxis biaurita (Lindl.) Kuntze
Terrestrial.
Central; 800–1400 m.
Distribution: India, Nepal, Bangladesh, China, Myanmar, Thailand, Laos.
Crepidium khasianum (Hook.f.) Szlach.
Microstylis khasiana Hook.f.
Malaxis khasiana (Hook.f.) Kuntze
Terrestrial.
Central; 1700–1800 m.
Distribution: Nepal, NE India, China, Myanmar, Thailand.
Crepidium purpureum (Lindl.) Szlach.
Microstylis purpurea Lindl.
Terrestrial.
Central; 800–1200 m.

Distribution: India, Nepal, Sri Lanka, China, SE Asia.

Cryptochilus Wall.

Cryptochilus acuminatus (Griff.) Schuit., Y.P.Ng & H.A.Pedersen

Xiphosium acuminatum Griff.

Eria carinata Gibs. ex Lindl.

Pinalia carinata (Gibs. ex Lindl.) Kuntze

Epiphytic.

Central; 1000–2000 m.

Distribution: Nepal, NE India, Bhutan, Thailand.

Cryptochilus luteus Lindl.

Epiphytic.

West, Central, East: 1200–2300 m.

Distribution: India, Nepal, Bhutan, Bangladesh, China, SE Asia.

Cryptochilus sanguineus Wall.

Epiphytic.

Central, East; 1000–2400 m.

Distribution: Nepal, NE India, Bhutan, China, Myanmar, Thailand.

Cryptochilus strictus (Lindl.) Schuit., Y.P.Ng & H.A.Pedersen

Eria stricta Lindl.

Mycaranthes stricta (Lindl.) Lindl.

Pinalia stricta (Lindl.) Kuntze

Epiphytic.

East; 300–1800 m.

Distribution: Nepal, NE India, Bhutan, Bangladesh, China, Myanmar, Thailand, Vietnam.

Cylindrolobus Blume

Cylindrolobus biflorus (Griff.) Rauschert

Eria biflora Griff.

Epiphytic.

Central; 300–500 m.

Distribution: Nepal, NE India, Bhutan, SE Asia.

Cymbidium Sw.

Cymbidium aloifolium (L.) Sw.

Epidendrum aloifolium L.

Cymbidium pendulum (Roxb.) Sw.

Epiphytic.

West, Central, East; 300–2400 m.

Distribution: India, Nepal, Bhutan, Bangladesh, Sri Lanka, China, SE Asia.

Use: Orn. Med.—Rhizome paste for fracture and dislocated bones.

Cymbidium crassifolium Herb.

Cymbidium mannii Rchb.f.

Epiphytic.

Central, East; 900–1500 m.

Distribution: India, Nepal, Bhutan, Bangladesh, China, SE Asia.

Cymbidium cyperifolium Wall. ex Lindl.

Epiphytic.
Central; 1500–2500 m.
Distribution: Nepal, NE India, Bhutan, Bangladesh, China, SE Asia.
Cymbidium devonianum Paxton
Epiphytic.
East; 1500–1800 m.
Distribution: Nepal, NE India, Bhutan, China, Myanmar, Thailand, Vietnam.
Use: Orn. Med.—Root paste applied to treat boils.
Cymbidium eburneum Lindl.
Epiphytic.
Central; 1000–1500 m.
Distribution: Nepal, NE India, Bhutan, China, Myanmar, Vietnam.
Cymbidium elegans Lindl.
    Cymbidium longifolium D.Don
Epiphytic.
Central; 1500–2800 m.
Distribution: Nepal, NE India, Bhutan, China, Myanmar.
Use: Orn. Med.—Leaves in demulcent.
Cymbidium erythraeum Lindl.
Epiphytic.
Central, East; 1500–2400 m.
Distribution: India, Nepal, Bhutan, China, Myanmar, Vietnam.
Cymbidium gammieanum King & Pantl.
Epiphytic.
Central; 1500–2400 m.
Distribution: Nepal, NE India.
Cymbidium hookerianum Rchb.f.
Epiphytic.
East; 1500–2600 m.
Distribution: Nepal, NE India, Bhutan, China, Myanmar, Thailand, Vietnam.
Use: Orn.
Cymbidium iridioides D.Don
Epiphytic.
Central, East; 1200–2800 m.
Distribution: India, Nepal, Bhutan, China, Myanmar, Vietnam.
Use: Orn., Med.—Leaf juice applied in deep wound to stop bleeding.
Cymbidium lancifolium Hook.
Terrestrial.
Central; 1500–2400 m.
Distribution: India, Nepal, Bhutan, Bangladesh, E & SE Asia, New Guinea.
Use: Orn.
Cymbilabia D.K.Liu & Ming H.Li
Cymbilabia undulata (Lindl.) D.K.Liu & Ming HLi
    Vanda undulata Lindl.
    Vandopsis undulata (Lindl.) J J Sm.

Epiphytic.
West, Central, East; 200–2400 m.
Distribution: India, Nepal, Bhutan, China, Myanmar, Vietnam.
Cypripedium L.
Cypripedium cordigerum D.Don
Terrestrial.
West, Central; 2100–4000 m.
Distribution: Pakistan, India, Nepal, Bhutan, China (Tibet).
Use: Orn. Med.—Root in tonic.
Cypripedium elegans Reichbenb.f.
Terrestrial.
Central; 3300–4200 m.
Distribution: India, Nepal, Bhutan, China (Tibet).
Use: Orn. Med.—Root in tonic.
Cypripedium himalaicum Rolfe
Terrestrial.
West, Central, East; 2800–4800 m.
Distribution: India, Nepal, Bhutan, China (Tibet).
Use: Orn. Med.—Plant for chest disorder and cough.
Cyrtosia Blume
Cyrtosia lindleyana Hook.f. & Thomson
Galeola lindleyana (Hook.f. & Thomson) Rchb.f.
Mycotrophic.
West, East; 1300–2800 m.
Distribution: Nepal, NE India, Bhutan, China, Vietnam, Indonesia.

Dactylorhiza **Necker ex Nevski**
Dactylorhiza hatagirea (D.Don) Soo
Orchis hatagirea D.Don
Terrestrial.
West, Central, East; 2800–4000 m.
Distribution: Pakistan, India, Nepal, Bhutan, China (Tibet).
Use: Med.—Tuber for tonic, wound healing.
Dendrobium Sw.
Dendrobium amoenum Wall. ex Lindl.
Epiphytic.
Central, East; 200–2900 m.
Distribution: India, Nepal, Bhutan.
Use: Orn. Med.—Stem for tonic.
Dendrobium amplum Lindl.
Epigeneium amplum (Lindl.) Summerh.
Epiphytic.
Central, East; 1300–2100 m.
Distribution: Nepal, NE India, Bhutan, Bangladesh, China, Myanmar, Thailand, Laos, Vietnam.

Use: Orn.
Dendrobium anceps Sw.
Aporum anceps (Sw.) Lindl.
Epiphytic.
East; 200–1400 m.
Distribution: Nepal, NE India, Bhutan, SE Asia.
Dendrobium aphyllum (Roxb.) C.E.C.Fisch.
Limodorum aphyllum Roxb.
Epiphytic.
Central, East; 200–1500 m.
Distribution: India, Nepal, Bhutan, China, SE Asia.
Use: Orn.
Dendrobium bicameratum Lindl.
Epiphytic.
West, Central; 1400–2400 m.
Distribution: India, Nepal, Bhutan, Myanmar.
Dendrobium chrysanthum Wall. ex Lindl.
Epiphytic.
Central, East; 1300–2000 m.
Distribution: India, Nepal, Bhutan, China, Myanmar, Thailand, Laos, Vietnam.
Dendrobium chryseum Rolfe
Epiphytic.
Central; 1200–2400 m.
Distribution: Nepal, NE India, China, Myanmar.
Use: Orn.
Dendrobium crepidatum Lindl. & Paxton
Epiphytic.
Central; 600–1400 m.
Distribution: India, Nepal, China, Myanmar, Thailand, Laos, Vietnam.
Use: Orn. Med.—Stem in fracture.
Dendrobium darjeelingense Pradhan
Epiphytic.
Central; 1500–2000 m.
Distribution: Nepal, NE India.
Dendrobium denneanum Kerr
Epiphytic.
Central; 600–2500 m.
Distribution: India, Nepal.
Dendrobium densiflorum Lindl.
Epiphytic.
Central, East; 300–2900 m.
Distribution: India, Nepal, Bhutan, China, Myanmar, Thailand.
Use: Orn. Med.—Pseudobulb in tonic, boils and pimples.
Dendrobium denudans D.Don
. Epiphytic.

Central, East; 1000–2200 m.
Distribution: India, Nepal, Bhutan.
Dendrobium eriiflorum Griff.
Epiphytic.
West, Central, East; 1400–2100 m.
Distribution: India, Nepal, Bhutan, Myanmar.
Use: Orn. Med.—Plant in tonic.
Dendrobium fimbriatum Hook.
Epiphytic.
Central, East; 200–2100 m.
Distribution: India, Nepal, Bhutan, Bangladesh, China, SE Asia.
Use: Orn. Med.—Plant in nervous debility.
Dendrobium formosum Roxb. ex Lindl.
Epiphytic.
Central, East; 500–1500 m.
Distribution: Nepal, NE India, Bangladesh, Sri Lanka, Myanmar, Thailand.
Use: Orn.
Dendrobium fugax Rchb.f.
Flickingeria fugax (Rchb.f.) Seidenf.
Epiphytic.
Central, East; 500–2000 m.
Distribution: Nepal, NE India, Bhutan, Myanmar, Thailand.
Use: Med.—Plant for tonic and stem in asthma, bronchitis, throat trouble and fever.
Dendrobium fuscescens Griff.
Epigeneium fuscescens (Griff.) Summerh.
Epiphytic.
East; 600–1800 m.
Distribution: Nepal, NE India, Bhutan, Bangladesh, China, Myanmar, Laos.
Use: Orn.
Dendrobium gibsonii Paxton
Epiphytic.
East; 900–2000 m.
Distribution: Nepal, NE India, Bhutan, China, Myanmar, Thailand, Laos, Vietnam.
Dendrobium heterocarpum Wall. ex Lindl.
Epiphytic.
Central, East; 1000–2200 m.
Distribution: India, Nepal, Bhutan, Sri Lanka, China, SE Asia.
Use: Orn. Med.—Stem paste in fracture and dislocated bone.
Dendrobium hookerinum Lindl.
Epiphytic.
East; 1800–1900 m.
Distribution: Nepal, NE India, Bhutan, Bangladesh, China (Tibet), Myanmar.
Use: Orn.

Dendrobium longicornu Lindl.
Epiphytic.
Central, East; 1300–3000 m.
Distribution: India, Nepal, Bhutan, China, Myanmar.
Use: Orn. Med.—Plant juice to relieve fever.
Dendrobium moniliforme (L.) Sw.
Epidendrum moniliforme L.
Dendrobium candidum Wall. ex Lindl.
Epiphytic.
Central, East; 1500–2700 m.
Distribution: India, Nepal, Bhutan, China, Korea, Japan, Myanmar, Vietnam.
Dendrobium monticola P.F.Hunt & Summerh.
Dendrobium alpestre Royle
Epiphytic.
Central; 1500–2700 m.
Distribution: India, Nepal.
Use: Orn. Med.—Pseudobulbs in boils and pimples.
Dendrobium moschatum (Banks) Sw.
Epidendrum moschatum Banks
Epiphytic.
Central, East; 200–1200 m.
Distribution: India, Nepal, Bhutan, Bangladesh, China, Myanmar, Thailand, Laos, Vietnam.
Use: Orn. Med.—Pseudobulb n fracture and dislocated bone.
Dendrobium nobile Lindl.
Epiphytic.
East; 300–2000 m.
Distribution: India, Nepal, Bhutan, China, Myanmar, Thailand.
Use: Orn. Med.—Stem in tonic.
Dendrobium plicatile Lindl.
Epiphytic.
East; 300–2400 m.
Distribution: India, Nepal, China, SE Asia.
Dendrobium polyanthum Wall. ex Lindl.
Dendrobium primulinum Lindl.
Epiphytic.
West, Central; 400–1400 m.
Distribution: India, Nepal, Myanmar.
Dendrobium porphyrochilum Lindl.
Epiphytic.
Central, East; 1800–2600 m.
Distribution: India, Nepal, China, North Myanmar, Thailand.
Dendrobium rotundatum (Lindl.) Hook.f.
Sarcopodium rotundatum Lindl.
Epiphytic.

Central; 1300–2500 m.
Distribution: Nepal, NE India, China, Myanmar. Use: Orn.
Dendrobium transparens Wall. ex Lindl.
Epiphytic.
Central; 700–2000 m.
Distribution: India, Nepal, Bhutan, Bangladesh, Myanmar.
Use: Orn., Med.—Pseudobulb paste and stem in fracture and dislocated bone.
Dendrolirium Blume
Dendrolirium kamlangensis (A.N.Rao) A.N.Rao
Eria kamlangensis A.N.Rao
Eria nepalensis Bajrach. & K.K.Shrestha
Epiphytic.
Central; 500 m.
Distribution: Nepal, NE India (Arunachal Pradesh).
Dendrolirium lasiopetalum (Willd.) S.C.Chen & J. J. Wood
Aerides lasiopetala Willd.
Eria lasiopetala (Willd.) Ormerod
Epiphytic.
West, Central, East; 500–1600 m.
Distribution: India, Nepal, Bhutan, Bangladesh, China, SE Asia.
Didymoplexis Griff.
Didymoplexis pallens Griff.
Mycotrophic.
East; 150 m.
Distribution: Afghanistan, Nepal, NE India, Bhutan, Bangladesh, China, SEAsia, New Guinea, Australia.
Dienia Lindl.
Dienia cylindrostachya Lindl.
Malaxis cylindrostachya (Lindl.) Kuntze
Terrestrial.
West, Central, East; 2000–3500 m.
Distribution: India, Nepal, Bhutan, China (Tibet).
Med.—Pseudobulb for tonic.
Dienia ophrydis (J.Koenig) Ormerod & Seidenf.
Epidendrum ophrydis J.Koenig
Malaxis latifolia Sm.
Terrestrial.
Central; 900–2100 m.
Distribution: India, Nepal, Bhutan, Sri Lanka, China, Japan, SE Asia, New Guinea, Australia.
Diplomeris D.Don
Diplomeris hirsuta (Lindl.) Lindl.
Diplochilos hirsutus Lindl.
Terrestrial.
Central; 100–1300 m.
Distribution: Nepal, NE India, China.

Epipactis **Zinn.**
Epipactis helleborine (L.) Crantz
Serapias helleborine L.
Terrestrial.
West, Central, East; 1500–3700 m.
Distribution: N Africa, Europe, Pakistan, India, Nepal, Bhutan, China, Myanmar.
Use: Med.—Tubers to treat headache and stomachache.
Epipactis royleana Lindl.
Terrestrial.
West, Central; 1400–3700 m.
Distribution: C Asia, Afghanistan, Pakistan, India, Nepal, Bhutan, China (Tibet).
Epipactis veratrifolia Boiss. & Hohen.
Terrestrial.
West, Central; 700–1900 m.
Distribution: Africa, C Asia, Afghanistan, Pakistan, India, Nepal, China.
Epipogium S.G.Gmelin ex Borkh.
Epipogium aphyllum Sw.
Mycotrophic.
West; 2500–3600 m.
Distribution: Europe, W Asia, India, Nepal, Bhutan, China, Japan.
Epipogium japonicum Makino
Mycotrophic.
Central; 2300 m.
Distribution: Nepal, NE India, China, Japan.
Epipogium roseum (D.Don) Lindl.
Limodorum roseum D.Don
Mycotrophic.
Central; 1000–2000 m.
Distribution: W Africa, India, Nepal, Sri Lanka, SE Asia, Australia.
Eria Lindl.
Eria clausa King & Pantl.
Epiphytic.
Central; 1100–2800 m.
Distribution: Nepal, NE India, Bhutan, China, Myanmar, Thailand, Vietnam.
Eria coronaria (Lindl.) Rchb.f.
Coelogyne coronaria Lindl.
Trichosma suavis Lindl.
Epiphytic.
Central, East; 1500–2300 m.
Distribution: India, Nepal, Bhutan, China, Thailand, Vietnam.
Eulophia R.Br. ex Lindl.
Eulophia bicallosa (D.Don) P.F.Hunt & Summerh.
Bletia bicallosa D.Don
Terrestrial.
Central; 200–300 m.

Distribution: India, Nepal, Bangladesh, China, SE Asia, New Guinea, Australia.

Eulophia dabia (D.Don) Hochr.

Bletia dabia D.Don

Terrestrial.

West, Central; 300–1300 m.

Distribution: C Asia, Afghanistan, Pakistan, India, Nepal, Bhutan, Bangladesh, China, Myanmar.

Use: Med.—Rhizome in tonic and aphrodisiac and tubers in cough and cold.

Eulophia explanata Lindl.

Terrestrial.

Central; 800–1000 m.

Distribution: India, Nepal.

Eulophia flava (Lindl.) Hook.f.

Cyrtopera flava Lindl.

Terrestrial.

Central; 100–2600 m.

Distribution: India, Nepal, China, Myanmar, Cambodia, Thailand, Laos, Vietnam.

Eulophia graminea Lindl.

Terrestrial.

Central; 900–1100 m.

Distribution: Pakistan, India, Nepal, Bhutan, China, Japan, SE Asia.

Eulophia herbacea Lindl.

Terrestrial.

West; 800–1400 m.

Distribution: India, Nepal, Bangladesh, China, Myanmar, Thailand, Laos.

Eulophia mackinnonii Duthie

Terrestrial.

Central; 800–1500 m.

Distribution: India, Nepal, Bangladesh.

Eulophia nuda Lindl.

Eulophia spectabilis (Dennst.) Suresh

Terrestrial.

Central, East; 400–2000 m.

Distribution: India, Nepal, Bhutan, Bangladesh, Sri Lanka, China, SE Asia, New Guinea.

Use: Med.—Tubers for bronchitis.

Eulophia obtusa (Lindl.) Hook.f.

Cyrtopera obtusa Lindl.

Terrestrial.

Central; 200–1500 m.

Distribution: India, Nepal, Bangladesh.

Eulophia picta (R.Br.) Ormerod

Cymbidium pictum R.Br.

Geodorum densiflorum (Lam.) Schltr.

Limodorum densiflorum Lam.
Geodorum pallidum D.Don
Terrestrial.
Central; 150–1400 m.
Distribution: India, Nepal, Bhutan, Bangladesh, Sri Lanka, China, SE Asia, New Guinea, Australia.

Galearis **Rafin.**
Galearis roborowskii (Maxim.) S.C.Chen, P.J.Cribb & S.W.Gale
Orchis roborowskii Maxim.
Galearis stracheyi (Hook.f.) P.F.Hunt
Terrestrial.
West, Central; 2000–4800 m.
Distribution: India, Nepal, Bhutan, China.
Use: Med.—Tubers as tonic.
Galearis spathulata (Lindl.) P.F.Hunt
Gymnadenia spathulata Lindl.
Aorchis spathulata (Lindl.) Vermeulen
Terrestrial.
West, Central, East. 3000–4500 m.
Distribution: India, Nepal, Bhutan, China.
Gastrochilus D.Don
Gastrochilus affinis (King & Pantl.) Schltr.
Saccolabium affine King & Pantl.
Gastrochilus nepalensis Raskoti
Epiphytic.
Central; 2300–2500 m.
Distribution: India, Nepal, Bhutan, China.
Gastrochilus calceolaris (Buch.-Ham. ex Sm.) D.Don
Aerides calceolaris Buch.-Ham. ex Sm.
Epiphytic.
Central, East; 1000–2300 m.
Distribution: India, Nepal, Bhutan, China, SE Asia.
Use: Orn.
Gastrochilus dasypogon (Sm.) Kuntze
Aerides dasypogon Sm.
Gastrochilus acutifolius (Lindl.) Kuntze
Saccolabium acutifolium Lindl.
Epiphytic.
Central, East; 1200–2100 m.
Distribution: India, Nepal, Bhutan, Vietnam.
Use: Orn.
Gastrochilus distichus (Lindl.) Kuntze
Saccolabium distichum Lindl.
Epiphytic.

Central, East; 1700–2900 m.

Distribution: Nepal, NE India (Sikkim), Bhutan, China, Myanmar, Laos, Vietnam.

Gastrochilus obliquus (Lindl.) Kuntze

Saccolabium obliquum Lindl.

Gastrochilus bigibbus (Rchb.f. ex Hook.f.) Kuntze

Saccolabium dasypogon (Sm.) Lindl.

Epiphytic.

Central, East; 100–1000 m.

Distribution: Nepal, NE India, Bhutan, China, Myanmar, Thailand, Vietnam, Laos.

Gastrochilus pseudodistichus (King & Pantl.) Schltr.

Saccolabium pseudodistichum King & Pantl.

Epiphytic.

Central; 1900–2900 m.

Distribution: Nepal, NE India, Bhutan, China.

Gastrodia R.Br.

Gastrodia falconeri D.L.Jones & M.A.Clements

Gamoplexis orobanchoides Falc.

Gastrodia orobanchoides (Falc.) Hook.f.

Mycotrophic.

Central; 2300–3000 m.

Distribution: Pakistan, India, Nepal.

Goodyera R.Br.

Goodyera biflora (Lindl.) Hook.f.

Georchis biflora Lindl.

Terrestrial.

West, Central; 1900–2200 m.

Distribution: India, Nepal, E Asia, Vietnam.

Goodyera foliosa (Lindl.) Benth. ex C.B.Clarke

Georchis foliosa Lindl.

Terrestrial.

Central; 2100–2300 m.

Distribution: India, Nepal, Bhutan, E Asia, Myanmar, Vietnam.

Goodyera fusca (Lindl.) Hook.f.

Aetheria fusca Lindl.

Terrestrial.

Central, East; 3200–4700 m.

Distribution: India, Nepal, Bhutan, China, Myanmar.

Goodyera hemsleyana King & Pantl.

Terrestrial.

Central; 2000–2200 m.

Distribution: Nepal, NE India, Bhutan, China, Myanmar.

Goodyera henryi Rolfe

Terrestrial.

Central; 2400 m.
Distribution: Nepal, E Asia.
Goodyera marginata Lindl.
Terrestrial.
West, Central; 2500 m.
Distribution: India, Nepal, China.
Goodyera nankoensis Fukuy.
Terrestrial.
East; 2000 m.
Distribution: Nepal, China.
Goodyera procera (Ker-Gawl.) Hook.
Neottia procera Ker-Gawl.
Terrestrial.
Central; 900–2000 m.
Distribution: India, Nepal, Bhutan, Sri Lanka, China, Japan, SE Asia.
Goodyera recurva Lindl.
Epiphytic.
Central; 1400–2800 m.
Distribution: Nepal, NE India, Bhutan, China.
Goodyera repens (L.) R.Br.
Satyrium repens L.
Terrestrial.
West, Central, East; 1000–4200 m.
Distribution: Europe, India, Nepal, Bhutan, China, Japan, SE Asia, N America.
Use: Med.—Plant extract taken as a blood purifier.
Goodyera schlechtendaliana Rchb.f.
Terrestrial.
Central; 1500–2500 m.
Distribution: India, Nepal, Bhutan, E & SE Asia.
Goodyera viridiflora (Blume) Lindl. ex D. Dietr.
Neottia viridiflora Blume
Terrestrial.
Central; 1300–2000 m.
Distribution: India, Nepal, China, SE Asia, New Guinea, Australia.
Goodyera vittata (Lindl.) Benth. ex Hook.f.
Georchis vittata Lindl.
Terrestrial.
East; 2000–3100 m.
Distribution: Nepal, NE India, Bhutan, China.
Gymnadenia R. Br.
Gymnadenia orchidis Lindl.
Terrestrial.
West, Central, East; 3000–4700 m.
Distribution: Pakistan, India, Nepal, Bhutan, China, Myanmar.
Use: Med.—Pseudobulb to treat cuts and wounds.

Habenaria **Willd.**

Habenaria aitchisonii Rchb.f.

Habenaria diceras Schltr.

Terrestrial.

West, Central, East; 2400–4000 m.

Distribution: Afghanistan, Pakistan, India, Nepal, Bhutan, China, Myanmar.

Habenaria arictina Hook.f.

Terrestrial.

West, Central, East; 1400–3000 m.

Distribution: India, Nepal, Bhutan, Bangladesh, China, Myanmar, Vietnam.

Habenaria commelinifolia (Roxb.) Wall. ex Lindl.

Orchis commelinifolia Roxb.

Terrestrial.

West, Central; 300–3000 m.

Distribution: India, Nepal, China (Yunnan), Myanmar, Thailand, Vietnam.

Use: Med.—Plant as salep.

Habenaria dentata (Sw.) Schltr.

Orchis dentata Sw.

Terrestrial.

Central; 500–2400 m.

Distribution: India, Nepal, Bhutan, China, S Korea, Japan, SE Asia.

Habenaria digitata Lindl.

Terrestrial.

Central, East; 200–1500 m.

Distribution: Pakistan, India, Nepal, Myanmar, Cambodia, Laos.

Habenaria diphylla (Nimmo) Dalzell

Liparis diphyllos Nimmo

Terrestrial.

Central, East; 800–1900 m.

Distribution: India, Nepal, China (Yunnan).

Habenaria furcifera Lindl.

Terrestrial.

West; 150–800 m.

Distribution: Pakistan, India, Nepal, Bangladesh, China, Myanmar, Thailand.

Habenaria intermedia D.Don

Terrestrial.

West, Central; 900–3300 m.

Distribution: Pakistan, India, Nepal, China (Tibet), Myanmar.

Use: Med.—Tuber paste in cough, asthma.

Habenaria malleifera Hook.f.

Terrestrial.

East; 1500–1800 m.

Distribution: Nepal, NE India, China, Myanmar, Vietnam.

Habenaria marginata Colebr.

Terrestrial.

West, Central; 500–1500 m.
Distribution: Pakistan, India, Nepal, Bhutan, Bangladesh, China, Myanmar, Thailand.
Use: Med.—Plant extract in flatulence, wound and tonic.
Habenaria palpensis Raskoti
Terrestrial.
Central; 1500 m.
Distribution: Endemic to Nepal.
Habenaria pantlingiana Kraenzl.
Terrestrial.
East; 2100–2200 m.
Distribution: India, Nepal, China, Myanmar, Thailand, Vietnam.
Habenaria pectinata D.Don
Terrestrial.
West, Central, East; 900–3500 m.
Distribution: Pakistan, India, Nepal, China (Yunnan), Myanmar.
Use: Orn. Med.—Tuber against arthritis.
Habenaria plantaginea Lindl.
Terrestrial.
West, Central, East; 200–1600 m.
Distribution: India, Nepal, Sri Lanka, Bangladesh.
Habenaria plantaginea Lindl. var. **nepalensis** (Kolan.) Rajbh. & Rita Chhetri
Habenaria nepalensis Kolan.
Terrestrial.
West, Central; 300–1000 m.
Distribution: Endemic to Nepal.
Habenaria pubescens Lindl.
Habenaria pubescens var. *nepalensis* Raskoti
Terrestrial.
Central; 600–1200 m.
Distribution: India, Nepal.
Habenaria reniformis (D.Don) Hook.f.
Listera reniformis D.Don
Habenaria wallichii (Kolan., Kras & Szlach.) J.M.H.Shaw
Habenella wallichii Kolan., Kras & Szlach.
Terrestrial.
Central; 400–1900 m.
Distribution: India, Nepal, China, SE Asia.
Habenaria sandiegoensis Raskoti
Terrestrial.
East; 1600 m.
Distribution: Endemic to Nepal.
Habenaria stenopetala Lindl.
Terrestrial.
West, Central; 1000–2500 m.

Distribution: India, Nepal, Bhutan, Bangladesh, China, Japan, Thailand, Vietnam, Philippines, New Guinea.

Hemipilia Lindl.

Hemipilia calcicola (W.W.Sm.) Y.Tang & H.Peng

Gymnadenia calcicola W.W.Sm.

Neottianthe cucullata (L.) Schltr. var. *calcicola* (W.W.Sm.) Soo

Ponerorchis cucullata (L.) X.H.Jin, Schuit. & W.T.Jin var. *calcicola* (W.W. Sm.) X.H.Jin, Schuit. & W.T.Jin

Terrestrial.

West, Central; 3700–5000 m.

Distribution: India, Nepal, Bhutan, China.

Use: Med.—Rhizome for tonic.

Hemipilia chusua (D.Don) Y.Tang & H.Peng

Orchis chusua D.Don

Chusua pauciflora (Lindl.) P.F.Hunt

Orchis chusua D.Don forma *parva* Kitam.

Ponerorchis chusua (D.Don) Soo

Terrestrial.

West, Central, East; 2400–4900 m.

Distribution: India, Nepal, Bhutan, China, Korea, Myanmar.

Hemipilia cordifolia Lindl.

Terrestrial.

West, Central; 1800–2600 m.

Distribution: India, Nepal, Bhutan, China, Myanmar.

Hemipilia cucullata (L.) Y.Tang, H.Peng & T.Yukawa

Orchis cucullata L.

Neottianthe cucullata (L.) Schltr.

Ponerorchis cucullata (L.) X.H.Jin, Schuit. & W.T.Jin

Terrestrial.

West, Central, East; 2900–4100 m.

Distribution: E Europe, India, Nepal, Bhutan, E Asia.

Hemipilia secundiflora (Hook.f.) Schltr.

Peristylus secundiflorus Kraenzl.

Neottianthe secundiflora (Kraenzl.) Schltr.

Ponerorchis secundiflora (Kraenzl.) X.H.Jin, Schuit. & W.T.Jin

Terrestrial.

West, Central, East; 2200–4200 m.

Distribution: India, Nepal, Bhutan, China.

Herminium R.Br.

Herminium alaschanicum Maxim.

Terrestrial.

West, Central; 1800–4500 m.

Distribution: Nepal, China, Mongolia.

Herminium albomarginatum (King) X.H.Jin, Schuit., Raskoti & Lu Q.Huang

Habenaria albomarginata King

Terrestrial.
Central, East; 3400–4400 m.
Distribution: Nepal, NE India (Sikkim), China.
Herminium chloranthum Tang & F.T.Wang
Terrestrial.
Central; 2500–4100 m.
Distribution: Nepal, China.
Herminium clavigerum (Lindl.) X.H.Jin, Schuit., Raskoti & Lu Q. Huang
Platanthera clavigera Lindl.
Terrestrial.
West, Central, East; 2300–3400 m.
Distribution: India, Nepal, Bhutan, China.
Herminium edgeworthii (Hook.f. ex Collett) X. H.Jin, Schuit., Raskoti & Lu Q. Huang
Habenaria edgeworthii Hook.f. ex Collett
Platanthera edgeworthii (Hook.f. ex Collett) R.K.Gupta
Terrestrial.
West, Central; 1200–3600 m.
Distribution: Pakistan, India, Nepal, Bhutan, China, Myanmar.
Use: Med.—Leaves and roots as blood purifier.
Herminium elisabethae (Duthie) Tang & F.T.Wang
Habenaria elisabethae Duthie
Peristylus elisabethae (Duthie) R.K.Gupta
Terrestrial.
West, Central, East; 2100–4100 m.
Distribution: India, Nepal, Bhutan, China, Myanmar.
Herminium fallax (Lindl.) Hook.f.
Peristylus fallax Lindl.
Terrestrial.
West, Central, East; 2000–4000 m.
Distribution: India, Nepal, Bhutan, China.
Herminium fimbriatum (Raskoti) X.H.Jin, Schuit., Raskoti & Lu Q.Huang
Bhutanthera fimbriata Raskoti
Terrestrial.
Central; 3600–3800 m.
Distribution: Endemic to Nepal.
Herminium gracile King & Pantl.
Terrestrial.
Central; 3300–4200 m.
Distribution: Nepal, NEIndia, China.
Herminium handelii X.H.Jin, Schuit., Raskoti & Lu Q.Huang
Terrestrial.
Central; 4200–4300 m.
Distribution: Nepal, NE India, Bhutan, China.
Herminium hongdeyuanii Raskoti

Terrestrial.
Central; 2200 m.
Distribution: Endemic to Nepal.
Herminium jaffreyanum King & Pantl.
Terrestrial.
East; 2300–3300 m.
Distribution: Nepal, NE India.
Herminium josephi Rchb.f.
 Herminium duthiei Hook.f.
 Peristylus duthiei (Hook.f.) Deva & H.B.Naithani
Terrestrial.
West, Central, East; 2400–4600 m.
Distribution: India, Nepal, Bhutan, China.
Herminium kamengense A.N.Rao
Terrestrial.
East; 2200–2500 m.
Distribution: Nepal, NE India, China.
Herminium lanceum (Thunb. ex Sw.) J.Vuijk
 Ophrys lancea Thunb. ex Sw.
Terrestrial.
West, Central, East; 700–4700 m.
Distribution: India, Nepal, Bhutan, China, Japan, SE Asia.
Use: Med.—Plant extract in suppressed urination.
Herminium latilabre (Lindl.) X.H.Jin, Schuit., Raskoti & Lu Q. Huang
 Platanthera latilabris Lindl.
Terrestrial.
West, Central; 900–3700 m.
Distribution: Pakistan, India, Nepal, Bhutan, China, Myanmar.
Herminium longilobatum S.N.Hegde & A.N.Rao
Terrestrial.
East; 2500–2800 m.
Distribution: Nepal, N. E. India, China.
Herminium mackinnonii Duthie
Terrestrial.
Central, East; 2100–2700 m.
Distribution: India, Nepal.
Herminium macrophyllum (D.Don) Dandy
 Neottia macrophylla D.Don
Terrestrial.
West, Central, East; 3100–4700 m.
Distribution: Pakistan, India, Nepal, Bhutan, China.
Herminium mannii (Rchb.f.) Tang & F.T.Wang
 Coeloglossum mannii Rchb.f.
 Peristylus mannii (Rchb.f.) Mukerjee
Terrestrial.

Central, East; 1700–3400 m.
Distribution: India, Nepal, China, Myanmar, Thailand.
Herminium monophyllum (D.Don) P.F.Hunt & Summerhayes
Neottia monophylla D.Don
Terrestrial.
West, Central; 1000–4200 m.
Distribution: India, Nepal, Bhutan.
Herminium monorchis (L.) R.Br.
Ophrys monorchis L.
Terrestrial.
West, Central, East; 2200–4600 m.
Distribution: Europe, C & W Asia, Afghanistan, Pakistan, India, Nepal, E Asia.
Use: Med.—Root in tonic.
Herminium pugioniforme Lindl. ex Hook.f.
Terrestrial.
Central; 2700–5200 m.
Distribution: India, Nepal, Bhutan, China.
Herminium quinquelobum King & Pantl.
Terrestrial.
Central, East; 1000–3000 m.
Distribution: Nepal, NE India, Bhutan, China.
Herminium souliei (Finet) Rolfe
Terrestrial.
Central, East; 1400–4200 m.
Distribution: Nepal, China.
Herpysma Lindl.
Herpysma longicaulis Lindl.
Terrestrial.
Central; 900–1700 m.
Distribution: Nepal, NE India, Bhutan, China, SE Asia.

Lecanorchis **Blume**
Lecanorchis japonica Blume
Mycotrophic .
Central; 1000–1400 m.
Distribution: Nepal, NE India, E Asia.
Liparis L.C.Rich.
Liparis bootanensis Griff.
Epiphytic.
Central; 1400–2500 m.
Distribution: Nepal, NE India, Bhutan, China, Japan, SE Asia.
Liparis campylostalix Rchb.f.
Leptorkis campylostalix (Rchb.f.) Kuntze;
Terrestrial.
West; 2800–3000 m.

Distribution: Nepal, E Asia, Vietnam.
Liparis cathcartii Hook.f.
Leptorkis cathcartii (Hook.f.) Kuntze
Terrestrial.
West, Central; 400–3100 m.
Distribution: Nepal, NE India, Bhutan, China, Myanmar.
Liparis cespitosa (Lam.) Lindl.
    Epidendrum cespitosa Lam.
Epiphytic.
Central, East; 500–2500 m.
Distribution: Africa, India, Nepal, Bhutan, China to the Pacific.
Liparis cordifolia Hook.f.
Terrestrial.
Central; 1400–2000 m.
Distribution: India, Nepal, Bhutan, China, Vietnam.
Liparis deflexa Hook.f.
Terrestrial.
Central, East; 400–1400.
Distribution: India, Nepal, Myanmar.
Liparis duthiei Hook.f.
    Leptorkis duthiei (Hook.f.) Kuntze
Epiphytic.
Central; 500–1000 m.
Distribution: Nepal, NE India.
Liparis elliptica Wight
    Leptorkis elliptica (Wight) Kuntze;
Terrestrial.
Central; 1400–1700 m.
Distribution: India, Nepal, Sri Lanka, China, Japan, SE Asia.
Liparis ferruginea Lindl.
Terrestrial.
Central; 400–800 m.
Distribution: Nepal, India, China, SE Asia.
Liparis glossula Rchb.f.
Terrestrial.
West, Central; 1800–4300 m.
Distribution: Nepal, NE India, China.
Liparis langtangensis Raskoti & Ale
Terrestrial.
Central; 3700–3900 m.
Distribution: Endemic to Nepal.
Liparis nervosa (Thunb.) Lindl.
    Ophrys nervosa Thunb.
    Liparis bituberculata (Hook.) Lindl.
Terrestrial.

West, Central, East; 400–2800 m.
Distribution: Old and New World tropics.
Use: Med.—Tuber to treat stomachache.
Liparis nervosa var. **khasiana** (Hook.f.) P.K.Sarkar
    Liparis bituberculata var. *khasiana* Hook.f.
Terrestrial.
Central.
Distribution: Nepal, NE India.
Liparis odorata (Willd.) Lindl.
    Malaxis odorata Willd.
Terrestrial.
Central; 900–2000 m.
Distribution: India, Nepal, China, Japan, Myanmar, Thailand, Laos, Vietnam.
Liparis olivacea Lindl.
Terrestrial.
Central.
Distribution: Nepal, NE India.
Liparis perpusilla Hook.f.
    Liparis togashii Tuyama
Epiphytic.
Central, East; 2500–2900 m.
Distribution: Nepal, NE India, Bhutan, China (Yunnan).
Liparis petiolata (D.Don) P.F.Hunt & Summerh.
    Acianthus petiolatus D.Don
Terrestrial.
Central; 1300–2800 m.
Distribution: Nepal, NE India, Bhutan, China, Thailand, Vietnam.
Liparis plantaginea Lindl.
    Leptorkis plantaginea (Lindl.) Kuntze
Terrestrial.
Central; 700–900 m.
Distribution: Nepal, NE India, Bhutan, Myanmar, Thailand, Vietnam.
Liparis platyrachis Hook.f.
    Leptorkis platyrachis (Hook.f.) Kuntze
Epiphytic .
Central; 1400–1600 m.
Distribution: India, Nepal, Bhutan, China.
Liparis pygmaea King & Pantl.
Terrestrial.
Central; 3000–3800 m.
Distribution: Nepal, NE India, China, Vietnam.
Liparis resupinata Ridl.
Epiphytic.
Central, East; 1300–2600 m.
Distribution: India, Nepal, Bhutan, China.

Liparis rostrata Rchb.f.
Terrestrial.
West; 2000–3000 m.
Distribution: India, Nepal, China.
Liparis somae Hayata
Liparis sikkimensis Lucksom & S Kumar
Epiphytic.
Central; 500–1300 m.
Distribution: Nepal, NE India, China.
Liparis stricklandiana Rchb.f.
Epiphytic.
Central; 600–1200 m.
Distribution: Nepal, NE India, Bhutan, China, Myanmar, Vietnam.
Liparis viridiflora (Blume) Lindl.
Malaxis viridiflora Blume
Epiphytic.
Central, East; 500–2500 m.
Distribution: India, Nepal, Bhutan, Sri Lanka, China, SE Asia.
Luisia Gaudich.
Luisia inconspicua (Hook.f.) King & Pantl.
Saccolabium inconspicuum Hook.f.
Gastrochilus inconspicuus (Hook.f.) Kuntze
Epiphytic.
Central; 700–1500 m.
Distribution: India, Nepal, Bhutan, Bangladesh, China.
Luisia trichorrhiza (Hook.) Blume
Vanda trichorrhiza Hook.
Epiphytic.
East; 1000–1300 m.
Distribution: India, Nepal, Bhutan, Bangladesh, Myanmar, Thailand.
Use: Orn. Med.—Leaf paste applied externally on muscular pain.
Luisia zeylanica Lindl.
Epiphytic.
West, Central, East; 300–1400 m.
Distribution: India, Nepal, Bhutan, Bangladesh, Sri Lanka.
Use: Orn. Med.—Leaf juice to treat wounds and boils.

Malaxis **Sol. ex Sw.**
Malaxis dolpensis M.R.Shrestha, L.R.Shakya & S.K.Ghimire
Terrestrial.
West; 4000–4200 m.
Distribution: Endemic to Nepal.
Malaxis monophyllos (L.) Sw.
Ophrys monophyllos L.
Terrestrial.

Central, East; 2600–4300 m.
Distribution: Europe, India, Nepal, E Asia, N America.
Malaxis monophyllos (L.) Sw. var. **obtusa** H.Tsukaya & H.Ohha
Terrestrial.
East; 3300–3400 m.
Distribution: Endemic to Nepal.
Malaxis muscifera (Lindl.) Kuntze
Dienia muscifera Lindl.
Terrestrial.
West, Central, East; 2000–4100 m.
Distribution: Pakistan, India, Nepal, Bhutan, China, Myanmar.
Use: Med.—Stem base in dysentery and fever.
Malaxis tamurensis Tuyama
Terrestrial.
East; 1200 m.
Distribution: Endemic to Nepal.
Mycaranthes Blume
Mycaranthes floribunda (D.Don) S.C.Chen & J.J.Wood
Dendrobium floribundum D.Don
Eria paniculata Lindl.
Epiphytic.
Central, East; 600–1800 m.
Distribution: Nepal, NE India, Bhutan, Bangladesh, China, SE Asia.

Neottia **Guett.**
Neottia acuminata Schltr.
Aphyllorchis parviflora King & Pantl.
Mycotrophic.
West, Central; 2500–4000 m.
Distribution: Nepal, NE India, Bhutan, E Asia.
Neottia chandrae Raskoti, J.J.Wood & R.Ale
Terrestrial.
Central, East; 3300–3800 m.
Distribution: India, Nepal.
Neottia listeroides Lindl.
Mycoheterotrophic.
West, Central; 2700–3200 m.
Distribution: Pakistan, India, Nepal, Bhutan, China.
Neottia longicaulis (King & Pantl.) Szlach.
Listera longicaulis King & Pantl.
Terrestrial.
Central; 2400–2500 m.
Distribution: India, Nepal, Bhutan, China.
Neottia mucronata (Panigrahi & J.J.Wood) Szlach.
Listera mucronata Panigrahi & J.J.Wood

Terrestrial.
Central; 2000–3500 m.
Distribution: Nepal, NE India, E Asia.
Neottia nepalensis (N.P.Balakr.) Szlach.
Listera nepalensis N.P.Balakr.
Terrestrial.
West, Central; 2900–4000 m.
Distribution: Endemic to Nepal.
Neottia pinetorum (Lindl.) Szlach.
Listera pinetorum Lindl.
Terrestrial.
Central; 2200–3100 m.
Distribution: India, Nepal, Bhutan, China, Japan, Myanmar.
Neottia tenii Schltr.
Mycotrophic.
Central; 2600 m.
Distribution: Nepal, China.
Neottia tenuis (Lindl.) Szlach.
Listera tenuis Lindl.
Terrestrial.
East; 3000–4000 m.
Distribution: Nepal, NE India, China (Tibet).
Nephelaphyllum Blume
Nephelaphyllum pulchrum Blume
Anoectochilus pulcher (Blume) Rollisson;
Terrestrial.
Central; 1000–1200 m.
Distribution: Nepal, NE India, SE Asia.
Nervilia Comm. ex Gaudich.
Nervilia concolor (Blume) Schltr.
Cordyla concolor Blume
Nervilia aragoana Gaudich.
Terrestrial.
Central; 500–1300 m.
Distribution: Pakistan, India, Nepal, Bhutan, Bangladesh, Sri Lanka, China, SE Asia, New Guinea, Australia.
Use: Med.—Plant in asthma and diarrhea.
Nervilia gammieana (Hook.f.) Pfitzer
Pogonia gammieana Hook.f.
Terrestrial.
Central, East; 400–800 m.
Distribution: Pakistan, India, Nepal, Bhutan.
Nervilia mackinnonii (Duthie) Schltr.
Pogonia mackinnonii Duthie
Terrestrial.

Central; 900–1400 m.
Distribution: India, Nepal, China, Myanmar, Vietnam.
Nervilia macroglossa (Hook.f.) Shcltr.
Poogonia macroglossa Hook.f.
Terrestrial.
Central; 600–1800 m.
Distribution: Nepal, NE India, China (Tibet), Myanmar.
Nervilia plicata (Andrews) Schltr.
Arethusa plicata Andrews
Terrestrial.
Central; 200–1300 m.
Distribution: Pakistan, India, Nepal, Bhutan, Bangladesh, Sri Lanka, China, SE Asia, New Guinea, Australia.
Nervila simplex (Thouars) Schltr.
Arethusa simplex Thouars
Terrestrial.
East; 200–1600 m.
Distribution: Africa, India, Nepal, Bhutan, China, SE Asia, New Guinea

Oberonia **Lindl.**
Oberonia acaulis Griff.
Oberonia myriantha Lindl.
Epiphytic.
Central, East; 150–2300 m.
Distribution: India, Nepal, Bhutan, China, Myanmar, Thailand, Vietnam.
Oberonia anthropophora Lindl.
Oberonia falcata King & Pantl.
Epiphytic.
Central, East; 1200–2400 m.
Distribution: India, Nepal, Bhutan, China, SE Asia.
Oberonia brachystachys Lindl.
Oberonia demissa Lindl.
Oberonia parvula King & Pantl.
Oberonia recurva Lindl.
Epiphytic
Central; 100–1500 m.
Distribution: India, Nepal, Sri Lanka, Myanmar, Thailand.
Oberonia caulescens Wall. ex Lindl.
Oberonia longilabris King & Pantl.
Epiphytic.
Central, East; 1200–2500 m.
Distribution: India, Nepal, Bhutan, China (Yunnan).
Use: Med.—Tuber in liver ailments.
Oberonia cavaleriei Finet
Epiphytic.

Central; 1000–2000 m.
Distribution: India, Nepal, China, Myanmar, Thailand, Laos, Vietnam.
Oberonia emarginata King & Pantl.
Oberonia micrantha King & Pantl.
Epiphytic.
Central, East; 1600–2700 m.
Distribution: Nepal, NE India, Bhutan, Thailand, Vietnam.
Oberonia ensiformis (Sm.) Lindl.
Malaxis ensiformis Sm.
Oberonia iridifolia Lindl.
Epiphytic.
Central, East; 600–2900 m.
Distribution: India, Nepal, China, Myanmar, Thailand, Laos, Vietnam.
Oberonia falconeri Hook.f.
Epiphytic.
West, Central, East; 200–1000 m.
Distribution: India, Nepal, China, SE Asia.
Oberonia jenkinsiana Griff. ex Lindl.
Epiphytic.
Central; 400–1400 m.
Distribution: Nepal, NE India, China, Myanmar, Thailand.
Oberonia lobulata King & Pantl.
Epiphytic.
Central; 300 m.
Distribution: Nepal, NE India.
Oberonia longilabris King & Pantl.
Terrestrial.
Central; 1200–1500 m.
Distribution: Nepal, NE India.
Oberonia mucronata (D.Don) Ormerod & Seidenf.
Stelis mucronata D.Don
Cymbidium iridifolium Roxb.
Terrestrial.
Central; 500–1800 m.
Distribution: India, Nepal, Bhutan, China, SE Asia.
Oberonia obcordata Lindl.
Epiphytic.
Central, East; 1100–1900 m.
Distribution: Nepal, NE India, China, Myanmar, Thailand, Vietnam.
Oberonia pachyphylla King & Pantl.
Epiphytic.
East; 200–800 m.
Distribution: Nepal, NE India, Thailand, Vietnam.
Oberonia pachyrachis Rchb.f. ex Hook.f.
Epiphytic.

West, Central, East; 600–1900 m.
Distribution: India, Nepal, Bhutan, China, Myanmar, Thailand, Vietnam.
Oberonia prainiana King & Pantl.
Epiphytic.
Central; 200–1500 m.
Distribution: India, Nepal, Myanmar, Thailand, Indonesia.
Oberonia pyrulifera Lindl.
Epiphytic.
Central; 300–2700 m.
Distribution: India, Nepal, Bhutan, China, Myanmar, Thailand.
Oberonia rufilabris Lindl.
Oberonia nepalensis L.R.Shakya & R.P.Chaudhary
Oberonia pantlingiana L.R.Shakya & R.P.Chaudhary
Oberonia rufilabris Lindl. var. *nepalensis* (L.R.Shakya & R.P.Chaudhary) Rajbh. & R.Chhetri
Epiphytic.
Central; 300–1500 m.
Distribution: Nepal, NE India, Bangladesh, China, SE Asia.
Odontochilus Blume
Odontochilus crispus (Lindl.) Hook.f.
Anoectochilus crispus Lindl.
Terrestrial.
Central; 400–2400 m.
Distribution: Nepal, NE India, Bhutan, China, Myanmar.
Odontochilus elwesii C.B.Clarke ex Hook.f.
Terrestrial.
Central; 1800–2300 m.
Distribution: Nepal, NE India, Bhutan, China, Myanmar, Thailand, Laos, Vietnam.
Odontochilus lanceolatus (Lindl.) Blume
Anoectochilus lanceolatus Lindl.
Terrestrial.
Central; 1600–1900 m.
Distribution: Nepal, NE India, Bhutan, Bangladesh, China, Myanmar, Thailand, Vietnam.
Odontochilus nandae Raskoti & H. Kurzweil
Terrestrial.
Central; 2400 m.
Distribution: Endemic to Nepal.
Odontochilus pumilus Hook.f.
Myrmechis pumila (Hook.f.) Tang & Wang
Terrestrial.
Central; 2500–3200 m.
Distribution: Nepal, NE India, Bhutan, China, Myanmar, Thailand.
Oreorchis Lindl.

Oreorchis foliosa (Lindl.) Lindl.
Corallorhiza foliosa Lindl.
Terrestrial.
West, Central; 2500–4500 m.
Distribution: India, Nepal, Bhutan, China, Myanmar.
Oreorchis indica (Lindl.) Hook.f.
Corallorhiza indica Lindl.
Oreorchis foliosa (Lindl.) Lindl. var. *indica* (Lindl.) N.Pearce & P.J.Cribb
Terrestrial.
Central, East; 2700–3800 m.
Distribution: India, Nepal, Bhutan, China, Japan, Myanmar.
Oreorchis micrantha Lindl.
Terrestrial.
Central, East; 2300–3400 m.
Distribution: India, Nepal, Bhutan, China, Myanmar.
Oreorchis nepalensis N.Pearce & P.J.Cribb
Terrestrial.
Central, 2700–3100 m.
Distribution: Nepal, China (Tibet).
Oreorchis porphyranthes Tuyama
Terrestrial.
Central; 2700–3800 m.
Distribution: Endemic to Nepal.

Pachystoma **Blume**
Pachystoma pubescens Blume
Terrestrial.
Central; 300–1400 m.
Distribution: India, Nepal, Bhutan, Bangladesh, China, SE Asia, New Guinea, Australia.
Papilionanthe Schltr.
Papilionanthe teres (Roxb.) Schltr.
Dendrobium teres Roxb.
Vanda teres Lindl.
Epiphytic.
Central, East; 200–2100 m.
Distribution: India, Nepal, Bhutan, Bangladesh, China, Myanmar, Thailand, Laos, Vietnam.
Use: Orn. Med.—Plant paste applied to treat dislocated bones.
Papilionanthe uniflora (Lindl.) Garay
Mesoclastes uniflora Lindl.
Aerides uniflora (Lindl.) Summerh.
Epiphytic.
Central, East; 1500–2100 m.
Distribution: Nepal, NE India, Bhutan.

Use: Orn.

Pecteilis Raf.

Pecteilis gigantea (Sm.) Raf.

Orchis gigantea Sm.

Terrestrial.

Central; 900–1800 m.

Distribution: Pakistan, India, Nepal, Bhutan, Myanmar.

Pecteilis susannae (L.) Raf.

Orchis susannae L.

Terrestrial.

West, Central, East; 150–2000 m.

Distribution: Nepal, NE India, China, SE Asia, New Guinea.

Pecteilis triflora (D.Don) Tang & F.T.Wang

Habenaria triflora D.Don

Terrestrial.

West, Central, East; 400–1200 m.

Distribution: N. W. India (Kumaun), Nepal, Bangladesh.

Pelatantheria Ridley

Pelatantheria insectifera (Rchb.f.) Ridl.

Sarcanthus insectifer Rchb.f.

Epiphytic.

Central, East; 150–1000 m.

Distribution: India, Nepal, Bangladesh, Myanmar, Thailand, Laos, Vietnam.

Peristylus Blume

Peristylus affinis (D.Don) Seidenf.

Habenaria affinis D.Don

Gymnadenia affinis (D.Don) Rchb.f.

Habenaria goodyeroides var. *affinis* (D.Don) King & Pantl.

Peristylus goodyeroides var. *affinis* (D.Don) C.M.Cooke

Terrestrial.

Central; 3100–3200 m.

Distribution: India, Nepal, China, SE Asia.

Peristylus biermannianus (King & Pantl.) X.H.Jin, Schuit. & W.T.Jin

Habenaria biermanniana King & Pantl.

Platanthera biermanniana (King & Pantl.) Kraenzl.

Terrestrial.

East; 2100–3200 m.

Distribution: Nepal, NE India.

Peristylus calcaratus (Rolfe) S.Y.Hu

Glossula calcarata Rolfe

Terrestrial.

Central; 1900–2000 m.

Distribution: Nepal, China, Vietnam.

Peristylus constrictus (Lindl.) Lindl.

Herminium constrictum Lindl.

Terrestrial.
Shady place, open ground; 300–2300 m.
Distribution: India, Nepal, Bhutan, Bangladesh, China (Yunnan), SE Asia.
Peristylus densus (Lindl.) Santapau & Kapadia
Coeloglossum densum Lindl.
Terrestrial.
Central, East; 2100–2800 m.
Distribution: Nepal, NE India, Bangladesh, China, Myanmar, Thailand, Vietnam.
Peristylus goodyeroides (D.Don) Lindl.
Habenaria goodyeroides D.Don
Terrestrial.
West, Central, East; 200–3600 m.
Distribution: India, Nepal, Bhutan, Bangladesh, China, SE Asia, New Guinea.
Peristylus hamiltonianus (Lindl.) Lindl.
Herminium hamiltonianum Lindl.
Terrestrial.
Central; 1500–2500 m.
Distribution: Nepal, NE India, Thailand.
Peristylus intrudens (Ames) Ormerod
Habenaria intrudens Ames
Terrestrial.
Central; 800–900 m.
Distribution: Nepal, NE India, China, Myanmar, Thailand, Philippines.
Peristylus lacertifer (Lindl.) Sm.
Coeloglossum lacertiferum Lindl.
Terrestrial.
Central; 200–3200 m.
Distribution: Nepal, NE India, Bhutan, China, Japan, SE Asia.
Peristylus lawii Wight
Terrestrial.
West, Central; 400–1000 m.
Distribution: India, Nepal, Myanmar.
Peristylus parishii Rchb.f.
Terrestrial.
Central; 600–2000 m.
Distribution: India, Nepal, China, Myanmar, Thailand, Vietnam.
Peristylus prainii (Hook.f.) Kraenzlin
Habenaria prainii Hook.f.
Terrestrial.
West, Central; 1000–2100 m.
Distribution: Nepal, NE India, Bhutan, Myanmar, Thailand, Vietnam.
Peristylus tipulifer (Parish & Rchb.f.) Mukerjee
Habenaria tipulifera Parish & Rchb.f.
Terrestrial.
Central, East; 1700–2400 m.

Distribution: Nepal, NE India, China, Myanmar, Thailand.

Phalaenopsis Blume

Phalaenopsis difformis (Wall. ex Lindl.) Kocyan & Schlt.

Aerides difforme Wall. ex Lindl.

Ornithochilus difformis (Wall. ex Lindl.) Schltr.

Epiphytic.

Central, East; 1100–1800 m.

Distribution: India, Nepal, Bhutan, China, SE Asia.

Use: Orn.

Phalaenopsis mannii Rchb.f.

Epiphytic.

Central, East; 150–600 m.

Distribution: Nepal, NE India, Bhutan, China (Yunnan), Myanmar, Vietnam.

Use: Orn.

Phalaenopsis parishii Rchb.f.

Doritis parishii (Rchb.f.) T.Yukawa & K.Kita

Phalaenopsis decumbens (Griff.) Holttum

Terrestrial.

Central; 200–500 m.

Distribution: Nepal, NE India, Myanmar, Thailand, Cambodia, Vietnam.

Phalaenopsis taenialis (Lindl.) EA Christenson & Pradhan

Aerides taenialis Lindl.

Epiphytic.

West, Central; 1510–2300 m.

Distribution: India, Nepal, Bhutan, China, Myanmar, Thailand, Laos, Vietnam.

Use: Orn.

Phreatia Lindl.

Phreatia elegans Lindl.

Epiphytic.

Central; 1600 m.

Distribution: India, Nepal, Sri Lanka.

Pinalia Lindl.

Pinalia acervata (Lindl.) Kuntze

Eria acervata Lindl.

Epiphytic.

East; 300–3000 m.

Distribution: Nepal, NE India, Bhutan, China, SE Asia.

Pinalia amica (Rchb.f.) Kuntze

Eria amica Rchb.f.

Epiphytic.

Central; 600–2100 m.

Distribution: Nepal, NE India, Bhutan, China, Myanmar, Thailand, Vietnam.

Pinalia annapurnensis (L.R.Shakya & M.R.Shrestha) Schuit., Y.P.Ng & H.A. Pedersen

Eria annapurnensis L.R.Shakya & M.R.Shrestha

Epiphytic.
Central; 2000 m.
Distribution: Endemic to Nepal.
Pinalia apertiflora (Summerh.) A.N.Rao
Eria apertiflora Summerh.
Epiphytic.
West, Central; 150–1600 m.
Distribution: Nepal, NE India, Myanmar, Thailand, Laos, Vietnam.
Pinalia baniae (Bajrach., L.R.Shakya & Chettri) Schuit., Y.P.Ng & H.A. Pedersen
Eria baniaii Bajrach., L.R.Shakya & Chettri
Epiphytic.
Central; 1600 m.
Distribution: Endemic to Nepal.
Pinalia bipunctata (Lindl.) Kuntze
Eria bipunctata Lindl.
Epiphytic.
Central; 1000–1700 m.
Distribution: Nepal, NE India, China (Yunnan), Veitnam.
Pinalia bractescens (Lindl.) Kuntze
Eria bractescens Lindl.
Epiphytic.
Central, East; 300–2000 m.
Distribution: Nepal, NE India, Bhutan, Bangladesh, SE Asia, New Guinea.
Pinalia concolor (E.C.Parish & Rchb.f.) Kuntze
Eria concolor E.C.Parish & Rchb.f.
Epiphytic.
East; 500 m.
Distribution: Nepal, Myanmar.
Pinalia excavata (Lindl.) Kuntze
Eria excavata Lindl.
Epiphytic.
Central, East; 500–2400 m.
Distribution: Nepal, NE India, Bhutan, China (Tibet), Thailand, Vietnam.
Pinalia graminifolia (Lindl.) Kuntze
Eria graminifolia Lindl.
Epiphytic.
Central, East; 1500–2500 m.
Distribution: Nepal, NE India, Bhutan, China, Myanmar.
Pinalia leucantha Kuntze
Eria alba Lindl.
Epiphytic.
Central, East; 1400–2500 m.
Distribution: India, Nepal, Bhutan.
Pinalia obesa (Lindl.) Kuntze

Eria obesa Lindl.
Epiphytic.
East; 500 m.
Distribution: Nepal, NE India, Bangladesh, Myanmar, Thailand.
Pinalia pokharensis (Bajrach., Subedi & K.K.Shrestha) Schuit., Y.P.Ng & H.A. Pedersen
Eria pokharensis Bajrach., Subedi & K.K Shrestha
Epiphytic.
Central; 900–1000 m.
Distribution: Endemic to Nepal.
Pinalia spicata (D.Don) S.C.Chen & J.J.Wood
Octomeria spicata D.Don
Eria spicata (D.Don) Hand.-Mazz.
Epiphytic.
West, Central; 900–2200 m.
Distribution: India, Nepal, Bhutan, Bangladesh, China, Myanmar, Thailand, Laos, Vietnam.
Use: Orn., Med.—Stem paste to reduce stomachache and applied externally to reduce headache.
Platanthera Rich.
Platanthera bakeriana (King & Pantl.) Kraenzl.
Habenaria bakeriana King & Pantl.
Terrestrial.
Central, East; 2500–4500 m.
Distribution: Nepal, NE India, Bhutan, China, Myanmar.
Platanthera carnosilabris (Tang & F.T.Wang) X.H.Jin, Schuit. & W.T.Jin
Herminium carnosilabris Tang & F.T.Wang
Terrestrial.
East; 3000–3400 m.
Distribution: Nepal, China.
Platanthera dambungensis K.Prasad
Terrestrial.
East; 2100–3800 m.
Distribution: Nepal, NE India.
Platanthera dulongensis X.H.Jin & Efimov
Epiphytic.
East; 2800 m.
Distribution: Nepal, NE India, China, Myanmar.
Platanthera japonica (Thunb.) Lindl.
Orchis japonica Thunb.
Platanthera arcuata Lindl.
Terrestrial.
West, Central; 2400–3100 m.
Distribution: India, Nepal, Bhutan, China.
Platanthera leptocaulon (Hook.f.) Soo

Habenaria leptocaulon Hook.f.
Terrestrial.
West, Central, East; 2400–4300 m.
Distribution: Nepal, NE India, Bhutan, China, Myanmar.
Platanthera nematocaulon (Hook.f.) Kraenzlin
Habenaria nematocaulon Hook.f.
Platanthera juncea (King & Pantl.) Kraenzlin
Peristylus nematocaulon (Hook.f.) M.L.Banerji & P.Pradhan
Terrestrial.
Central, East; 2700–3600 m.
Distribution: Nepal, NE India, Bhutan, China (Tibet), Myanmar.
Platanthera pachycaulon (Hook.f.) Soo
Habenaria pachycaulon Hook.f.
Platanthera exelliana Soo
Terrestrial.
Central, East; 2400–4200 m.
Distribution: Nepal, NE India, Bhutan, China.
Platanthera sikkimensis (Hook.f.) Kraenzlin
Habenaria sikkimensis Hook.f.
Terrestrial.
East; 2500–4000 m.
Distribution: Nepal, NE India, China.
Use: Med.—Tuber for tonic.
Platanthera stenantha (Hook.f.) Soo
Habenaria stenantha Hook.f.
Terrestrial.
Central, East; 1500–3600 m.
Distribution: Nepal, NE India, Bhutan, China, Myanmar.
Platanthera stenoglossa Hayata
Terrestrial.
Central; 3200 m.
Distribution: Nepal, China, Japan.
Platanthera superantha (J.J.Wood) X.H.Jin, Schuit., Raskoti & Lu Q.Huang
Peristylus superanthus J.J.Wood
Terrestrial.
East; 3300–3500 m.
Distribution: Nepal, NE India, Myanmar.
Platanthera urceolata (C.B.Clarke) Bateman
Habenaria urceolata C.B.Clarke
Diphylax urceolata (C.B.Clarke) Hook.f.
Terrestrial.
Central, East; 3000–4200 m.
Distribution: India, Nepal, Bhutan, China, Myanmar.
Pleione D.Don
Pleione coronaria P.J.Cribb & C.Z.Tang

Epiphytic.
Central; 2500–3000 m.
Distribution: Endemic to Nepal.
Pleione hookeriana (Lindl.) Rollisson
Coelogyne hookeriana Lindl.
Epiphytic.
Central, East; 2200–3600 m.
Distribution: India, Nepal, Bhutan, China, Myanmar, Thailand, Laos, Vietnam.
Pleione humilis (Sm.) D.Don
Epidendrum humile Sm.
Epiphytic.
West, Central, East; 1800–3000 m.
Distribution: Nepal, NE India, Bhutan, China (Tibet), Myanmar.
Use: Orn. Med.—Pseudobulb paste in cut and wounds.
Pleione maculata (Lindl.) Lindl. & Paxton
Coelogyne maculata Lindl.
Epiphytic.
Central; 1400–2700 m.
Distribution: Nepal, NE India, Bhutan, Bangladesh, China, Thailand.
Use: Orn. Med.—Rhizome for liver and stomach ailments.
Pleione praecox (Sm.) D.Don
Epidendrum praecox Sm.
Epiphytic.
Central, East; 1500–2500 m.
Distribution: Nepal, India, Bhutan, China, Myanmar, Thailand.
Use: Orn. Med. - Pseudobulb paste in cut and wounds.
Podochilus Blume
Podochilus cultratus Lindl.
Epiphytic.
Central, East; 400–1000 m.
Distribution: Nepal, NE India, Bhutan, China, Myanmar, Thailand, Vietnam.
Porpax Lindl.
Porpax elwesii (Rchb.f.) Rolfe
Eria elwesii Rchb.f.
Epiphytic.
Central, East; 800–1800 m.
Distribution: India, Nepal, China (Yunnan), Thailand, Vietnam, Malaysia.
Porpax extinctoria (Lindl.) Schuit., Y.P.Ng & H.A.Pedersen
Dendrobium extinctorium Lindl.
Conchidium extinctorium (Lindl.) Y.P.Ng. & P.J.Cribb
Eria extinctoria (Lindl.) Oliv.
Epiphytic.
Central; 400–600 m.
Distribution: Nepal, Myanmar.
Porpax meirax (E.C.Parish & Rchb.f.) King & Pantl.

Cryptochilus meirax E.C.Parish & Rchb.f.
Epiphytic.
East; 1400–1800 m.
Distribution: Nepal, NE India, Myanmar, Malayasia.
Porpax parviflora (D.Don) Ormerod & Kurzweil
Dendrobium parviflorum D.Don
Conchidium muscicola (Lindl.) Rauschert
Dendrobium muscicola Lindl.
Eria muscicola (Lindl.) Lindl.
Porpax muscicola (Lindl.) Schuit., Y.P.Ng & H.A.Pedersen
Epiphytic.
Central, East; 1200–1800 m.
Distribution: India, Nepal, Bhutan, China, SE Asia.
Use: Med.—Plant in cardiac, respiratory and nervous disorder.
Porpax pusilla (Griff.) Schuit. Y.P.Ng & H.A.Pederson
Conchidium pusillum Griff.
Eria pusilla (Griff.) Lindl.
Pinalia pusilla (Griff.) Kuntze
Terrestrial.
Central; 1300–1700 m.
Distribution: Nepal, NE India, China, Myanmar, Thailand, Vietnam.
Pteroceras Hassk.
Pteroceras teres (Blume) Holttum
Dendrocolla teres Blume
Epiphytic.
East; 200–1000 m.
Distribution: India, Nepal, Bangladesh, China, SE Asia.

Rhomboda **Lindl.**
Rhomboda abbreviata (Lindl.) Ormerod
Hetaeria abbreviata Lindl.
Odontochilus abbreviatus (Lindl.) Tang & Wang
Terrestrial
Central; 500–1600 m.
Distribution: Nepal, NE India, China, Myanmar, Thailand.
Rhynchostylis Blume
Rhynchostylis retusa (L.) Blume
Epidendrum retusum L.
Epiphytic.
West, Central, East; 300–1800 m.
Distribution: India, Nepal, Bhutan, Sri Lanka, China, SE Asia.
Use: Orn. Med.—Leaves to treat rheumatism and root juice applied to cuts and wounds.
Risleya King & Pantl.
Risleya atropurpurea King & Pant.

Mycotrophic.
East; 3400 m.
Distribution: Nepal, NE India (Sikkim), Bhutan, China, Myanmar.
Robiquetia Gaudich.
Robiquetia succisa (Lindl.) Seidenf. & Garay
Sarcanthus succisus Lindl.
Epiphytic.
East; 400–1200 m.
Distribution: Nepal, NE India, Bhutan, China, SE Asia.

Satyrium **Sw.**
Satyrium nepalense D.Don
Terrestrial.
West, Central, East; 600–4600 m.
Distribution: Pakistan, India, Nepal, Bhutan, Sri Lanka, China, Myanmar, Thailand.
Use: Orn. Med.—Tuber as tonic.
Satyrium nepalense D.Don var. **ciliatum** (Lindl.) Hook.f.
Satyrium ciliatum Lindl.
Terrestrial.
Central, East; 1800–4300 m.
Distribution: India, Nepal, Bhutan, China, Thailand.
Schoenorchis Reinw. ex Blume
Schoenorchis gemmata (Lindl.) J.J.Sm.
Saccolabium gemmata Lindl.
Epiphytic.
East; 800–1800 m.
Distribution: Nepal, NE India, Bhutan, China, SE Asia.
Smitinandia Holttum
Smitinandia micrantha (Lindl.) Holttum
Saccolabium micranthum Lindl.
Epiphytic.
Central, East; 500–1400 m.
Distribution: India, Nepal, Bhutan, Bangladesh, China (Yunnan), SE Asia.
Use: Med.—Root powder as tonic.
Spathoglottis Blume
Spathoglottis ixioides (D.Don) Lindl.
Cymbidium ixioides D.Don
Terrestrial.
Central, East; 1500–4600 m.
Distribution: Nepal, NE India, Bhutan, China (Tibet).
Use: Orn.
Spiranthes Rich.
Spiranthes australis (R.Br.) Lindl.
Neottia australis R.Br.

Terrestrial.
West, Central, East; 150–4600 m.
Distribution: Russia, S Asia, China, Japan, SE Asia, New Guinea, Australia.
Use: Med.—Tubers as tonic.
Spiranthes flexuosa (Sm.) Lindl.
Neottia flexuosa Sm.
Ophrys spiralis L.
Terrestrial.
Central; 900 m.
Distribution: Afghanistan, Pakistan, India, Nepal, Sri Lanka, China, Japan, SE Asia.

Taeniophyllum **Blume**
Taeniophyllum scaberulum Hook.f.
Epiphytic.
Central; 600–1500 m.
Distribution: India, Nepal.
Tainia Blume
Tainia minor Hook.f.
Terrestrial.
East; 1500–2300 m.
Distribution: Nepal, NE India, China, Myanmar.
Thelasis Blume
Thelasis longifolia Hook.f.
Epiphytic.
Central, East; 800–1300 m.
Distribution: India, Nepal.
Thelasis pygmaea (Griff.) Lindl.
Euproboscis pygmaea Griff.
Epiphytic.
Central; 800–1500 m.
Distribution: India, Nepal, China, SE Asia, New Guinea.
Thunia Rchb.f.
Thunia alba (Lindl.) Rchb.f.
Phaius albus Lindl.
Epiphytic.
West, Central, East; 500–1900 m.
Distribution: India, Nepal, Bhutan, Bangladesh, China, SE Asia.
Use: Orn. Med.—Plant paste applied to set dislocated bones.
Thunia alba (Lindl.) Rchb.f. var. **bracteata** (Roxb.) N.Pearce & P.J.Cribb
Limodorum bracteatum Roxb.
Epiphytic.
Central; 600–1600 m.
Distribution: India, Nepal, Bhutan, Myanmar, Thailand.
Tipularia Nutt.

Tipularia josephi Rchb.f. ex Lindl.
Terrestrial.
Central, East; 2000–3100 m.
Distribution: India, Nepal, Bhutan, China, Myanmar.
Trichotosia Blume
Trichotosia dasyphylla (C.S.P.Parish & Rchb.f.) Kraenzlin
Eria dasyphylla C.S.P.Parish & Rchb.f.
Epiphytic.
Central; 800–1600 m.
Distribution: Nepal, NE India, China, SE Asia.
Tropidia Lindl.
Tropidia angulosa (Lindl.) Blume
Cnemidia angulosa Lindl.
Terrestrial
Central; 100–200 m.
Distribution: India, Nepal, Bhutan, Bangladesh, China, SE Asia.
Tropidia curculigoides Lindl.
Terrestrial.
Central, East; 200–1000 m.
Distribution: India, Nepal, Bhutan, China, SE Asia.

Uncifera **Lindl.**
Uncifera acuminata Lindl.
Saccolabium acuminatum (Lindl.) Hook.f.
Epiphytic.
East; 1200–1600 m.
Distribution: Nepal, NE India, Bhutan, Bangladesh, China, Vietnam.
Uncifera lancifolia (King & Pantl.) Schltr.
Saccolabium lancifolium King & Pantl.
Epiphytic.
Central; 2000 m.
Distribution: Nepal, NE India, China (Tibet), Vietnam.
Uncifera obtusifolia Lindl.
Saccolabium obtusifolium (Lindl.) Hook.f.
Epiphytic.
Central; 900–1100 m.
Distribution: Nepal, NE India, Bhutan, Myanmar, Thailand, Vietnam.

Vanda **Jones ex R.Br.**
Vanda alpina (Lindl.) Lindl.
Luisia alpina Lindl.
Trudelia alpina (Lindl.) L.A.Garay
Epiphytic.
West, Central; 1100–1800 m.
Distribution: India, Nepal, Bhutan, China (Yunnan), Vietnam.

Vanda amplullacea (Roxb.) L.M.Gardiner
Aerides ampullacea Roxb.
Ascocentrum ampullaceum (Roxb.) Schltr.
Epiphytic.
Central, East; 100–1500 m.
Distribution: India, Nepal, Bhutan, Bangladesh, China, Myanmar, Thailand, Laos, Vietnam.
Use: Orn.
Vanda bicolor Griff.
Epiphytic.
Central; 700 m.
Distribution: Nepal, NE India, China (Tibet), Myanmar.
Vanda cristata Lindl.
Trudelia cristata (Wall. ex Lindl.) Senghas
Epiphytic.
West, Central, East; 1000–2300 m.
Distribution: India, Nepal, Bhutan, China, Vietnam.
Use: Orn. Med.—Root powder applied to cuts and wounds.
Vanda griffithii Lindl.
Luisia griffithii (Lindl.) Kraenzl.
Trudelia griffithii (Lindl.) Garay
Epiphytic.
Central; 1500 m.
Distribution: Nepal, NE India.
Vanda parviflora Lindl.
Vanda testacea var. *parviflora* (Lindl.) M.R.Almeida
Epiphytic.
East; 200–2200 m.
Distribution: India, Nepal, Sri Lanka.
Vanda tessellata (Roxb.) Hook. ex G.Don
Epidendrum tessellatum Roxb.
Vanda roxburghii R.Br.
Epiphytic.
Central; 200–600 m.
Distribution: India, Nepal, Sri Lanka, Myanmar, Vietnam.
Use: Orn. Med.—Root in rheumatism.
Vanda testacea (Lindl.) Rchb.f.
Aerides testaceum Lindl.
Epiphytic.
Central, East; 200–2000 m.
Distribution: India, Nepal, Bhutan, Sri Lanka, Myanmar, Thailand.
Use: Orn. Med.—Leaves as antiviral agent.

Zeuxine **Lindl.**
Zeuxine affinis (Lindl.) Benth. ex Hook.f.
Monochilus affinis Lindl.
Zeuxine grandis Seidenf.

Terrestrial.
Central; 700–2200 m.
Distribution: India, Nepal, Bhutan, Bangladesh, China, SE Asia.
Zeuxine flava (Wall. ex Lindl.) Trimen
Monochilus flavus Wall. ex Lindl.
Terrestrial.
Central, East; 1200–2300 m.
Distribution: India, Nepal, Bhutan, China (Yunnan), SE Asia.
Zeuxine goodyeroides Lindl.
Terrestrial.
Central; 1200–3100 m.
Distribution: Nepal, NE India, Bhutan, China, Myanmar, Vietnam.
Zeuxine integrilabella C.S.Leou
Hetaeria integrilabella (C.S.Leou) S.S.Ying
Epiphytic.
Central; 2100 m.
Distribution: Nepal, China.
Zeuxine lindleyana A.N.Rao
Epiphytic.
Central; 200–800 m.
Distribution: India, Nepal.
Zeuxine longilabris (Lindl.) Trimen
Monochilus longilabris Lindl.
Terrestrial.
Central; 200–800 m.
Distribution: Nepal, NE India, Bangladesh, Sri Lanka, Myanmar, Cambodia.
Zeuxine membranacea Lindl.
Terrestrial.
Central; 400–700 m.
Distribution: Nepal, NE India, Bhutan, China (Hong Kong), Myanmar, Thailand, Cambodia, Vietnam.
Zeuxine nervosa (Wall. ex Lindl.) Benth. ex Trimen
Monochilus nervosus Wall. ex Lindl.
Terrestrial.
Central; 200–1200 m.
Distribution: Nepal, NE India, Bhutan, Bangladesh, China, Thailand, Philippines.
Zeuxine reflexa King & Pantl.
Terrestrial.
East; 400–1400 m.
Distribution: Nepal, NE India, Bhutan, China, Thailand.
Zeuxine strateumatica (L.) Schltr.
Terrestrial.
West, Central, East; 150–1200 m.
Distribution: SW, S & SE Asia, New Guinea.
Use: Med.—Tuber as tonic.

References

Acharya KP, Rokaya MB (2010) Medicinal orchids of Nepal: are they well protected? OurNature (Biratnagar, Nepal) 8:82–91
Acharya KP, Vetaas OR, Birks HJB (2011) Orchid species richness along Himalayan elevational gradients. J Biogeogr 38(9):1821–1833
Bajracharya DM (2010) Nepalese orchids. MD Bajracharya, Kathmandu, Nepal
Bajracharya DM (2011) Orchid flora of Raja Rani (Morang district) and adjoining areas and their conservation. Kriti Publication, Kirtipur, Nepal
Banerji ML (1978) Orchids of Nepal. Bishen Singh Mahendra Pal Singh, India
Banerji ML, Pradhan P (1984) Orchids of Nepal Himalaya. J. Cramer, Germany
Bashyal R, Paudel K, Hinsley A, Phelps J (2023) Making sense of domestic wildlife and CITES legislation: the example of Nepal's orchids. Biol Conserv 280:109951
Bhandari P, Shrestha K, Kc A, Shrestha KK (2020a) Three new records of orchids from Nepal. J Jap Bot 95(1):41–46
Bhandari P, Shakya LR, Chaudhary RP (2019a) *Zeuxine lindleyana* (Orchidaceae)—a new record for the flora of Nepal. J Jap Bot 94(1):45–46
Bhandari P, Uprety Y, Shakya LR (2019b) *Platanthera carnosilabris* (Orchidaceae), an addition to the flora of Nepal. J Jap Bot 94(1):42–44
Bhandari P, Shrestha K, Subedi CK (2020b) Orchids of Panchase forest, central Nepal: a checklist. J Plant Resources (Nepal) 18(1):143–156
Bhardowaj B, Rai KR, Raskoti BB (2015) Inventory of orchids in the Hanspokhari area of Kolbung VDC, Ilam district, eastern Nepal. MIOS J 16(7):18–24
de Milleville R, Shrestha TB (2004) Nepal orchids in pictures. Malla Prakashan, Nepal
Don D (1825) Prodromus Florae Nepalensis, London
Elliott A, Hyam R, Watson M, Wrankmore E, Hartley H, Krieger J, Gandhi K, Acuna R, Almeida RFD, Amorim G, Anderson G, Andrella GC, Anguiano M, Antonio-domingues H, Ard WH, Atkins H, Atwood JJ, Aubriot X, Zizka G (2025) World flora online plant list June 2025 (2025–06) (datas set). The World Flora Online Consortium, Zenodo. https://doi.org/10.5281/zenodo.15704590
Gupta S, Ghimire SK, Parmar G (2025) *Tropidia angulosa* (Orchidaceae)—a new record for the flora of Nepal. J Jap Bot 100(1):84–85
Hara H, Stearn WT, Williams LHJ (1978) An enumeration of the flowering plants of Nepal, vol 1. British Museum, Natural History, London
Joshi PR, Thapa BB, Chand K, Maharjan L, Paudel MR, Pant B (2024) Biological activities of protocorms and stems extracts of Dendrobium transparens. Trends Phytochem Res 2(2):122
Kafle S, Pathak HN (2023) Diversity of orchids in Shiureni forest, Aadhkhola rural municipality, Syangja, Nepal. Him Biodiv (Pokhara, Nepal) 9:22–28
Kafle R, Magar MST, Parmar G (2024) *Calanthe longipes* Hook. f., a new record of Orchidaceae for the flora of Nepal. Banko Janakari (J Forestry Information for Nepal) 34(1):51–56
Kafle R, Raskoti BB, Parmar G (2025) Dendrobium hookerianum (Orchidaceae)—a new official record for the flora of Nepal. J Jap Bot 100(4):347–349
Kurzweil H, Raskoti BB, Witthana S, Pingyot T (2017) *Habenaria iyoensis* (Orchidaceae), a new record for Thailand and Nepal. Thai Forest Bull Bot 45(2):108–112
Lamichhane D (2014) Trade of orchids and conservation efforts in Nepal. In: Jha PK et al (eds) Proceedings of National workshop on NTFP/MAPs sector action plan development: orchid. Department of Plant Resources, Ministry of Forests and Soil Conservation and Central Department of Botany, Tribhuvan University, Nepal, pp 49–55
Pant B (2014) Medicinal orchids and their research in Nepal. In: Jha PK et al (eds) Proceedings of National Workshop on NTFP/MAPs sector action plan development: Orchid. Department of Plant Resources, Ministry of Forests and Soil Conservation and Central Department of Botany, Tribhuvan University, Nepal, pp 44–48

Pant B, Raskoti BB (2013) Medicinal orchids of Nepal. Himalayan Map House (P.) Ltd., Kathmandu, Nepal

Pant B, Poudel MR, Chand MB, Wagner SH (2016) A treasure trove of orchids in central Nepal. Kathmandu, Nepal, Central Department of Botany, Tribhuvan University

Pant B, Poudel MR, Chand MB, Pradhan S, Malla BB, Raskoti BB (2018) Orchid diversity in two community forests of Makwanpur district, central Nepal. J Threatened Taxa 10:12523–12530

Pant B, Pradhan S, Paudel MR, Shah S, Pandey S, Joshi PR (2019) Various culture techniques for the mass propagation of medicinal orchids from Nepal. Acta Horticult 1262:109–124

Pant B, Chand K, Paudel MR et al (2021a) Micropropagation, the antioxidant and anticancer activity of pineapple orchid: Dendrobium densiflorum Lindl. J Plant Biochem Biotechnol. https://doi.org/10.1007/s13562-021-00692-y

Pant B, Paudel MR, Joshi PR (2021b) Orchids as potential sources of anticancer agents: our experience. Annapurna J Health Sci 1(1):42–51. https://www.ajhs.org.np/ajhs/index.php/ajhs/article/view/17

Pant B, Chand K, Paudel MR et al (2022) Micropropagation, antioxidant and anticancer activity of pineapple orchid: Dendrobium densiflorum Lindl. J Plant Biochem Biotechnol 31:399–409. https://doi.org/10.1007/s13562-021-00692-y

Parmar G, Kafle R, Lamichhane D, Magar MST (2024) Catalogue of plants of National Botanical Garden, 2024. National Botanical Garden, Department of Plant Resources, Lalitpur, Nepal

Paudel HR (2020) The first record of the genus *Tropidia* Lindl. (Orchidaceae) from Nepal. Intern J Appl Chem Biol Sci 1(3):53–57

Paudel MR, Bhattarai HD, Pant B (2020) Traditionally used medicinal Dendrobium: a promising source of active anticancer constituents. In: Merillon JM, Kodja H (eds) Orchids phytochemistry, biology and hortculture. Springer Nature, Switzerland. https://doi.org/10.1007/978-3-030-11257-8_16-1

Pun U (2013) Orchids. In: Jha PK et al (eds) Nepalpedia series no. 2: biological diversity and conservation. Nepal Academy of Science and Technology, Lalitpur, Nepal, pp 427–431

Pun U (2014) Orchid farming potentiality and commercialization (aesthetic and medicinal orchids) in Nepal. In: Jha PK et al (eds) Proceedings of National Workshop on NTFP/MAPs sector action plan development: orchid, Department of Plant Resources, Ministry of Forests and Soil Conservation and Central Department of Botany, Tribhuvan University, Nepal, pp 41–43

Rajbhandari KR (2014) Orchids of Nepal: status, threat and conservation. In: Jha PK et al (eds) Proceedings of national workshop on NTFP/MAPs sector action plan development: orchid. Department of Plant Resources, Ministry of Forests and Soil Conservation and Central Department of Botany, Tribhuvan University, Nepal, pp 1–40

Rajbhandari KR (2015) A handbook of the orchids of Nepal. Department of Plant Resources, Kathmandu, Nepal

Rajbhandari KR, Bhattarai S (2001) Beautiful orchids of Nepal. Authors, Kathmandu, Nepal

Rajbhandari KR, Dahal S (2004) Orchids of Nepal: a checklist. Bot Orient (Central Department of Botany, Tribhuvan University, Nepal) 4(1):89–106

Rajbhandari KR, Pandey TR (2024a) An introduction of the orchids of Nepal, vol 1. The Gardens and Water Parks (Lalitpur Metropolitan City, Lalitpur, Nepal), pp 97–104

Rajbhandari KR, Pandey TR (2024b) Know our Nepali Dendrobium orchids, vol 2. The Gardens and Water Parks (Lalitpur Metropolitan City, Lalitpur, Nepal), pp 72–75

Rajbhandari KR, Rai SK (2017) A handbook of the flowering plants of Nepal, vol 1. Department of Plant Resources, Kathmandu, Nepal

Rajbhandari KR, Bhattarai S, Joshi R (2000) Orchid diversity of Nepal and their conservation need. In: Proceedings of the 8th international workshop of BIO-REFOR, Kathmandu, Nepal, November 28–December 2, 1999, Biotechnology applications for reforestation and biodiversity conservation, 2000, pp 249–252

Rajbhandari KR, Bhatt GD, Chhetri R, Pandey TR, Thapa TK, Budhathoki KB (2024a) A checklist of the seed plants of Khaptad National Park, Nepal. Plant Research Centre, Department of Plant Resources, Dhangadhi, Kailali, Nepal

Rajbhandari KR, Rai SK, Chhetri R (2024b) A handbook of the flowering plants of Nepal, vol 5. National Herbarium and Plant Laboratories, Department of Plant Resources, Kathmandu, Nepal

Raskoti BB (2009) The orchids of Nepal. BB Raskoti and R Ale, Kathmandu, Nepal

Raskoti BB, Ale R (2019a) New species of orchids and notes on Orchidaceae of Nepal. Phytotaxa 394(4):257–266

Raskoti BB, Ale R (2019b) Molecular phylogeny and morphology reveal a new epiphytic species of Habenaria (Orchidaceae, Orchideae, Orchidinae) from Nepal. PLoS ONE 14(10):e223355. https://doi.org/10.1371/journal.pone.0223355. 12 pp

Raskoti BB, Ale R (2023) A new variety of *Habenaria* (Orchidaceae, Orchideae; Habenariinae) and new records of orchids from Nepal. Taiwania 68(1):122–128

Rokaya MB, Raskoti BB, Timsina B, Muenzbergova Z (2013) An annotated checklist of the orchids of Nepal. Nordic J Bot 31:511–550

Samuel AH (2022) Orchids of Godavari and surrounding areas. International Centre for Integrated Mountain Development, Kathmandu, Nepal

Samuel AH (2023) *Lecanorchis japonica* (Orchidaceae)—a new generic record for the orchid flora of Nepal. J Jap Bot 98(4):205–207

Shahi S, Budhathoki KB, Bhatt GD (2025) Catalogue of plants of Dewahariya Botanical Garden, Kailali. Plant Research Centre, Department of Plant Resources, Kailali, Nepal

Shakya LR, Bajracharya DM (2013) Survey of the orchids of Shivapuri National Park in Kathmandu (Nepal) and their conservation. Pleione 7(1):39–45

Subedi A, Subedi N, Chaudhary RP (2007) Panchase forest: an extraordinary place for wild orchids in Nepal. Pleione 1:23–31

Subedi CK, Bhandari P, Thapa S, Pandey M, Gurung J, Shakya LR, Chaudhary RP (2018) *Cephalanthera erecta* var. *oblanceolata* (Orchidaceae)—a new record for the flora of Nepal. J Jap Bot 93(4):287–290

Teoh ES (2019) Medicinal orchids of Nepal and Bhutan. In: Orchids as Aphrodisiac, medicine or food. Springer, Cham. https://doi.org/10.1007/978-3-030-18255-7_13

Tiruwa BL, Neupane BD, Kadariya R, Pokhrel CP, Pant B (2024a) Diversity of orchids in terms of their distribution, uses and conservation in Annapurna Conservation Area of Nepal. Am J Plant Sci 15:422–440

Tiruwa BL, Neupane BD, Karki S, Sharma S, Paudel RP, Kadariya R (2024b) A pictorial handbook of orchids of the Annapurna Conservation Area. National Trust for Nature Conservation—Annapurna Conservation Area Project, Pokhara, Nepal

White K, Sharma B (2000) Wild orchids of Nepal, guide to the Himalayan orchids of Tribhuvan Rajpath and Chitwan Jungle. White Lotus, Bangkok, Thailand

4 Taxonomy, Ecology and Floral Biology of Forest-Based Orchids in Nainital and Surrounding Areas of Uttarakhand, India

Vasundhra Lodhiyal, Jeewan Singh Jalal, Renu Suyal, L. M. Tewari, N. C. Pandey, Disha Upreti, Vishal Bisht, and Lata Nitwal

4.1 Introduction

Nainital, one of the most prominent hill stations of Uttarakhand, occupies a distinct place both in Indian mythology and modern geographical literature. The region is mentioned in the *Manas Khanda* of the *Skanda Purana* as *Tri-Rishi Sarovar*, a sacred lake created by the sages Atri, Pulastya, and Pulaha, who, finding no water for sustenance during their pilgrimage, invoked waters from *Mansarovar* in Tibet. A second cultural reference identifies Nainital as one of the sixty-four *Shakti Peeths*, consecrated by the fall of Sati's eyes (*nain*) while her body was being carried by Lord Shiva, thus giving the site its name "Naini-tal" or "Lake of the Eye". The revered Naina Devi temple on the northern shore of Naini Lake remains a site of Shakti worship today (Atkinson 1882; Joshi and Tiwari 2010). The colonial discourse around Nainital begins with the British merchant P. Barron, whose observations were published in the *Calcutta Englishman* (1841) and *Agra Akhbar*. His writings popularized Nainital's scenic charm to the wider public. Subsequently, Barron constructed his residential lodge in 1842, catalyzing the development of Nainital as a health resort and summer retreat for Europeans in colonial India. Geographically, Nainital is situated between 29°19′–29°28′ N latitude and 79°22′–79°38′ E longitude in the Kumaun Lesser Himalaya, at an altitude of 1938 m. Naini Lake, a tectonic origin water body, is enclosed by seven surrounding ridges: Naina (2611 m), Alma (2230 m), Lariakanta (2481 m), Sher-ka-danda (2402 m), Deopata

V. Lodhiyal · L. M. Tewari (✉) · N. C. Pandey · D. Upreti · V. Bisht · L. Nitwal
Department of Botany, D.S.B. Campus, Kumaun University, Nainital, Uttarakhand, India

J. S. Jalal
Central National Herbarium, Howarh, Botanical Survey of India, Kolkata, West Bengal, India

R. Suyal
Ecology Development and Research, Centre for Ecology Development and Research (CEDAR), Dehradun, Uttarakhand, India

A. Husen, M. Pant (eds.), *Forest-based Orchids*,
https://doi.org/10.1007/978-981-92-0096-2_4

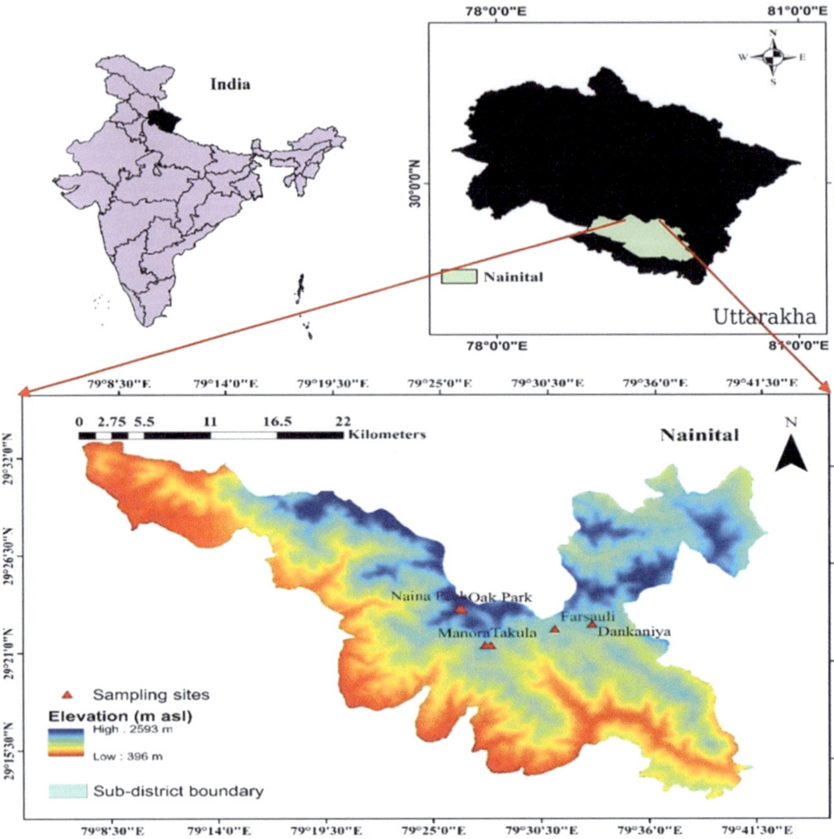

Fig. 4.1 Map of Nainital and it surrounding areas

(2434 m), Handibandi (2180 m), and Ayarpatta (2352 m). Other notable lakes in the vicinity include Bhimtal, Naukuchiatal, Sattal, Khurpatal, Sukhatal, and Sariatal, collectively earning the Kumaun region its epithet of "*Lake District of India*" (Rawat 2005) (Fig. 4.1).

The orchid family (*Orchidaceae*) represents one of the largest flowering plant families globally, comprising ~28,000 species across ~850 genera (Chase et al. 2015; Lal et al. 2024). The name *Orchis* was first introduced by the Greek philosopher Theophrastus (370–285 BC) in *Enquiry into Plants*, in reference to paired tuberous roots. In Indian Vedic literature, orchids were referred to as *Vanda*, a genus that today remains one of the most prized ornamental groups (Misra 2007). Orchids embody remarkable diversity in size, morphology, growth habit, and pollination biology, often displaying specialized co-evolutionary relationships with pollinators and mycorrhizal fungi (Arditti 1992; Dressler 1993) (Fig. 4.2).

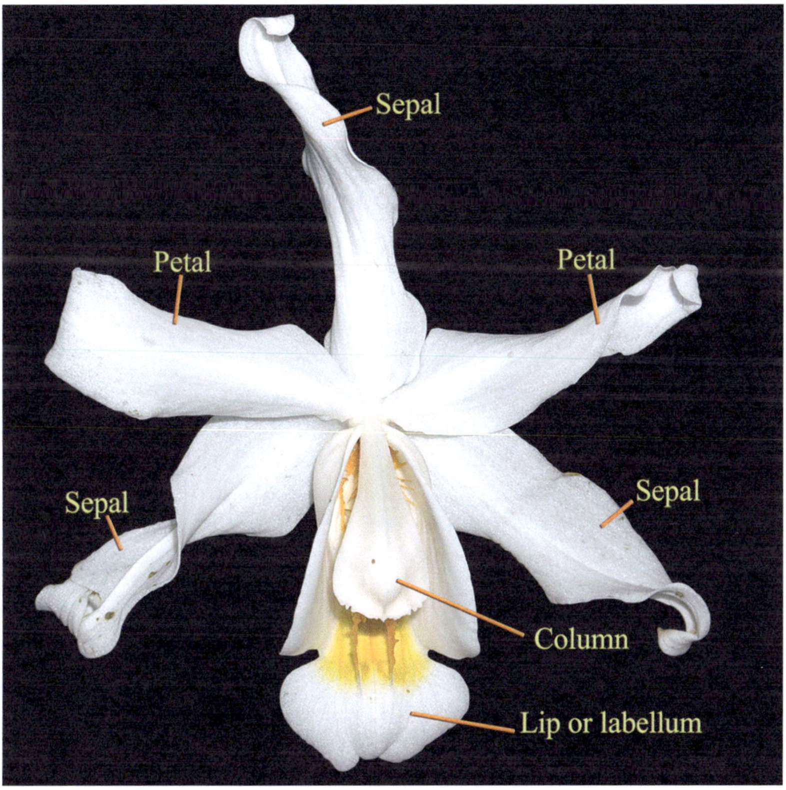

Fig. 4.2 An orchid flowers representing all the floral parts of orchids

Globally, orchids occur in nearly all ecosystems, with the highest species richness in tropical America, Indo-Malayan regions, and the eastern Himalaya. Approximately, 72% of orchid species are epiphytes, predominantly in moist tropical forests, while terrestrial orchids are more common in temperate and alpine zones (Rasmussen 1995; Cribb and Govaerts 2005). In India, orchids are exceptionally diverse, with ~1331 species belonging to 184 genera, accounting for ~7.6% of the angiosperm flora (Misra 2007; Jalal and Jayanthi). Their distribution is concentrated in the Eastern Himalaya and Northeast India (notably Arunachal Pradesh, Sikkim, and Meghalaya), Western Ghats, and to a lesser extent, the Western Himalaya. Within the Western Himalaya, the Kumaun region, encompassing Nainital, represents a transition zone where both terrestrial and epiphytic forms coexist due to its climatic and altitudinal gradients (Deva and Naithani 1986; Pangtey et al. 1991a). The growth forms of orchids in Nainital reflects their ecological versatility. Based on growth habits, they may be classified as terrestrial (e.g., genera *Habenaria*, *Peristylus*), mycoheterotrophic (e.g., *Neottialisteroides*, *Gastrodia*), or epiphytic/lithophytic (e.g., *Aerides*, *Dendrobium*, *Coelogyne*, *Vanda*). Terrestrial orchids often develop rhizomes, tubers, or corms, enabling persistence in seasonal droughts. Epiphytic orchids, predominantly inhabiting canopy trees and moss-laden rocks, evolve specialized aerial roots with velamen for rapid water absorption from mist and rainfall (Rasmussen 1995). Sympodial and monopodial growth patterns further differentiate orchids. Sympodial orchids generate successive shoots and often pseudobulbs (e.g., *Bulbophyllum*, *Pholidota*), whereas monopodial orchids (e.g., *Vanda*, *Phalaenopsis*) elongate from a singular terminal growth axis. The striking floral adaptations of orchids, particularly the zygomorphic flowers with a highly modified labellum acting as a pollinator landing platform, the column (fusion of androecium and gynoecium), and the formation of pollinia instead of powdery pollen, represent advanced reproductive innovations (Dressler 1993; Chase et al. 2015).

Botanical exploration in Kumaun began with the collections of Richard Blinworth for Dr. Nathaniel Wallich (1815–1820), followed by J.F. Royle (1833–1839), and most comprehensively, Sir Richard Strachey. Duthie (1906a) later consolidated their work in his *Catalogue of Plants of Kumaun*, reporting 13 orchid species. Subsequent floristic inventories, including those by Hooker (1890), Champion (1921), Gupta (1968), Vij et al. (1983a), Deva and Naithani (1986), Pangtey and Samant (1991), and Uniyal et al. (1997, 2007), progressively enriched the checklist of orchids in Nainital. The present synthesis reveals the presence of 43 genera and 85 species, distributed across three subfamilies—Cypripedioideae (1 genus/1 species), Orchidoideae (16 genera/32 species), and Epidendroideae (26 genera/52 species). Among these, *Habenaria* (9 spp.), *Dendrobium* (6 spp.), and *Peristylus* (5 spp.) represent the most diverse genera. Epiphytic orchids (*Aerides*, *Dendrobium*, *Phalaenopsis*, *Vanda*) dominate the canopy habitats, whereas terrestrial (*Habenaria*, *Platanthera*, *Neottia*) and mycoheterotrophic (*Gastrodia*) forms occupy shaded forest understoreys and grassy slopes. Orchids are not merely ornamental "flagship plants," but ecologically they act as indicators of ecosystem health due to their sensitivity to microclimatic and edaphic conditions (Swarts and Dixon 2009). Their obligate dependence on pollinator guilds and fungal symbionts underlines their keystone role in Himalayan Forest ecosystems. In local ethnomedicine, orchid tubers form part of *Ashtavarga*—a vital group of Ayurvedic tonics (Chauhan 1996). Internationally, orchids play a major role in the floriculture industry, with genera like *Phalaenopsis*, *Dendrobium*, and *Cymbidium* forming high-value commodities (De and Vij 1992). Yet, habitat perturbations, deforestation, over-collection, and climate change remain significant threats to their survival in Himalayan landscapes (Jalal and Jayanthi 2012).

There are various forest habitats and vegetation zones in Nainital and its surroundings where Orchids commonly occur. In Nainital, the forest habitats and vegetation zones are strongly structured by altitude and aspect, creating a mosaic of ecological communities that range from xeric chir-pine forests at lower elevations to mesic oak, deodar, and cypress forests at higher zones, interspersed with secondary grassy slopes shaped by disturbance. Chir-pine (*Pinus roxburghii* Sarg.) forests represent one of the most distinctive vegetation types of the Western Himalaya, typically occupying exposed south-facing slopes between 1600–2000 m elevation (Banday et al. 2017). These forests form extensive pure stands characterized by minimal species diversity and open canopy structure, with occasional emergent individuals of *Quercus leucotrichophora* and other broadleaved associates occurring sporadically (Lodhiyal and Lodhiyal 2012). The edaphic preferences of chir-pine are fundamentally distinct from those of oak forests, being restricted to xerophytic conditions such as ridge crests, south-facing slopes, and well-drained rocky substrates (Jackson 1994). The dominance of *P. roxburghii* in these habitats is attributed to its exceptional fire adaptation mechanisms, including thick protective bark, vigorous coppicing ability when young, and rapid establishment following disturbance (Sigdel et al. 2018). Historical fire regimes have been instrumental in maintaining the open character of these forests, preventing establishment of fire-sensitive species and promoting grass-dominated understoreys. The ground layer is typically covered by a persistent litter of fallen needles, which creates allelopathic conditions that inhibit herbaceous development. Chir-Pine and Banj-Oak Mixed Forests (1200–1500 m) form ecologically significant transition zones where *P. roxburghii* and *Q. leucotrichophora* achieve co-dominance. This vegetation type represents a steep ecological gradient where chir-pine dominates xeric south-facing slopes with graminoid understoreys, while banj-oak occupies mesic north-facing aspects supporting diverse herbaceous communities (Champion and Seth 1968). Banj-Oak Forests (1400–2400 m) dominated by *Quercus leucotrichophora* A. Camus constitute the most widespread forest type in the study region (Singh and Singh 1987; Lodhiyal and Lodhiyal 2012). These forests demonstrate remarkable ecological amplitude, extending from moist ravines at 1100 m to exposed ridges at 2400 m elevation. The structural characteristics of banj-oak forests vary dramatically with aspect and moisture availability—dense, closed-canopy forests with rich understoreys occur on north-facing slopes, while open, grass-dominated stands characterize south-facing exposures (Rawat and Pangtey 1987). The fire-resistant qualities of banj-oak, combined with its prolific coppicing ability, enable persistence across diverse habitat conditions (Savard et al. 2000). In mesic habitats, these forests support exceptional floristic diversity, particularly of cryptogams (ferns, mosses, and liverworts), and often exhibit gregarious stands of *Sinarundinaria falcata* (ringal bamboo) (Singh and Singh 1987). Telunj-Oak Forests (2200–2500 m) dominated by *Quercus floribunda* Lindl. occupy intermediate elevations between banj-oak and kharsu-oak zones. These forests achieve optimal development on deep, calcareous soils and represent transitional communities with moderate species diversity (Champion and Seth 1968). Kharsu-Oak Forests (2500–2611 m) constitute the uppermost broad-leaved forest zone, dominated by *Quercus semecarpifolia* Sm. This vegetation type occurs on both north and south aspects, exhibiting a more open structure with scattered subdominant tree species and variable shrub density (Singh and Singh 1987). Deodar Forests (2200–2600 m) dominated by *Cedrus deodara* (Roxb. ex D. Don) G. Don represent climatically controlled vegetation confined to north-west facing slopes of Naina Peak (Prahlad et al. 2018). These forests exhibit exceptional productivity with dense canopies that create persistently moist, shaded ground conditions supporting luxuriant cryptogamic communities (Champion and Seth

1968). The species demonstrates remarkable ecological adaptability, capable of establishment on precipitous rocky substrates with minimal soil development. Cypress Forests (2400–2600 m) dominated by *Cupressus torulosa* D.Don occur as highly localized communities restricted to south-west slopes of Naina Peak. This vegetation type occupies extremely specialized habitats, including rock crevices and precipitous slopes with minimal soil development, representing one of the most habitat-restricted forest communities in the region. Grassy Slopes represent secondary vegetation resulting from forest degradation through fire, grazing, and anthropogenic disturbances. These communities are predominantly restricted to south-facing exposures where intense solar radiation and moisture stress preclude tree establishment. Despite their degraded status, these habitats provide critical refugia for terrestrial orchids, particularly in microhabitats with enhanced moisture retention (Jalal and Jayanthi 2012).

Systematic botanical exploration of the Western Himalaya commenced with Major General Thomas Hardwicke's pioneering collections in 1796, marking the beginning of European scientific interest in Himalayan flora (Hooker 1906). In the Kumaun region specifically, Richard Blinworth (1815–1820) initiated comprehensive plant collections under the auspices of Dr. Nathaniel Wallich, Superintendent of the Saharanpur Botanical Garden (Duthie 1906a). J.F. Royle's phytogeographical investigations (1833–1839) culminated in his seminal work *Illustrations of the Botany and other Branches of Natural History of the Himalayan Mountains and of the Flora of Kashmir*, establishing foundational knowledge of Himalayan biogeography (Royle 1833–1839). The most comprehensive early collections were undertaken by Sir Richard Strachey, whose systematic inventories formed the basis for all subsequent floristic studies in Kumaun (Atkinson 1882). Their original catalogue, published in Atkinson's *Gazetteer of Himalayan Districts of the North West Provinces and Oudh* (1882), represented the first authoritative documentation of the region's flora. The appointment of J.F. Duthie as Director of Saharanpur Botanical Garden in 1876 marked a transformative period in Himalayan orchid research (Duthie 1906a). Duthie's revised catalogue (1906b) documented 13 orchid species in 11 genera from Nainital and adjacent areas, while his comprehensive orchid flora (Duthie 1906a) catalogued 173 orchid species in 45 genera from the entire Western Himalaya. This work established Duthie as the pre-eminent authority on Western Himalayan orchids and provided the taxonomic framework for all subsequent investigations. Concurrent with Duthie's work, Colonel R.E. Davidson's collections from Nainital vicinity contributed numerous type specimens to the Saharanpur herbarium, several of which were described as new species (Hooker 1890). Sir J.D. Hooker's *Flora of British India* (1890) further consolidated knowledge of Kumaun orchids within the broader Indian floristic context. Post-independence orchid research intensified with establishment of the Northern Circle of Botanical Survey of India at Dehradun (Jain 1956). Seidenfaden and Arora (1982a) enumerated 250 species from the Western Himalaya, adding 77 species to Duthie's original inventory and describing two new species. The definitive modern treatment by Deva and Naithani (1986) provided an illustrated monograph of 72 genera and 239 species, representing the most comprehensive systematic account of Western Himalayan orchids (Kumar and Manilal 1994). Subsequent investigations by Pangtey and Samant (1991) recorded 36 genera and 70 species from Nainital based on fresh collections, while Pangtey et al. (1991b) documented 61 genera and 193 species from the broader Kumaun region. Recent contributions by Jalal et al. (2005, 2009a) and Jalal and Pangtey (2010a) have added several new records, including the significant rediscovery of *Gastrodia falconeri* after nearly a century.

4.2 The Orchid Flora of Nainital, Uttarakhand

The orchid flora of Nainital, comprising 44 genera and over 80 species, represents significant floristic richness in the western Himalayas, reflecting tropical, subtropical, and temperate influences (Arora 1982; Deva and Naithani 1986; Pangtey and Samant 1991). This diversity is closely linked to the region's topography, climate, and habitat heterogeneity, providing niches from moist riverine valleys to shaded forest floors and open grasslands on ridges (Vij et al. 1983a; Jyoti and Tewari 2020). The altitudinal gradient (500–3000 m) facilitates the coexistence of orchids adapted to lowland tropical and temperate climates (Cribb and Govaerts 2005; Chase et al. 2015).

4.2.1 Life Forms and Habitat Specialization

Orchids in Nainital occur as epiphytic, terrestrial, and lithophytic life forms, each showing distinctive ecological preferences. Epiphytes such as *Acampe*, *Aerides*, *Vanda*, *Phalaenopsis*, *Rhynchostylis*, and *Thunia* dominate lower to mid-elevations (500–1800 m), growing on various host trees where factors like bark texture and canopy structure influence colonization (Pangtey et al. 1991b; Jyoti and Tewari 2020). Adaptations including fleshy leaves, thick aerial roots, and water-storage tissues allow survival amid varying humidity and nutrient availability in the canopy (Duthie 1906a; Arditti 1992).

Terrestrial orchids like *Platanthera*, *Spiranthes*, *Satyrium*, *Zeuxine*, *and Pholidota* are found mainly at higher elevations (1500–2600 m), preferring moist, shaded slopes and moss-dominated forest floors (Pangtey and Samant 1991; Rasmussen 1995). These species depend on soil moisture and organic content, emphasizing the importance of intact forests (Rawat 2005). Lithophytes such as some Thunia and *Smitinandia* grow on rocky riverine surfaces, showing adaptations for nutrient-poor substrates (Seidenfaden and Arora 1982a).

4.2.2 Morphology and Reproductive Adaptations

Nainital orchids exhibit diverse flower morphologies associated with their pollination syndromes. For example, *Platanthera edgeworthii* and *Satyrium nepalense* have elongated spurs adapted to long-tongued moths, while the spirally arranged flowers of *Spiranthes sinensis* favour small insect pollinators and potential self-pollination (Jalal et al. 2009b; Mahapatra et al. 2021). Epiphytes like *Vanda cristata* display large, colorful flowers attracting bees and butterflies, whereas cryptic forest-floor orchids such as *Zeuxine flava* rely on less conspicuous insect visitors (Pangtey and Samant 1987; Rasmussen 1995).

Leaf morphology also reflects ecophysiological adaptation; epiphytes possess narrow, leathery leaves to reduce water loss, whereas terrestrials tend to have broader, fleshy leaves optimizing photosynthesis under shade (Pangtey et al. 1991a; Arditti 1992). Pseudobulbs and tubers provide nutrient and water storage, aiding survival in drought or low activity periods (Rasmussen 1995).

4.2.3 Phenology and Flowering Dynamics

Flowering times are staggered from April (*Zeuxine flava*) to November (*Spiranthes sinensis*), indicating temporal niche differentiation that reduces interspecific pollinator competition and maintains nectar availability (Pellissier et al. 2010). Early bloomers such as *Vanda tessellata* and *Acampe praemorsa* align with peak pollinator activity in warm valleys, while late-flowering species like *Platanthera japonica* reproduce in cooler high-altitude zones (Vij et al. 1983b; Jalal et al. 2005).

4.2.4 Altitudinal and Microhabitat Distribution

Altitudinal patterns define orchid distribution, with epiphytes concentrated at lower humid elevations and terrestrials dominant at higher cooler zones (Seidenfaden and Arora 1982b; Chase et al. 2015). Deceptive orchids are more frequent at lower altitudes where pollinator diversity is greater, supporting ecological strategies shaped by altitude (Pellissier et al. 2010). Microhabitat specializations such as *Smitinandia micrantha*'s restriction to riverine forests and *Thunia alba*'s dual lithophytic and epiphytic habits showcase ecological plasticity within the region (Joshi et al. 2009).

4.2.5 Biogeography and Floristic Affinities

The Nainital flora spans Indo-Malayan tropical and temperate Himalayan elements. Genera like *Acampe* and *Phalaenopsis* reflect tropical Indo-Malayan origins, while *Platanthera* and *Zeuxine* align with temperate Himalayan lineages (Chase et al. 2015; Givnish et al. 2016). Some taxa have Southeast Asian distributions, suggesting historical dispersal and ecological connectivity (Seidenfaden and Arora 1982b; Cribb and Govaerts 2005).

4.2.6 Species Interactions and Ecological Roles

Orchids engage in complex networks with pollinators, mycorrhizal fungi, and host trees. Mycorrhizal fungi are crucial for germination and nutrient uptake, especially in terrestrial species (Waud et al. 2014; McCormick et al. 2004). Pollination modes vary widely, with entomophily predominating, along with self-pollination and hawkmoth specialization, reinforcing species coexistence (Swarts and Dixon 2009; Mahapatra et al. 2021). Orchids contribute to local biodiversity by providing nectar and pollen resources (Pangtey and Samant 1991).

4.2.7 Orchid Diversity in the Region

Describe below are 43 genera and 85 species documented in the Nainital region, distributed across three subfamilies: Cypripedioideae (1 genus, 1 species), Orchidoideae (16 genera, 32 species), and Epidendroideae (26 genera, 52 species). The generic diversity reflects the biogeographical position of Nainital as a transition zone between tropical and temperate orchid floras. Species-rich genera include *Habenaria* (9 spp.), *Dendrobium* (6 spp.), *Peristylus* (5 spp.), *Eria*, *Liparis*, *Nervilia*, and *Platanthera* (4 spp. each) (Pangtey and Samant 1991). This distribution pattern aligns with broader Himalayan trends where terrestrial genera (*Habenaria*, *Platanthera*) achieve maximum diversity in temperate zones, while epiphytic genera (*Dendrobium*, *Aerides*, *Bulbophyllum*) are more diverse in subtropical elevations (Deva and Naithani 1986). Epiphytic orchids comprise 15 genera including *Aerides*, *Bulbophyllum*, *Coelogyne*, *Dendrobium*, *Eria*, *Oberonia*, *Phalaenopsis*, *Pholidota*, *Rhynchostylis*, *Thunia*, and *Vanda* (Vij et al. 1983b). These taxa exhibit pronounced host specificity, with *Quercus leucotrichophora* supporting the greatest diversity (6 species), followed by *Sapium insigne* (5 species) and *Syzygium operculatum* (4 species) (Pangtey and Samant 1991). Mycoheterotrophic orchids are represented by four species across three genera: *Cymbidium macrorhizon*, two *Neottia* species, and *Gastrodia falconeri*. These saprophytic taxa occupy specialized ecological niches in humus-rich soils and represent some of the most conservation-sensitive elements of the orchid flora. The remarkable orchid diversity of Nainital reflects the region's position at the confluence of biogeographical zones, supporting both tropical epiphytic elements and temperate terrestrial assemblages within a relatively compact area. However, several historically recorded species remain unconfirmed, highlighting the urgent need for comprehensive resurveys and conservation prioritization (Uniyal et al. 2007).

Orchids in Nainital represent an important component of the region's floral diversity. They occur across a range of habitats and elevations, thriving in both shaded forest floors and moist slopes, and contribute significantly to the ecological richness of the area. The list of orchids in Nainital and its surrounding areas is as follows:

1. ACAMPE **Lindl.**
Acampe (from Greek *akampas* = rigid) refers to the brittle flowers. Two species are known from Uttarakhand; one is recorded in Nainital.

1.1. ***Acampe papillosa*** Lindl.
Epiphytic, up to 40 cm tall, stem stout and sheathed. Leaves 7–10 × 1–2 cm, conduplicate, fleshy, bilobed at apex. Inflorescence umbellate, 4–8-flowered. Flowers are yellow with brown blotches. Lip white with transverse bars; spur half as long as ovary, hairy within. Capsule fusiform.

Flowering: October–November.
Habitat: Moist-deciduous and riverine forests up to 1300 m, often on *Shorea robusta* and *Toona ciliata*; also on boulders near streams.
Distribution in Nainital: Dogaon, Jeolikote, Kathgodam, Ranibagh.
General Distribution: India (NE and central states, Uttarakhand, WB), Bangladesh, Bhutan, China, Indo-China, Myanmar, Nepal, Thailand.
References: Chowdhery and Agrawala (2013).

2. AERIDES **Lour.**
Aerides (from Greek *aer* = air, *eides* = child) refers to "children of the air" or air-plants. Two species occur in Uttarakhand, also found in Nainital.

2.1. ***Aerides multiflora*** (Roxb.)
Epiphytic/lithophytic, stem 10–25 cm. Leaves conduplicate, 10–15 cm, bifid at apex, often reddish. Inflorescence 10–30 cm, racemose, many-flowered, longer than leaves. Flowers pinkish-white, petals longer than sepals, lip twice as long as sepals with a short spur. Capsule ovoid.

Flowering: April–May.
Habitat: Riverine forests up to 1500 m, on *Mangifera indica*, *Toona ciliata*, *Mallotus philippensis*.
Distribution in Nainital: Kathgodam, Ranibagh, Dogaon, Jeolikote.
General Distribution: India (NE, central, and Himalayan states, AandN Islands), Bangladesh, Bhutan, Indo-China, Myanmar, Nepal, Thailand.
References: Roxburgh (1820), Chowdhery and Agrawala (2013).

2.2. ***Aerides odorata*** (Lour.)
Epiphytic, stem 20–30 cm, woody. Leaves 10–30 × 4–5 cm, thick, bifid at apex. Inflorescence 10–30 cm, racemose, many-flowered, fragrant. Flowers white with purple blotches, petals oblong, lip 3-lobed with long horn-shaped spur (10 mm). Capsule ovoid.

Flowering: May–July.
Habitat: Dense riverine forests up to 1500 m, on *Mangifera indica* and *Toona ciliata*.

Distribution in Nainital: Haldwani, Ranibagh, Jeolikote, Dogaon.
General Distribution: India (NE, central, southern states, AandN Islands), Bangladesh, Bhutan, Indo-China, Indonesia, Malaysia, Myanmar, Nepal, Philippines, Thailand.
References: Loureiro (1790), Chowdhery et al. (2010)

3. ANDROCORYS **Schlt.**
Androcorys (Greek *andrus* = anther, *corys* = hood) refers to the hood formed by dorsal sepal and petals. Three species occur in Uttarakhand; one in Nainital.

3.1. ***Androcorys monophylla*** (D.Don)
Terrestrial/lithophytic, up to 12 cm tall. Leaf solitary, 6–10 × 1 cm. Inflorescence lax, 3–10 cm spike with minute yellowish-green flowers (2–3 mm). Bracts subulate. Lip ovate, ciliate margins, base pale yellow, apex bright yellow and reflexed.

Flowering: July–August.
Habitat: Common in shaded moist places, 800–2400 m, in groups on rock crevices and grassy slopes.
Distribution in Nainital: Naina Peak, Kilbury, Snow View, Lariakanta, Tiffin Top, Bhowali road, Dogaon.
General Distribution: India (Himalayan states), Bhutan, Nepal.
References: Agrawala and Chowdhery (2010).

4. BRACHYCORYTHIS **Lindl.**
Brachycorythis (Greek *brachys* = short, *korys* = helmet) refers to the cucullate labellum resembling a helmet. One species in Uttarakhand and Nainital.

4.1. ***Brachycorythis obcordata*** (Lindl. ex Wall.)
Terrestrial, 10–20 cm tall. Tubers globose; stem slender, purple-tinged. Leaves 2–5 cm, elliptic-lanceolate, with purple midrib. Inflorescence 5–15 cm, lax raceme, 12–15-flowered. Flowers pale-purple. Lip obcordate, puberulous at base; spur stout, half as long as ovary.

Flowering: July–August.
Habitat: Chir-pine and banj-oak forests, grassy moist slopes, 1000–1900 m.
Distribution in Nainital: Above Kainchi, Bhimtal, Jeolikote, Ratighat road, Patuwadangar.
General Distribution: India (NE Himalaya, Uttarakhand, Himachal), Bangladesh, Bhutan, Nepal.
References: Chowdhery and Agrawala (2013).

5. BULBOPHYLLUM **Thouars**
Bulbophyllum (Greek *bulbos* = bulb, *phyllon* = leaf) refers to leaves with pseudobulbs. Twelve species occur in Uttarakhand; three in Nainital.

5.1. ***Bulbophyllum cariniflorum*** (Rchb.f.)
Epiphytic. Pseudobulbs ovoid, grooved with age; leaves 2 per pseudobulb, 9–16 cm. Inflorescence spike 8–10 cm, many-flowered. Flowers yellowish-green; lip mobile, tongue-shaped, orange.

Flowering: July–August.
Habitat: Rare, 1000–1500 m, riverine forests on *Quercus leucotrichophora* and *Pinus roxburghii.*
Distribution in Nainital: Bajoon, Patuwadangar.
General Distribution: India (NE Himalaya, Uttarakhand), Bhutan, Nepal, Myanmar.
References: Deva and Naithani (1986).

5.2. ***Bulbophyllum polyrrhizum*** Lindl.
Epiphytic, up to 10 cm. Pseudobulbs ovoid, solitary leaf 5–7 cm, appearing after flowering. Inflorescence raceme, 6-flowered, 8–10 cm. Flowers yellowish-green, appearing before leaves; lip papillate, reflexed.

Flowering: March–April.
Habitat: Infrequent, at elevations between 1000–1500 m, on *Quercus* and *Pinus* trunks.
Distribution in Nainital: Baldiyakhan, Patuwadangar.
General Distribution: India (NE Himalaya, Odisha, Uttarakhand, WB), Nepal.
References: Deva and Naithani (1986)

5.3. ***Bulbophyllum umbellatum*** Lindl.
Epiphytic, 10–15 cm. Pseudobulbs contiguous; leaf solitary, 9–15 cm, thick. Inflorescence umbel, 2–4 flowers. Flowers cream with red spots. Lip oblong, grooved; column short, winged.

Flowering: April–May.
Habitat: Rare, 1200–1400 m, on *Quercus* and *Pinus* in mixed forests.
Distribution in Nainital: Bajoon, Patuwadangar.
General Distribution: India (NE Himalaya, Uttarakhand), Bhutan, China, Nepal, Myanmar, SE Asia.
References: Deva and Naithani (1986).

6. CEPHALANTHERA **Rich.**
Cephalanthera (Greek *cephale* = head, *anthera* = anther) refers to the anther position. About 27 species worldwide; one in Uttarakhand and Nainital.

6.1. ***Cephalanthera longifolia*** L.
Terrestrial, up to 40 cm. Leaves lanceolate, 5–12 cm, folded. Inflorescence raceme, few to many-flowered. Flowers white, 1.5 cm, lip with orange-yellow crests.

Flowering: May–July.
Habitat: Rare, 1800–2500 m, shaded moist forest floors and grassy slopes.
Distribution in Nainital: Lariakanta, Naina Peak, Snow View.
General Distribution: India (Himalayan states, NE), Africa, Bhutan, China, Nepal, Europe, Myanmar, Sri Lanka, Thailand.
References: Deva and Naithani (1986).

7. CHEIROSTYLIS **Blume**
Cheirostylis (Greek *cheir* = hand, *stylis* = style) refers to the lobed column resembling a hand. One species in Uttarakhand and Nainital.

7.1. ***Cheirostylis griffithii*** Lindl.
Terrestrial, 8–12 cm. Stem decumbent below. Leaves ovate, wither at flowering. Inflorescence 6–10 cm, 1–3-flowered. Flowers white, 1.5 cm, sepals connate into pubescent tube; lip 3-lobed with comb-like calli.

Flowering: March–April.
Habitat: Very rare, ~1400 m, in *Pinus roxburghii* and *Quercus* mixed forests.
Distribution in Nainital: Bajoon.
General Distribution: India (NE Himalaya, Uttarakhand), Bhutan, Myanmar, Nepal, Thailand.
References: Deva and Naithani (1986).

8. COELOGYNE **Lindl.**
8.1. ***Coelogyne cristata*** Lindl.
Epiphytic or lithophytic, 10–20 cm. Pseudobulbs ovoid–oblong, 4–6 cm, bearing 2 linear-oblong leaves, 10–30 × 1–3 cm, acute. Inflorescence drooping, 10–30 cm, 5–8-flowered. Flowers large, 6–8 cm, pure white, fragrant; sepals 3.5–5.5 cm; petals undulate, 3–5 cm; lip 3-lobed with 5 fimbriate lamellae; column long, hooded at apex.

Flowering: February–March.
Habitat: Frequent but scattered, 1200–1500 m, on mossy trunks, branches, and rocks.
Distribution in Nainital: Bajoon, Mangoli, Kainchi temple, Patuwadangar.
General Distribution: India (Arunachal Pradesh, Meghalaya, Sikkim, Uttarakhand), Bhutan, China, Nepal.
References: Hooker (1890), King and Pantling (1898), Duthie (1906a), Deva and Naithani (1986), Pangtey et al. (1991a), Uniyal et al. (2007), Jalal et al. (2011).

9. CREPIDIUM **Blume**
9.1. ***Crepidium acuminatum*** D.Don
Terrestrial, up to 30 cm. Pseudobulbs ~3 cm. Leaves 3 or more, ovate-lanceolate, 10–20 cm, acute or acuminate, membranous. Inflorescence longer than leaves,

many-flowered. Flowers yellowish-green, tinged purple; sepals oblong; petals linear, longer; lip ovate, bilobulate, with small auricles; column fleshy with rounded arms.

Flowering: July–August.

Habitat: Moist shaded forest floor, 1200–2400 m, often in groups of 20–40.

Distribution in Nainital: Badapather, Hanumangari, Kilbury, Naina Peak, Patuwadangar, Snow View, Tiffin Top.

General Distribution: India (Arunachal Pradesh, Assam, Himachal Pradesh, Manipur, Meghalaya, Mizoram, Nagaland, Sikkim, Tripura, Uttarakhand), Bangladesh, Bhutan, China, Myanmar, Nepal, Thailand.

References: Gupta (1968).

9.2. ***Crepidium purpureum*** Lindl.

Terrestrial, 20–30 cm. Pseudobulbs 7–10 cm. Leaves 3, petioled, ovate to suborbicular, 7–10 × 2–5 cm, sheathing at base. Inflorescence raceme 9–25 cm, many-flowered. Flowers yellow, 6–10 mm; sepals oblong; petals linear, equal to sepals; lip concave, deeply bilobed.

Flowering: July–August.

Habitat: Subtropical forests, 900–1200 m, shaded, humus-rich soils, among mosses, ferns, and liverworts.

Distribution in Nainital: Dogaon, Hanumangari, Patuwadangar.

General Distribution: India (Arunachal Pradesh, Jharkhand, Karnataka, Manipur, Mizoram, Odisha, Sikkim, Uttarakhand, West Bengal), China, Philippines, Sri Lanka, Thailand, Vietnam.

References: Jalal et al. (2011).

10. CYMBIDIUM **Swartz**

10.1. ***Cymbidium macrorhizon*** Lindl.

Mycoheterotrophic, 15–20 cm. Rhizome scaly, fleshy, branched; stem underground. Inflorescence erect, 10–15 cm, 2–5-flowered. Flowers 4 cm, white or yellowish, streaked purple; sepals linear-lanceolate, spreading; petals erect, lanceolate; lip subovate, 3-lobed, papillate, with 2 lamellae; column arcuate, narrowly winged; pollinia 4.

Flowering: June–July.

Habitat: Rare, 800–1500 m, in damp shaded humus of sal, chir-pine, and riverine forests.

Distribution in Nainital: Sattal.

General Distribution: India (Arunachal Pradesh, Assam, Himachal Pradesh, Madhya Pradesh, Meghalaya, Mizoram, Nagaland, Sikkim, Uttarakhand), Bhutan, China (Yunnan), Laos, Myanmar, Nepal, Pakistan, Thailand, Vietnam.

References: Hooker (1890), Duthie (1906a), Deva and Naithani (1986), Pangtey et al. (1991b), Uniyal et al. (2007), Jalal et al. (2011), Chowdhery (2009).

11. DENDROBIUM **Sw.**

A large genus of epiphytic orchids distributed from tropical Asia to Australia, with about 1500 species worldwide. In India c.135 species occur, six in Uttarakhand, all represented in Nainital.

11.1. ***Dendrobium amoenum*** Wall. ex Lindl.

Epiphytic or rarely lithophytic; pseudobulbs slender, pendulous, 20–70 cm, leaves linear-lanceolate, 5–11 cm; flowers white with violet tips, fragrant, borne in pairs on leafless pseudobulbs; lip 3-lobed, orbicular apical lobe hairy, spur swollen.

Flowering: May–June.
Habitat: Common epiphyte on oak trunks and treetops between 800- and 1600-m elevations; rarely lithophytic.
Distribution in Nainital: Bajoon, Baldiyakhan, Jeolikote, Mangoli, Patuwadangar, Sattal, Ratighat.
General Distribution: India (Assam, Arunachal Pradesh, Meghalaya, Sikkim, Uttarakhand), Bangladesh, Bhutan, Myanmar, Nepal.

11.2. ***Dendrobium bicameratum*** Lindl.

Epiphytic; pseudobulbs clavate or fusiform, erect, 6–10 cm; leaves elliptic-oblong, acute, up to 10 × 1.2 cm; flowers fleshy, yellow with red markings, in dense capitate racemes; lip concave, 3-lobed with two large oblong nectar cells.

Flowering: July–August.
Habitat: Infrequent, on mossy branches and pine trunks in open sites between 1200–1500 m elevation.
Distribution in Nainital: Bajoon, Khurpatal, Patuwadangar.
General Distribution: India (Arunachal Pradesh, Jharkhand, Manipur, Meghalaya, Mizoram, Nagaland, Odisha, Sikkim, Uttarakhand, West Bengal), Bhutan, Nepal, Myanmar.

11.3. ***Dendrobium denudans*** D.Don

Epiphytic; pseudobulbs 7–20 cm, terete, tapering; leaves 5–10 cm, deciduous at flowering; inflorescence slender, many-flowered; flowers white or pale green with reddish veins; lip short, decurved, 3-lobed with radiating red lines, margin serrate.

Flowering: September–October.
Habitat: Rare, on mossy trunks in exposed sunlight between 1200–1500 m elevations.
Distribution in Nainital: Bhimtal, Sattal.
General Distribution: India (Assam, Arunachal Pradesh, Meghalaya, Mizoram, Sikkim, Uttarakhand), Bhutan, Myanmar, Nepal, Thailand.
References: Jalal et al. (2011).

11.4. ***Dendrobium fimbriatum*** Hook.

Epiphytic; pseudobulbs 20–35 cm, erect, terete, sulcate; leaves several, linear-lanceolate, 7–14 cm, acuminate; flowers golden yellow, 4–6 cm across, borne in few-flowered racemes near apex; petals broader than sepals; lip orbicular, fringed, often with dark brown patch at centre.

Flowering: May–June.

Habitat: Rare, epiphytic on moss-laden tree trunks and branches in mixed forests around 1500 m elevation.

Distribution in Nainital: Sattal.

General Distribution: India (Arunachal Pradesh, Assam, Meghalaya, Nagaland, Sikkim, Uttarakhand), Bangladesh, Bhutan, China (Yunnan), Laos, Myanmar, Nepal, Thailand, Vietnam.

References: Hooker (1823), Uniyal et al. (2007).

11.5. ***Dendrobium primulinum*** Lindl.

Epiphytic; pseudobulbs pendulous, 15–25 cm; leaves broadly oblong, ~10 cm, coriaceous; flowers solitary or paired on leafless pseudobulbs, 5–6 cm across, white to pale rose; sepals lanceolate, petals narrower; lip suborbicular, ciliate, yellow with purple streaks, base convolute.

Flowering: March–April.

Habitat: Frequent pendent epiphyte on medium-sized trees in warm open sites between 600–1400 m elevations.

Distribution in Nainital: Bhujiaghat, Dogaon, Jeolikote, Patuwandagar, Ranibagh.

General Distribution: India (Andaman and Nicobar Islands, Arunachal Pradesh, Mizoram, Sikkim, Uttarakhand), Bhutan, China, Malaya, Myanmar, Nepal, Thailand.

11.6. ***Dendrobium transparens*** Wall. ex Lindl.

Epiphytic; pseudobulbs slender, 30–60 cm, slightly thickened at nodes, subpendulous; leaves linear-lanceolate, 7–10 cm; flowers 2–3, fragrant, white tinged purplish-rose at tips; sepals and petals lanceolate, purple-tipped; lip oblong, obscurely 3-lobed, mid-lobe orbicular, white with large pink blotch; column with two horn-like processes.

Flowering: May–June.

Habitat: On older leafless pseudobulbs, rare in Nainital region.

Distribution in Nainital: Collected near Nainital by J.F. Duthie (1906a).

General Distribution: India (Arunachal Pradesh, Jharkhand, Manipur, Mizoram, Nagaland, Odisha, Sikkim, Tripura, Uttarakhand), Nepal.

12. DIENIA **Lindl.**

The genus name derives from Greek *dienos* (= 2 years old), in allusion to its flowering thought to occur in alternate years. It includes about six species distributed mainly in the tropics and subtropics of the Old and New World, with few extending into temperate Asia, Europe, and the Americas. In India four species occur, three in Uttarakhand, two in Nainital.

12.1. ***Dienia cylindrostachya*** Lindl.
Terrestrial herbs 10–20 cm tall; pseudobulbs small, conical; stem with 2–3 tubular sheaths; leaf solitary, 3–10 × 2–5 cm, elliptic, subacute, margin undulate, long tubular petiole. Inflorescence 5–6 cm, many-flowered raceme, flowers greenish yellow; sepals ovate, acuminate; petals linear-lanceolate; lip fleshy, broadly ovate, with beak-like apex, surface with raised central line and two concavities below the column.

Flowering: July–August.
Habitat: Temperate region between 2200–2600 m elevations, in shaded moist forest floor or grassy slopes.
Distribution in Nainital: Lariakanta, Naina Peak range, Snow View.
General Distribution: India (Arunachal Pradesh, Himachal Pradesh, Jammu and Kashmir, Sikkim, Uttarakhand, West Bengal), Bhutan, China, Nepal.
References: Uniyal et al. (2007).

12.2. ***Dienia muscifera*** Lindl.
Terrestrial, 15–25 cm tall; stem from ovoid pseudobulb, with 1–2 tubular sheaths; leaves usually 2, subopposite, unequal, ovate-rotund to lanceolate, 2–8 cm; inflorescence 7–22 cm, laxly flowered, exceeding leaves; flowers yellowish-green, ~3 mm across; sepals oblong-lanceolate, longer than petals; petals linear, shorter; lip fleshy, broadly ovate, acute, with beak, convex and thickened; column very short, fleshy.

Flowering: July–September.
Habitat: Rare ground orchid on grassy slopes around 2500 m elevation.
Distribution in Nainital: Naina Peak range.
General Distribution: India (Arunachal Pradesh, Himachal Pradesh, Jammu and Kashmir, Sikkim, Uttarakhand), Bhutan, Nepal, Pakistan.
References: Uniyal et al. (2007).

13. DITHRIX **(Hook.f.) Schltr. ex Brummitt**
The genus name is derived from Greek *di* (two) and *thrix* (hair). It has one species distributed from Afghanistan to the western Himalaya; in India restricted to Uttarakhand and recorded from Nainital.

13.1. ***Dithrix griffithii*** (Hook.f.) Ormerod and Gandhi
Terrestrial; stem 15–30 cm, slender, straight or flexuous, with few elliptic or linear-oblong leaves, 2.5–6 cm long; inflorescence spike 2–6 cm, few- to many-flowered; flowers very small, greenish-white, deflexed, tubular; lip 3–4 mm, 3-fid with divaricate lateral lobes, mid-lobe broader; spur c. 1 mm, slightly curved.

Flowering: March–May.
Habitat: Very rare; open grassy slopes between 1400–1600 m elevations.

Distribution in Nainital: On way to Ratighat.
General Distribution: India (Himachal Pradesh, Uttarakhand), Pakistan.

14. EPIPACTIS **L.**
Derived from the Greek *epipaktis*, an ancient name for a plant thought to curdle milk (*E. helleborine*). About 30 species occur across Europe, Asia, Africa, and North America; six species in India, three in Uttarakhand, one in Nainital.

14.1. ***Epipactis helleborine*** L.
Terrestrial up to 90 cm; rhizome short, woody; stem glabrous below, puberulous above; leaves ovate-lanceolate to broadly elliptic, 5–15 × 2–8 cm, many-nerved, margin ciliolate; inflorescence raceme up to 30 cm, many-flowered; flowers green flushed with red-violet, 1.5–2 cm across; lip shorter than petals, hypochile saccate, epichile cordate with erose margins and two basal calli.

Flowering: July–August.
Habitat: Frequent in oak forests, forest edges, grassy slopes between 1500–2600 m elevations.
Distribution in Nainital: Kilbury, Badapathar, Government House, Hanumangari, Naina Peak, Pangtey's Gorge, Patuwandagar, Snow View, Tiffon Top.
General Distribution: India (Arunachal Pradesh, Himachal Pradesh, Jammu and Kashmir, Sikkim, Uttarakhand, West Bengal), Africa, Afghanistan, Bhutan, China, Europe, Kazakhstan, Kyrgyzstan, Myanmar, Nepal, N. America, Pakistan, Russia, SW Asia, Tajikistan, Uzbekistan.
References: Linnaeus (1753).

15. ERIA **Lindl.**
The genus name is derived from Greek *erio-* (wool), referring to the woolly hairs on the inflorescence and flowers. It comprises about 500 species distributed across tropical Asia to New Guinea, Australia, and the Pacific Islands; 60 species occur in India, 10 in Uttarakhand, 4 in Nainital.

15.1. ***Eria alba*** Lindl.
Epiphytic; pseudobulbs ovoid, c. 3 cm, marked with vertical lines; leaves 3–5, oblanceolate, 15–20 × 1–3 cm, strongly nerved; inflorescence spike 4–6-flowered, pubescent; flowers white, fragrant; lip 3-lobed, tinged reddish-purple, mid-lobe yellow and red with two ridges.

Flowering: June–July.
Habitat: Rare; moss-covered banj-oak trees between 1400–1600 m elevations.
Distribution in Nainital: Naukuchiatal.
General Distribution: India (Arunachal Pradesh, Himachal Pradesh, Mizoram, Nagaland, Sikkim, Uttarakhand), Bhutan, Nepal.

15.2. ***Eria lasiopetala*** Willd.

Epiphytic; rhizome stout, clothed with tubular sheaths; pseudobulbs 7–10 cm, compressed, ovoid or clavate; leaves 3–5, lanceolate, 15–25 × 2–4 cm; raceme 7–12-flowered, stout; flowers greenish-yellow; sepals tomentose; lip 3-lobed, mid-lobe oblong, acute, deflexed, slightly 2-ridged.

Flowering: April–June.

Habitat: Rare; moist shaded sal and broad-leaved forests between 600–1500 m elevations.

Distribution in Nainital: Bajoon, Jeolikote, Patuwadangar.

General Distribution: India (Arunachal Pradesh, Assam, Bihar, Chhattisgarh, Manipur, Meghalaya, Mizoram, Odisha, Sikkim, Uttarakhand, West Bengal), Bangladesh, Bhutan, China, Indo-China, Myanmar, Nepal, Thailand.

15.3. ***Eria muscicola*** Lindl.

Epiphytic; pseudobulbs globose, 4–6 mm; leaves 2–4, oblanceolate, up to 3 cm; raceme 2–6-flowered; flowers yellowish-green; sepals lanceolate, lateral falcate; lip concave, erose at apex with two basal calli.

Flowering: August–September.

Habitat: Rare; epiphytic on moist shaded riverine forest trees around 1400 m elevation.

Distribution in Nainital: Sattal.

General Distribution: India (Andaman and Nicobar Islands, Arunachal Pradesh, Manipur, Meghalaya, Sikkim, Uttarakhand, West Bengal), Bhutan, China, Laos, Myanmar, Nepal, Sri Lanka, Vietnam.

15.4. ***Eria spicata*** D.Don

Epiphytic or lithophytic, 20–25 cm tall; pseudobulbs flattened, 4–12 cm; leaves 4–6, elliptic-lanceolate, 10–17 × 3–5 cm; raceme dense, many-flowered; flowers white to pale-yellow; lip obscurely 3-lobed, tinged yellow.

Flowering: August–September.

Habitat: Very rare; subtropical epiphyte between 600–1600 m elevations.

Distribution in Nainital: Patuwadangar.

General Distribution: India (Arunachal Pradesh, Assam, Manipur, Meghalaya, Mizoram, Sikkim, Uttarakhand), Bhutan, Cambodia, China, Laos, Myanmar, Nepal, Thailand, Vietnam.

16. EULOPHIA **R.Br. ex Lindl.**

The genus name is from Greek *eu-* (well-developed) and *lophus* (crest), referring to the prominent calluses on the lip. About 250 species are distributed in tropical and subtropical regions. Twenty-four species occur in India, eight in Uttarakhand, one in Nainital.

16.1. ***Eulophia herbacea*** Lindl.

Terrestrial, glabrous; pseudobulbs broadly ovate to conical; stem 10–15 cm; leaves 2–5, linear to elliptic-lanceolate, up to 30 cm, plicate; inflorescence 30–60 cm, erect, lax, with few basal sheaths; flowers appearing with leaves, pale green, fragrant, 4–5 cm across; sepals linear-lanceolate, acuminate, spreading; petals white, shorter; lip yellow, 3-lobed, mid-lobe oblong, rounded, finely fimbriate with lamellae; spur very short, geniculate; column stout, curved, slightly winged.

Flowering: June–July.
Habitat: Very rare terrestrial orchid around 600 m elevation; not recollected after initial report.
Distribution in Nainital: Reported from Ranibagh.
General Distribution: India (Chhattisgarh, Gujarat, Himachal Pradesh, Karnataka, Madhya Pradesh, Maharashtra, Manipur, Mizoram, Odisha, Sikkim, Tripura, Uttarakhand), Bangladesh, China, Laos, Myanmar, Nepal, Thailand.
References: Jalal et al. (2011).

17. GASTROCHILUS **D.Don**

The genus name is from Greek *gaster* (= belly) and *cheilos* (= lip), referring to the sac-like lip. It comprises about 50 species distributed in India, Nepal, Southeast Asia, Malaysia, and adjacent islands. Eighteen species occur in India, five in Uttarakhand, one in Nainital.

17.1. ***Gastrochilus calceolaris*** (Buch.-Ham. ex Sm.)

Epiphytic; stem very short, pendulous; leaves 6–8, linear-oblong, 10–30 × 2–3 cm, unequally bifid at apex; inflorescence corymbose to umbellate, many-flowered, shorter than leaves; flowers pale green speckled red-brown, waxy, faintly fragrant; sepals unequal, dorsal obovate-oblong, lateral falcate; petals oblong-obovate; lip adnate to lower column, base forming short yellow sac, apical lobe semi-circular, fleshy patch central; spur short, yellow; column very short, thick.

Flowering: April–May.
Habitat: Upper subtropical region between 1000–1200 m elevations, as semi-pendulous epiphyte on mossy trunks in partial shade.
Distribution in Nainital: Jeolikote.
General Distribution: India (Arunachal Pradesh, Assam, Manipur, Meghalaya, Mizoram, Nagaland, Sikkim, Tripura, Uttarakhand), Bhutan, China, Malaysia, Nepal, Thailand.
References: Jalal et al. (2011).

18. GASTRODIA **R. Br.**

18.1. ***Gastrodia falconeri*** D.L.Jones & M.A.Clem

Terrestrial, leafless, mycoheterotrophic. Rhizome tuberous, scaly, 5–6 cm, giving rise to erect, stout stems 30–90 cm, pale to yellowish-brown, smooth. Sheaths few, loose, the upper acuminate. Inflorescence a lax raceme 20–30 cm, many-flowered. Flowers numerous, reddish-brown tinged with yellow, 12 × 5 mm; sepals and petals united into a ventricose tube; lip longer than sepals, hypochile narrow, epichile ovate; column elongate, winged above.

Flowering: July–August.

Habitat: Extremely rare; damp, well-littered forest floors of *Quercus leucotrichophora* forest around 2300 m elevation.

Distribution in Nainital: Snow View (only two individuals recorded in 2009, recollected after 100 years since S.W. Seers' collection).

General Distribution: India (Himachal Pradesh, Jammu and Kashmir, Uttarakhand), Pakistan.

19. GOODYERA **L.**

19.1. ***Goodyera procera*** (Wall. ex Ker Gawl.)

Robust terrestrial herb up to 70 cm. Stem stout, glabrous, clothed by sheathing bases. Leaves 6–8, elliptic to ovate-elliptic, acute, 7–15 × 4–5 cm, long-petioled with broadly sheathing base. Inflorescence dense, spike-like, 8–30 cm, many-flowered. Flowers minute, white tinged green, fragrant; bracts ovate-acuminate, pubescent at margin, equalling ovary; sepals broadly ovate, obtuse, 2.5 mm; petals cuneate, erose-dentate; lip cymbiform, saccate, reflexed at apex with two oblong calli within; column short.

Flowering: May–June.

Habitat: Rare; shaded moist localities near perennial streams, often with *Eupatorium adenophorum* at elevation 600 m.

Distribution in Nainital: Bhujiaghat.

General Distribution: India (plains to NE Himalaya, central states, Uttarakhand), Bhutan, China, Myanmar, Nepal, Sri Lanka, Thailand.

19.2. ***Goodyera repens*** L.

Slender terrestrial herb, 15–30 cm. Rhizome creeping, stoloniferous. Stem erect, glabrous or glandular-pubescent. Leaves crowded at base, elliptic, 1–3 cm, dark green above with striking silver venation, petiole sheathing. Inflorescence raceme 2–5 cm, loosely many-flowered. Flowers white flushed pink, globular, at right angle to ovary; bracts lanceolate, equalling or longer than ovary; sepals ovate, acute, glabrous or pubescent; petals lanceolate, adhering to dorsal sepal; lip smaller than sepals, saccate at base, apical lobe short, ovate, slightly undulate; column short, without appendages.

Flowering: July–August.

Habitat: Frequent; moist damp shady slopes with mosses and liverworts between 2200–2400 m elevations.

Distribution in Nainital: Naina Peak, Rajbhawan, Snow View, trail to Tiffin Top.

General Distribution: India (Himalayan states from Uttarakhand to Arunachal Pradesh, Himachal Pradesh, Jammu and Kashmir), China, Myanmar, Nepal.

20. GYMNADENIA **R.Br.**

20.1. ***Gymnadenia orchidis*** Lindl.

Terrestrial, robust, 20–50 cm. Roots tuberous, palmately divided with elongate lobes. Stem clothed at base with tubular sheaths. Leaves 3–5, elliptic-oblong to linear-oblong, 5–12 cm, sessile. Inflorescence densely flowered, cylindric, up to 10 cm. Flowers purplish-red or pink, rarely whitish; bracts lanceolate, acuminate, basal exceeding ovary. Sepals broadly ovate, dorsal forming hood with petals, lateral reflexed; petals rhomboid-orbicular; lip adnate to column base, shallowly 3-lobed, smooth above; spur filiform, curved, slightly clavate, as long as or longer than ovary. Column short, broad.

Flowering: July–August.

Habitat: High-elevation moist grassy slopes between 2200–2600 m elevations; very rare, not recollected in recent surveys.

Distribution in Nainital: Reported from Naina Peak by J.F. Duthie and H.G. Champion.

General Distribution: India (Arunachal Pradesh, Himachal Pradesh, Jammu and Kashmir, Uttarakhand), Bhutan, Nepal, Pakistan, Tibet.

21. HABENARIA **Willd.**

21.1. ***Habenaria aitchisonii*** Rchb.f.

Terrestrial. Tuber 2 cm, narrow. Stem stout, to 40 cm. Leaves two, subopposite at base, sessile, orbicular to broadly ovate, cuspidate, 4–7 cm. Inflorescence spike 3–8 cm, 15–20-flowered. Flowers yellowish-green, erect, fragrant. Bracts lanceolate, shorter than ovary. Sepals ovate, dorsal erect and concave, lateral reflexed; petals lanceolate, auricled at base; lip as long as sepals, 3-lobed, lateral lobes filiform, mid-lobe acute, deflexed; spur variable but shorter than ovary.

Flowering: July–August.

Habitat: Moist shaded slopes with mosses and liverworts between 2000–2600 m elevations.

Distribution in Nainital: Ayapatta, Badapathar, below Naina Peak, Snow View.

General Distribution: India (Himachal Pradesh, Jammu and Kashmir, Sikkim, Uttarakhand), Afghanistan, Bhutan, China (Tibet), Nepal, Pakistan.

21.2. ***Habenaria digitata*** Lindl.

Terrestrial, 15–60 cm. Leaves 5–7, clustered near middle, lanceolate to ovate-lanceolate, 4–6 × 3–4 cm, margins undulate. Inflorescence 7–10 cm, lax, many-flowered. Flowers greenish-white, subsecund. Bracts narrowly lanceolate, longer than ovary. Sepals green, unequal, dorsal erect, concave, suborbicular, lateral longer, ovate, reflexed. Petals deeply 2-partite, linear, falcate, white at base with green tips. Lip 3-lobed, mid-lobe longer, linear; spur 10 mm, compressed, inflated towards apex, slightly shorter than ovary.

Flowering: July–August.
Habitat: Rare; shady moist sites around 1400 m elevation, not recollected recently.
Distribution in Nainital: Reported from Khurpatal (Champion).
General Distribution: Widespread in India (plains, peninsula, Himalaya), Bangladesh, Cambodia, Laos, Myanmar, Nepal, Pakistan.

21.3. ***Habenaria ensifolia*** Lindl.
Terrestrial, tuberous. Stem 60–70 cm, robust, leafy in lower half with long acute tubular sheaths. Leaves 7–15 cm, oblong or linear-lanceolate, acuminate, mostly in upper portion. Inflorescence raceme 10–20 cm. Flowers green, 3–4 cm across; bracts leaf-like, linear-lanceolate, longer than ovary. Sepals unequal, erect, tips recurved, lateral wider and subfalcate. Petals forming hood with dorsal sepal, falcate-oblong, ciliolate. Lip pale green, slightly longer than sepals, 3-lobed; side-lobes deeply pectinate, mid-lobe linear. Spur compressed, dilated towards apex, nearly equalling ovary.

Flowering: July–August.
Habitat: Rare; reported earlier but not recollected in recent surveys.
Distribution in Nainital: Nainital, Lariakanta (J.F. Duthie, H.G. Champion, T.A. Rao).
General Distribution: India (Arunachal Pradesh, Himachal Pradesh, Sikkim, Uttarakhand), Bhutan, Nepal.

21.4. ***Habenaria intermedia*** (D.Don)
Terrestrial. Tuber oblong, ~4 cm. Stem terete, 15–40 cm+, with 2–3 basal sheaths. Leaves scattered, 6–15 × 2–5 cm, ovate-oblong, acuminate, cordate to rounded at base. Inflorescence 2–6-flowered raceme. Flowers white or greenish-white, 5 cm across. Bracts leaf-like, lanceolate, equalling or exceeding ovary. Sepals persistent, ovate-lanceolate, reflexed tips, lateral ones spreading. Petals pure white, equalling dorsal sepal, strongly 5-nerved, margin ciliolate. Lip longer than sepals, 3-lobed; side-lobes deeply pectinate, mid-lobe linear (3–5 cm). Spur stout, exceeding ovary, tapering, curved.

Flowering: July–August.
Habitat: Frequent to common along forest edges and moist grassy slopes between 1500–2600 m elevations.
Distribution in Nainital: Badapather, Bajoon, Baldiyakhan, Bhowali, Camel's Back, Hanumangari, Kilbury, Lariakanta, Mangoli, Naina Peak range, Pangtey's Gorge, Snow View, Tiffin Top, Ratighat.
General Distribution: India (Himachal Pradesh, Jammu and Kashmir, Uttarakhand), China, Nepal, Pakistan.

21.5. ***Habenaria marginata*** (Colebr)
Terrestrial, tuberous. Tubers small, cylindric. Stem reduced, subscapigerous. Leaves few, radical, sessile, oblong or linear-oblong, 3–12 × 1–2.5 cm, with pale yellow

margin. Inflorescence spike 5–15 cm, densely flowered. Flowers greenish-yellow to orange, ~1.5 cm across. Bracts semi-amplexicaul, ovate-lanceolate, as long as ovary. Sepals unequal, pale green, dorsal erect, broadly ovate, lateral falcately lanceolate. Petals yellow with green tinge, falcate, forming hood with dorsal sepal. Lip yellow, fleshy, 3-partite; lateral lobes filiform, mid-lobe short, broader with reflexed margin. Spur stout, compressed, equalling or shorter than ovary.

Flowering: July–August.
Habitat: Infrequent but locally frequent in moist grassy slopes and forest edges between 800–1200 m elevations.
Distribution in Nainital: Dogaon, Birbhatti, Gaithia, Jeolikote.
General Distribution: India (widespread: Andhra Pradesh, Himachal Pradesh, Uttarakhand, Tamil Nadu, etc.), Bangladesh, Bhutan, China, Myanmar, Nepal, Thailand.

21.6. ***Habenaria pectinata*** (J.E. Sm.) D.Don
Terrestrial, up to 30 cm tall. Stem stout, quadrangular, winged at angles. Leaves 6–8 × 1–2 cm, ovate-lanceolate, acuminate, strongly 3-nerved. Inflorescence 8–12 cm raceme. Flowers greenish-white, ~2.5 cm across. Bracts leaf-like, longer than flowers. Sepals green, dorsal broadly lanceolate, obtuse, 3-nerved; lateral narrower, subfalcate, spreading, mucronate. Petals white, falcate-oblong, obtuse, forming hood with dorsal sepal. Lip pure white, longer than sepals, 3-partite; mid-lobe linear, longer than side lobes; side lobes deeply pectinate. Spur stout, longer than ovary, funnel-shaped at base, apex curved.

Flowering: July–August.
Habitat: Infrequent, open moist-shaded grassy slopes between 1200- and 1600-m elevations.
Distribution in Nainital: Bhowali, Hanumangari, Jeolikote, Patawandagar, Ratighat.
General Distribution: India (Arunachal Pradesh, Meghalaya, Sikkim, Uttarakhand), Bhutan, China, Nepal, Pakistan.
Key References: Don (1825), Jalal et al. (2011).

21.7. ***Habenaria plantaginea*** Lindl.
Terrestrial, 20–30 cm. Tubers ovoid-globose, 2 cm long. Stem glabrous, bracteate. Leaves 3–7, subradical, elliptic-oblong, 7.5–15 cm, clustered near base. Inflorescence 5–15 cm, lax-flowered spike. Bracts lanceolate, half as long as ovary. Flowers white, 1–1.5 cm across. Sepals subequal, dorsal ovate-oblong, lateral oblong, falcate, spreading. Petals linear-lanceolate, denticulate, pressed to dorsal sepal forming hood. Lip flabelliform, 3-lobed; side lobes semi-ovate, entire or denticulate; mid-lobe linear-oblong, shorter. Spur about equalling ovary.

Flowering: September–October.
Habitat: Infrequent, open grassy slopes between 600–1200 m elevations.

Distribution in Nainital: Amparao, Jeoligaon, above Ranibagh.
General Distribution: India (Himachal Pradesh, Jammu and Kashmir, Karnataka, Maharashtra, Sikkim, Uttarakhand), Bhutan, Myanmar, Nepal.
Key References: Lindley (1835), Jalal et al. Check List 4(3):308 (2011).

21.8. ***Habenaria pubescens*** Lindl.
Terrestrial, 30–70 cm, pubescent. Tubers 2, ellipsoidal. Leaves basal, elliptic lanceolate, acuminate, 7–9 nerved, grooved midrib, white undulate margin, pubescent. Inflorescence raceme, stalk pubescent with basal sterile bracts. Flowers greenish-yellow, perpendicular to axis. Bracts oblanceolate, shorter than ovary. Dorsal sepal erect, broadly ovate, acuminate, ciliate; lateral sepals reflexed, longer, ovate-obtuse. Petals falcate, acute, arching with dorsal sepal. Lip 3-lobed, longer than sepals; lateral lobes broader at middle, filiform at tip, crenate margin, distinct notch; mid-lobe shorter, rounded apex. Spur as long as ovary, ovary long-beaked.

Flowering: July–August.
Habitat: Rare; dry deciduous *Shorea robusta* forests on sandy soils between 600–1000 m elevations.
Distribution in Nainital: Chopra Village above Ranibagh.
General Distribution: India (Sikkim, Uttarakhand), Myanmar, Nepal.
Key References: Jalal et al. (2009b).

21.9. ***Habenaria stenopetala*** Lindl.
Terrestrial, 30–50 cm. Tubers oblong or narrowly ellipsoid, ~2.5 cm. Leaves 15–20 cm, on upper ⅔ of stem, oblong-lanceolate to ovate-lanceolate, acute, undulate, sheathing. Inflorescence dense raceme. Flowers pale green, spirally arranged. Bracts lanceolate or ovate-lanceolate, longer than flowers. Sepals subequal, aristate-tipped. Petals 2-partite, shorter than sepals, upper longer, lower reduced, bases prolonged downwards. Lip 10–15 mm, deeply 3-lobed; lateral lobes linear or subulate, shorter; mid-lobe linear or ligulate, broader. Spur slender, equal to or longer than ovary, slightly curved upwards.

Flowering: August–September.
Habitat: Very rare; moist-shaded grassy slopes, under shrubs between 1000–2000 m elevations.
Distribution in Nainital: Ghorakhal, below Nainital (C.W. Hope).
General Distribution: India (Arunachal Pradesh, Assam, Himachal Pradesh, Jammu and Kashmir, Karnataka, Maharashtra, Manipur, Meghalaya, Mizoram, Nagaland, Sikkim, Tripura, Uttarakhand), Bhutan, China, Nepal, Thailand.

22. HEMIPILIA **Lindl.**
22.1. ***Hemipilia cordifolia*** Lindl.
Terrestrial herbs, up to 20 cm tall. Tuber ovoid or subglobose. Leaf solitary, fleshy, orbicular-cordate, obtuse or acute, amplexicaul, 5–10 cm, many-nerved. Inflorescence a lax raceme, 3–5-flowered, 4–6 cm. Flowers purple, ~1.5 cm across. Bracts

lanceolate, acuminate, shorter than ovary. Sepals equal in length; dorsal sepal erect, lanceolate-ovate, obtuse; lateral sepals falcate, oblong, spreading. Petals slightly shorter, obliquely ovate, forming hood with dorsal sepal. Lip broadly 3-lobed; lateral lobes rounded, reflexed; mid-lobe orbicular, bluntly acuminate, pubescent. Spur trumpet-shaped, deflexed, shorter or equalling ovary, faintly 2-lobed or entire at tip.

Flowering: July–September.
Habitat: Rare but locally frequent on moist-shaded walls and rock crevices between 1800–2200 m elevations.
Distribution in Nainital: On the way to Ratighat.
General Distribution: India (Himachal Pradesh, Uttarakhand), Nepal.

23. HERMINIUM **Guett.**
23.1. ***Herminium lanceum*** Thunb. ex Sw.
Terrestrial, 10–60 cm tall. Tubers subglobose to ellipsoid. Stem with 2 basal tubular sheaths, bearing 2–4 alternate leaves. Leaves linear to lanceolate, 6–20 × 0.5–1.4 cm, acute or acuminate. Inflorescence 8–40 cm, dense, many-flowered spike. Flowers erect to horizontal, pale green to yellowish green. Bracts lanceolate, shorter than ovary, acuminate. Sepals subequal, obtuse, oblong; dorsal connivent with petals to form hood, lateral spreading. Petals linear, acute, 1-nerved, equal to sepals. Lip long rectangular, base slightly saccate, upper half deeply 3-lobed; lateral lobes linear, recurved, 2–5 mm; mid-lobe minute (0.5 mm long). Column very short (1 mm long).

Flowering: July–August.
Habitat: Common on open grassy slopes, moist-shaded forest edges between 1200–2600 m elevations.
Distribution in Nainital: Badapather, Bajoon, Bhowali, Ghorakhal, Hanumangari, Kilbury, Naina Peak range, Pangtey's Gorge, Patwadangar, Tiffin Top.
General Distribution: India (Assam, Himachal Pradesh, Jammu and Kashmir, Manipur, Meghalaya, Mizoram, Nagaland, Sikkim, Tripura, Uttarakhand), Bhutan, China, Malaya, Myanmar, Nepal, Pakistan, Thailand.

24. LIPARIS **Rich.**
24.1. ***Liparis deflexa*** Hook.f.
Terrestrial, 10–25 cm tall. Pseudobulbs tuberoides, 1–1.5 cm, covered in membranous sheaths. Stem slender, arising from pseudobulb. Leaves 2, subopposite, elliptic to ovate-lanceolate, 5–10 × 3–4 cm, subsessile, acuminate, sheathing at base. Inflorescence erect, 12–24 cm, 7–12-flowered. Flowers yellow, ~6 mm across. Bracts reflexed, lanceolate, half as long as ovary. Sepals unequal, obtuse, 3–5-nerved; dorsal acute; laterals oblong, reflexed. Petals narrowly oblong, 1-veined, reflexed, margins recurved. Lip orbicular-reniform, crenulate except near base, apex mucronate, basal calli two. Column slightly curved, with narrow rounded wings, base swollen.

Flowering: July–August.

Habitat: Rare terrestrial orchid of sandy soils in partial shade of sal forests between 600–1000 m elevations.

Distribution in Nainital: Chopra village, Ranibagh.

General Distribution: India (Andhra Pradesh, Arunachal Pradesh, Chhattisgarh, Goa, Karnataka, Kerala, Sikkim, Tamil Nadu, Uttarakhand), Indo-China, Myanmar.

Key References: Hooker (1890).

24.2. ***Liparis glossula*** Rchb.f.

Terrestrial, 10–20 cm tall. Pseudobulbs ovoid to oblong, 3–5 cm. Leaf solitary, oblong or linear-oblong, 8–10 × 3–4 cm, subacute. Inflorescence erect, racemose, longer than leaf, many-flowered. Flowers purplish red, ~1.5 cm across. Bracts linear, half to equal ovary length. Sepals linear-lanceolate, 3-veined, apex subacute, margins revolute. Petals filiform. Lip broadly oblong to subquadrate, 10 × 7–8 mm long, base slightly contracted, central band dark coloured, apex subtruncate and mucronate, margin minutely ciliate, callus absent. Column curved, 4–5 mm.

Flowering: August–September.

Habitat: Occasional terrestrial orchid of moist-shaded mossy sites between1600–2400 m elevations.

Distribution in Nainital: Naina Peak, Snow View, Zoo, Tiffin Top.

General Distribution: India (Himachal Pradesh, Sikkim, Uttarakhand), China, Nepal.

24.3. ***Liparis odorata*** (Willd.) Lindl.

Terrestrial, 15–25 cm tall. Pseudobulbs narrowly ovoid, 1–2 cm. Stem bulbous at base, 10–14 cm, sheathed. Leaves 2, elliptic-lanceolate, 8–12 × 2–2.7 cm, plicate, 5-nerved. Inflorescence 10–15 cm, racemose. Flowers pale yellow tinged reddish-purple, ~8 mm. Bracts ovate-lanceolate, shorter than ovary. Dorsal sepal linear, blunt; lateral sepals broadly ovate, 5-nerved. Petals spathulate, 3-veined. Lip concave, obovate, with 2 tooth-like calli at base, apex broad, emarginate, crenulate. Column erect, slightly hook-like.

Flowering: July–August.

Habitat: Occasional in moist-shaded evergreen forests between 1800–2200 m elevations.

Distribution in Nainital: Kailakhan.

General Distribution: India (widespread: Andhra Pradesh, Arunachal Pradesh, Assam, Goa, Jharkhand, Karnataka, Kerala, Maharashtra, Manipur, Meghalaya, Mizoram, Nagaland, Odisha, Sikkim, Tamil Nadu, Uttarakhand, West Bengal), Bhutan, China, Japan, Laos, Myanmar, Nepal, Pacific Islands, Thailand, Vietnam.

24.4. ***Liparis rostrata*** Rchb.f.

Terrestrial, 6–10 cm tall. Pseudobulbs ovoid, small, sheathed in white membranes. Leaves 2, opposite, 11 × 3–4 cm, petiolate at base. Inflorescence raceme, several-flowered. Flowers brownish-green, ~2 cm. Bracts small, broad based. Sepals narrowly lanceolate to spathulate, 3-nerved; dorsal with cordate base. Petals linear, as long as sepals, reflexed. Lip broadly cordate, apiculate, margin minutely crenulate, claw channelled above, no callus. Column slender, long, with two knob-like basal protuberances.

Flowering: July–August.

Habitat: Common terrestrial orchid of moist-shaded mossy habitats between 1500–2500 m elevations.

Distribution in Nainital: Barapather, Camel's Back, Deopatta, Lariakanta, Naina Peak, Rajbhawan, Snow View, Tiffin Top.

General Distribution: India (Himachal Pradesh, Jammu and Kashmir, Uttarakhand), China, Nepal, Pakistan.

25. LUISIA **Gaud.**

25.1. ***Luisia trichorrhiza*** (Hook.) Blume

Epiphytic, 15–25 cm tall. Stem stout, covered by leaf sheaths. Leaves fleshy, terete, 8–15 cm long. Inflorescence extra-axillary, 4–5-flowered, leaf-opposed. Flowers greenish-pink to purple, ~10 mm across. Bracts broadly ovate-triangular. Sepals subequal, lateral acute, petals oblong, obtuse. Lip 3-lobed, dark purple, 5–7 mm, longer than sepals; epichile broadly cordate, >5 mm broad; hypochile concave with erect rounded lobes. Column stout, 2 mm long.

Flowering: March–April.

Habitat: Epiphyte in moist-shaded *Shorea robusta* and broad-leaved forests between 600–1400 m elevations.

Distribution in Nainital: Ranibagh, Raatighat.

General Distribution: India (Andhra Pradesh, Arunachal Pradesh, Assam, Chhattisgarh, Jharkhand, Maharashtra, Meghalaya, Nagaland, Odisha, Sikkim, Tamil Nadu, Uttarakhand), Bangladesh, Bhutan, Myanmar, Nepal, Thailand.

25.2. ***Luisia zeylanica*** Lindl.

Epiphytic, 15–30 cm tall. Stem stout, covered by leaf sheaths. Leaves alternate, terete, 6–15 cm long. Inflorescence condensed raceme, 2–4-flowered. Flowers greenish-pink or purple, 5–7 mm across. Bracts ovate-lunate. Sepals subequal, dorsal obtuse, lateral keeled towards apex. Petals oblong, obtuse, slightly longer than sepals. Lip longer and broader than sepals, 3.5–5 mm, with maroon bands; 3-lobed, epichile<5 mm broad, rhomboid; hypochile auricular, shorter. Column very short.

Flowering: April–June.

Habitat: Epiphyte on tree trunks and branches between 600–800 m elevations.

Distribution in Nainital: Ranibagh.
General Distribution: India (Andaman and Nicobar Islands, Arunachal Pradesh, Goa, Kerala, Madhya Pradesh, Maharashtra, Meghalaya, Sikkim, Tamil Nadu, Uttarakhand), Nepal, Sri Lanka.

26. NEOTTIA **Guett.**
26.1. ***Neottia listeroides*** Lindl.
Mycoheterotrophic, 10–40 cm tall. Rhizome with crowded fleshy roots. Stem with 3–5 sheaths, brownish-olive green. Inflorescence raceme, 5–20 cm, many-flowered. Flowers brownish-olive green, 7–10 mm. Sepals subequal, elliptic-ovate; petals linear, blunt. Lip narrowly obovate-oblong, 5–7 mm, apex deeply 2-lobed; lobes ovate-lanceolate. Column curved, dilated at base and apex.

Flowering: July–September.
Habitat: Moist-shaded forested areas, especially banj-oak (*Quercus leucotrichophora*) between 1800–2400 m elevations, forming small colonies.
Distribution in Nainital: Badapathar, Karnataka, Kilbury, Lariakanta, Rajbhawan, Naina Peak, Snow View.
General Distribution: India (Arunachal Pradesh, Himachal Pradesh, Jammu and Kashmir, Sikkim, Uttarakhand), Myanmar, Nepal, SE Tibet.

26.2. ***Neottia microglottis*** (Duthie) Schltr.
Mycoheterotrophic, up to 30 cm tall. Rhizome thick, fibrous, pale brown. Stem stout, creamish-white, with 2–4 loose sheaths. Inflorescence raceme, 10–20 cm, glandular pubescent. Flowers pale green, 7 mm, crowded. Sepals ovate, hairy, 4 mm; petals linear-spathulate, reflexed. Lip linear-oblong, entire, margins undulate. Column erect, shorter than petals, dilated at base and apex.

Flowering: August–September.
Habitat: Extremely rare; last recorded from Deopatta, not recollected recently.
Distribution in Nainital: Deopatta (Champion).
General Distribution: India (Uttarakhand, endemic).
Key References: Pangtey et al. (1991a).

27. NERVILIA **Comm.** ***ex*** **Gaudich.**
27.1. ***Nervilia concolor*** (Blume) Schltr.
Terrestrial, 8–12 cm tall. Tuber whitish, subglobose to ovoid, 1–1.5 cm wide. Leaf solitary, appearing after flowering, 5–7 cm, broadly flabellate-cordate, cuspidate, many-nerved, margin undulate, often reddish-purple blotched. Inflorescence 6–10 cm, 4–6-flowered. Flowers pale green with cream, pendulous; lip greenish, ovate, trilobed, white tinged yellow, purple-veined; mid-lobe ovate, pubescent along nerves. Bracts deflexed, twice as long as ovary. Column 6–8 mm, widened towards apex.

Flowering: June–July.

Habitat: Subtropical regions between 600–1500 m elevations; gravely soil near water channels, shaded colonies with mosses and ferns.

Distribution in Nainital: Jeolikote–Dogaon, Ranibagh.

General Distribution: India (widespread, including AandN Islands, NE states, Himalaya, Peninsular India), Bangladesh, Bhutan, China, Myanmar, Nepal, Japan, SE Asia, Pacific Islands.

Key References: Uniyal et al. (2007).

27.2. ***Nervilia crociformis*** (Zoll. and Moritzi) Seidenf.

Terrestrial, 4–5 cm tall. Tubers subglobose to ovoid, ~12 mm diam. Leaf appearing post-flowering, orbicular-cordate, funnel-shaped, 2.5–5 cm, hairy dorsally. Inflorescence 3–6 cm, solitary-flowered. Flower faintly fragrant, solitary. Sepals narrowly oblong, petals linear-lanceolate, forward-directed. Lip as long as sepals, 3-lobed; side-lobes short; terminal lobe orbicular, margin fimbriate, disc with 7 glandular ridges, purple towards apex. Column clavate.

Flowering: May–June.

Habitat: Frequent in shaded evergreen forest floors, well-littered soil, up to 1500 m elevation.

Distribution in Nainital: Jeolikote, Patuwandagar, Sattal.

General Distribution: India (Assam, Karnataka, MP, Maharashtra, Manipur, Odisha, Sikkim, Tamil Nadu, Uttarakhand), Africa, China, SE Asia, New Guinea, Philippines, Australia.

Key References: Uniyal et al. (2007).

27.3. ***Nervilia gammieana*** (Hook.f.) Pfitzer

Terrestrial, 10–15 cm tall. Tubers globular, ~3 cm, warty. Leaf 10–12 × 5–7 cm, reniform, apiculate, many-nerved, margin undulate. Inflorescence 10–15 cm, stout, 5–8-flowered. Flowers drooping, pink with white streaks. Sepals and petals subequal, elliptic-oblanceolate. Lip yellowish-pink, elliptic-obovate, hypochile convolute, with 2 side-lobes +2 ridges; epichile crenulate, hairy above, with 3 central ridges. Column 10–15 mm, apex purple. Capsule 1.5 cm long.

Flowering: May–June.

Habitat: Subtropical Sal forests between 600–1600 m elevations, moist-shaded partially open floors; often in groups (18–20 plants).

Distribution in Nainital: Dogaon, Jeolikote, Patuwadagar, Ranibagh.

General Distribution: India (Arunachal Pradesh, Himachal Pradesh, JandK, Sikkim, Uttarakhand), Nepal, Pakistan.

Key References: Hooker (1872–1897), Uniyal et al. (2007).

27.4. ***Nervilia mackinnonii*** (Duthie) Schltr.

Terrestrial, 3–10 cm tall. Tubers globose, ~6 mm, whitish, warty. Leaf appearing after flowering, 2.5–4.5 × 3.7–5 cm, 7-sided to cordate, membranous, apex acute, petiole 4–8 cm, with basal sheaths. Inflorescence 4–6 cm, solitary-flowered. Flower white or light-green, flushed reddish-brown externally. Sepals longer, acute; petals slightly shorter. Lip oblong, 3-nerved, white with green base, purple-spotted, mid-lobe narrower and elongated. Column short.

Flowering: May–June.
Habitat: Rare, grassy moist shaded places, up to 2200 m elevation.
Distribution in Nainital: Above Zoo, Bajoon, Cheena Peak, Hanumangari, Kilbury, Pangote.
General Distribution: India (Uttarakhand, endemic focus), China, Myanmar, Nepal.
Key References: Duthie (1902), Uniyal et al. (2007).

28. OBERONIA **Lindl.**
28.1. ***Oberonia ensiformis*** (J.E.Sm.) Lindl.
Epiphytic, 15–20 cm tall. Leaves 4–5, linear-ensiform, 6–18 × 0.5–1.5 cm, falcate, coriaceous, acuminate. Inflorescence erect, shorter than leaves, densely flowered, with numerous minute appressed bracteoles. Flowers orange-yellow, turning brown with age. Bracts ovate, convolute, subacute. Sepals broadly ovate, subacute. Petals ovate, stiff, pubescent, ciliolate. Lip twice the size of sepals and petals, 2-lobed, pubescent, margins erose, disc with two globular calli at the junction of mid-lobe and lateral lobes; mid-lobe further 2-lobed, kidney-shaped.

Flowering: May–July.
Habitat: Epiphyte on tree trunks, occasionally lithophyte on rocks between 1000–1500 m elevations, warmer valleys.
Distribution in Nainital: Dogaon.
General Distribution: India (Arunachal Pradesh, Assam, Manipur, Mizoram, Nagaland, Sikkim, Tripura, Uttarakhand), Bhutan, Nepal, Thailand.
Key References: Uniyal et al. (2007).

28.2. ***Oberonia falconeri*** Hook.f.
Epiphytic, 4–8 cm tall. Leaves 4–5, overlapping at base, 2–6 × 1.5–2 cm, thick, broadly ensiform, acute or acuminate. Inflorescence terminal, equal to or longer than leaves, laxly many-flowered. Flowers greenish-yellow, small, spiral. Bracts longer than ovary, oblong-lanceolate, serrulate. Sepals oblong-elliptic, obtuse. Petals ovate, obtuse, spreading. Lip broadly oblong, twice as long as sepals; side-lobes small; mid-lobe with two incurved lobules, separated by a broad sinus; surface with scattered gland-like glistening blisters.

Flowering: September–October.
Habitat: Rare, epiphyte on tree trunks in open places between 600–1500 m elevation.
Distribution in Nainital: Kathgodam, Ranibagh.

General Distribution: India (Andhra Pradesh, Arunachal Pradesh, Assam, Bihar, Chhattisgarh, Jharkhand, Karnataka, Madhya Pradesh, Maharashtra, Meghalaya, Odisha, Sikkim, Uttarakhand, West Bengal), Bangladesh, China, Laos, Malaysia, Nepal, Thailand, Vietnam.
Key References: Uniyal et al. (2007).

28.3. ***Oberonia pachyrachis*** Rchb.f. ex Hook.f.
Epiphytic, 5–10 cm tall. Leaves narrowly ensiform, acuminate, slightly falcate, 5–8 × 1–1.5 cm. Inflorescence 4–6 cm, many-flowered; peduncle short, broad, compressed, adnate nearly to leaf apex. Flowers pale brown, minute (<1 mm), sunk in pits of floral axis; sepals and lip flat on surface. Bracts ovate-rotund, deeply serrate. Sepals elliptic, obtuse, glandular dorsally, lateral sepals mamillate at apex. Petals linear-oblong, obtuse. Lip orbicular-elliptic, broader than long, concave, entire to slightly erose, with deep basal depression beneath column.

Flowering: July–August.
Habitat: Locally frequent, epiphyte on tree trunks in mixed deciduous and riverine forests, sometimes lithophytic on boulders between 1000–1500 m elevations.
Distribution in Nainital: Bajoon, below Patwadangar, Jeolikote, Kathgodam, Ranibagh.
General Distribution: India (Arunachal Pradesh, Manipur, Meghalaya, Mizoram, Nagaland, Sikkim, Tripura, Uttarakhand, West Bengal), Bhutan, China, Myanmar, Nepal, Thailand, Vietnam.
Key References: Hooker (1872–1897), Uniyal et al. (2007).

29. OREORCHIS **Lindl.**
The genus *Oreorchis* (Greek: *oreos* = mountain, *orchis* = orchid) comprises c. 16 species, distributed in Bhutan, E and SW China, NE India, Japan, Korea, Myanmar, Nepal, and Russia. Four species occur in India and Uttarakhand, one in Nainital.

29.1. ***Oreorchis micrantha*** Lindl.
Terrestrial, 15–20 cm tall. Pseudobulbs ovoid-conic, 2 cm long, basal fibres present. Leaves 2, linear-lanceolate, 15–18 cm, 2–5-nerved. Inflorescence laxly racemose, 10–15 cm, arising from pseudobulb side. Flowers sessile, yellowish, lip white with purple spots. Sepals linear-lanceolate; petals shorter, sometimes purple spotted. Lip 3-lobed, lateral lobes falcate, hypochile with fleshy channelled callus, epichile bilobed with crumpled margin. Column curved, base concave.

Flowering: June–July.
Habitat: Very rare, in dense moist evergreen broad-leaved forests around 2200 m elevation.
Distribution in Nainital: Kilbury (near Pangtey's Gorge).

General Distribution: India (Arunachal Pradesh, Himachal Pradesh, Jammu and Kashmir, Sikkim, Uttarakhand); Bhutan, China (Yunnan), Myanmar, Nepal, Tibet.
Key References: Deva and Naithani (1986).

30. PACHYSTOMA **Blume**
The genus *Pachystoma* (Greek: *pachys* – thick, *stoma* – mouth) comprises c. 20 species distributed in tropical Asia, Malay Archipelago, New Guinea, N Australia, and New Caledonia. Two species occur in India, one in Uttarakhand, Nainital.

30.1. ***Pachystoma pubescens*** Blume
Terrestrial, 20–35 cm tall. Leaves grass-like, 10–35 cm, appearing after flowering. Inflorescence erect, 6–8-sheathed, 8–10-flowered. Flowers dull purplish to pinkish, glandular-puberulous, subsecund. Sepals oblong-lanceolate, 5-nerved, lateral sepals subsaccate; petals narrower, spathulate, beaked at tip. Lip 3-lobed, saccate at base, margins hairy, mid-lobe orbicular with 5 thick crested lamellae. Column puberulous, dilated at apex. Pollinia 8.

Flowering: March–April.
Habitat: Very rare, shaded grassy sites in plains and foothills up to 1200 m elevation.
Distribution in Nainital: Near Nainital.
General Distribution: India (A and N Islands, Arunachal Pradesh, Himachal Pradesh, Karnataka, Kerala, Madhya Pradesh, Maharashtra, Manipur, Meghalaya, Mizoram, Nagaland, Odisha, Punjab, Sikkim, Tamil Nadu, Uttar Pradesh, Uttarakhand, West Bengal); Australia, Bangladesh, Bhutan, Indo-China, Indonesia, Malaysia, Myanmar, Nepal, New Guinea, Philippines.
Key References: Deva and Naithani (1986).

31. PECTEILIS **Raf.**
The genus *Pecteilis* (Greek: *pecteilon* = comb, referring to the comb-like lobes of the lip) comprises c. 5 species, distributed in E and SE Asia, Himalayas, and China. Five species occur in India, two in Uttarakhand, one in Nainital.

31.1. ***Pecteilis gigantea*** (J.E. Sm.) Raf.
Terrestrial herb, 40–80 cm tall. Tubers elongate, 4–5 cm. Leaves mostly mid-stem, ovate-oblong to lanceolate, 6–20 × 2–8 cm; upper leaves bract-like. Inflorescence raceme 2–6-flowered, 15–20 cm. Flowers large (6–10 cm across), white tinged greenish-yellow, strongly fragrant. Sepals spreading, dorsal suborbicular, laterals oblong; petals linear-falcate. Lip 3-lobed, side lobes deeply pectinate with 10–16 filiform segments, mid-lobe tongue-like; spur 8–11 cm, obtuse. Column white.

Flowering: September–October.
Habitat: Scattered but locally frequent in open moist-shaded grassy slopes at elevation between 700–1000 m elevations.

Distribution in Nainital: Aamparao, Baldiyakhan, Bajoon, Dogaon.

General Distribution: India (Andhra Pradesh, Bihar, Chhattisgarh, Goa, Gujarat, Himachal Pradesh, JandK, Jharkhand, Karnataka, Kerala, Madhya Pradesh, Maharashtra, Nagaland, Odisha, Tamil Nadu, Uttarakhand), Myanmar, Pakistan.

Key References: Uniyal et al. (2007).

32. PERISTYLUS **Blume**

The genus *Peristylus* (Greek: *peri* = around, *stylus* = column) comprises c. 100 species, distributed in SE Asia to New Guinea, NE Australia, and SW Pacific. Thirty species occur in India, eight in Uttarakhand, five in Nainital.

32.1. ***Peristylus affinis*** (D.Don) Seidenf.

Terrestrial, 10–15 cm tall. Leaves clustered near lower stem, ovate to elliptic. Inflorescence lax, up to 15 cm. Flowers white; dorsal sepal hooded with petals; lip longer than petals, basal half concave forming a pouch, 3-lobed with broad mid-lobe. Spur globular.

Flowering: July–August.

Habitat: Rare, on moist shaded grassy slopes up to 1500 m elevation.

Distribution in Nainital: Baldiyakhan.

General Distribution: India (Arunachal Pradesh, Himachal Pradesh, Jharkhand, Manipur, Nagaland, Sikkim, Uttarakhand, W Bengal), China, Myanmar, Nepal, Thailand, Vietnam.

Key References: Uniyal et al. (2007).

32.2. ***Peristylus constrictus*** Lindl.

Terrestrial, up to 70 cm. Leaves 5–6, elliptic, clustered mid-stem. Inflorescence dense, cylindric, up to 30 cm, many-flowered. Flowers white, sweet-scented, spreading, 1.5 cm. Sepals oblong, obtuse; petals obliquely ovate-lanceolate; lip oblong, 3-lobed with linear lateral lobes. Spur globular.

Flowering: July–August.

Habitat: Frequent, in open grassy slopes up to 1200 m elevation.

Distribution in Nainital: Bhujiaghat, Dogaon, Jeolikote.

General Distribution: India (Arunachal Pradesh, Assam, Himachal Pradesh, JandK, Jharkhand, Madhya Pradesh, Maharashtra, Meghalaya, Nagaland, Odisha, Sikkim, UP, Uttarakhand, W Bengal), Bangladesh, Bhutan, China, Myanmar, Nepal, Thailand, Vietnam.

Key References: Uniyal et al. (2007).

32.3. ***Peristylus elisabethae*** (Duthie) R.K. Gupta

Terrestrial, 20–25 cm. Leaves 2–3, basal, lanceolate. Inflorescence lax, 15–20 cm, many-flowered. Flowers green, small, crowded, deflexed. Sepals connivent, dorsal ovate; petals oblong; lip longer than sepals, 3-lobed, lateral lobes linear, spreading, mid-lobe oblong. Spur ovate, longer than broad.

Flowering: July–August.
Habitat: Rare, on open mossy slopes around 2500 m elevation.
Distribution in Nainital: Naina Peak range, Ratighat, Snow View.
General Distribution: India (Himachal Pradesh, Sikkim, Uttarakhand, W Bengal), China, Myanmar, Nepal.
Key References: Duthie (1906a), Uniyal et al. (2007).

32.4. ***Peristylus fallax*** Lindl.
Terrestrial, 10–30 cm. Tubers small. Stem with 2 basal sheaths. Leaf solitary, 3–15 cm, narrowly elliptic-lanceolate, acute. Inflorescence lax–dense, 10–15 cm. Flowers green, small. Sepals equal, oblong, spreading; dorsal sepal hooded with petals; lateral sepals slightly falcate. Petals oblong, blunt. Lip fleshy, flat, 3-lobed, mid-lobe oblong, longer than side-lobes. Spur saccate, shorter than sepals. Column long.

Flowering: July–August.
Habitat: Rare, on moist shaded slopes at elevations between 2200–2500 m elevations.
Distribution in Nainital: Naina Peak range.
General Distribution: India (Arunachal Pradesh, Himachal Pradesh, Sikkim, Uttarakhand), Nepal, Tibet.
Key References: Uniyal et al. (2007).

32.5. ***Peristylus goodyeroides*** (D.Don) Lindl.
Terrestrial, 20–60 cm. Tubers ellipsoid, hairy. Leaves clustered in upper third of stem, ovate-elliptic, acute, sheathing at base. Inflorescence erect, 10–20 cm, lax. Flowers white or pale greenish-yellow, sweet-scented. Dorsal sepal broad, hooded with petals; lateral sepals oblong. Petals trapeziform, broader than sepals. Lip 3-lobed, basal part flat, with triangular nectary; mid-lobe broad, blunt; side-lobes equal to mid-lobe. Spur globular.

Flowering: July–August.
Habitat: Occasional, in grassy slopes and forest edges at elevations between 2000–2400 m elevations.
Distribution in Nainital: Lariakanta, below Cheena Peak, Kilbury, Snow View, Tiffin Top.
General Distribution: India (Andhra Pradesh, Arunachal Pradesh, Assam, Bihar, Chhattisgarh, Himachal Pradesh, Jharkhand, Karnataka, Kerala, Madhya Pradesh, Manipur, Mizoram, Odisha, Sikkim, Tamil Nadu, Tripura, Uttarakhand, W Bengal), Bangladesh, Bhutan, China, Indonesia, Malaysia, Myanmar, Nepal, New Guinea, Philippines, Sri Lanka, Thailand.
Key References: Uniyal et al. (2007).

33. PHALAENOPSIS **Blume**

From Greek *phalaina* (moth) and *opsis* (appearance), referring to the moth-like flowers. Comprises c. 40–45 species, mainly in SE Asia to New Guinea, most diverse in Indonesia and the Philippines. Four species in India, three in Uttarakhand, two in Nainital.

33.1. ***Phalaenopsis difformis*** (Wall. ex Lindl.)

Epiphytic, 15–30 cm. Stems sheathed. Leaves few, elliptic-oblong, coriaceous, 7–10 × 4–5 cm, apex obtuse. Inflorescence 15–25 cm, lax, many-flowered. Flowers small, greenish-yellow with brown bands, 1 cm. Sepals oblong-ovate, reflexed; petals smaller. Lip longer than sepals, with hypochile-bearing rounded lobes and horn-like spur, and epichile with 2 recurved pectinate lobules.

Flowering: June–July.

Habitat: Rare epiphyte, on tree trunks and branches of evergreen trees at elevations between 800–1600 m elevations.

Distribution in Nainital: Jeolikote (near Sunrise Hotel), Patuawadangar.

General Distribution: India (Arunachal Pradesh, Assam, Manipur, Meghalaya, Mizoram, Sikkim, Tripura, Uttarakhand), Indo-China, Myanmar, Nepal, Thailand.

33.2. ***Phalaenopsis taenialis*** Lindl.

Small epiphyte, 4–7 cm tall. Roots flattened, verrucose. Stem short, without pseudobulb. Leaves 1–2, sometimes absent, pendulous, narrowly elliptic-oblong, 6–13 × 1.5–2 cm. Inflorescence pendulous, longer than leaves, 6–8-flowered. Flowers pale pink, 2–2.5 cm across, widely opening. Sepals unequal, obtuse; petals obovate, rounded. Lip 3-lobed, with long subcylindric spur. Column stout, 5 mm, stigma broadly dilated; fruit cylindric, 5–6 cm.

Flowering: April–May.

Habitat: Very rare, on oak (*Quercus leucotrichophora*) trunks and branches at elevation of 1600 m.

Distribution in Nainital: Near Bhowali.

General Distribution: India (Arunachal Pradesh, Assam, Manipur, Meghalaya, Mizoram, Nagaland, Sikkim, Uttarakhand), Myanmar, Nepal.

34. PHOLIDOTA **Lindl.**

From Greek *pholidotos* = scaly, referring to the large inflorescence bracts. About 30 spp., distributed in SE Asia, Australia, New Guinea, Pacific islands. Twelve species in India; two in Uttarakhand and Nainital.

34.1. ***Pholidota articulate*** Lindl.

Epiphytic or lithophytic. Pseudobulbs 5–10 cm, stem-like, jointed, branching, with new growths near apex of old ones. Leaves 2, narrowly oblong, membranous, 5–10 × 2.5–4 cm. Inflorescence dense raceme, 5–7 cm, many-flowered. Flowers brownish-pink. Bracts large, convolute, caducous. Sepals orbicular to ovate; petals orbicular. Lip 3-lobed, hypochile with rounded lobes and 5 lamellate nerves, epichile broad, bilobulate. Column slightly winged at apex.

Flowering: July–August.
Habitat: Frequent in moist, shaded ravines and valleys at elevations between 1000–1600 m elevations.
Distribution in Nainital: Bajoon, Jeolikote, Khurpatal, Mangoli, Naukuchiatal, Patawandagar.
General Distribution: India (Arunachal Pradesh, Assam, Himachal Pradesh, Manipur, Meghalaya, Mizoram, Nagaland, Sikkim, Tripura, Uttarakhand, West Bengal), Bhutan, China, Indo-China, Indonesia, Malaysia, Myanmar, Nepal, Thailand.
Key References: Uniyal et al. (2007).

34.2. ***Pholidota imbricata*** Lindl.
Epiphyte or lithophyte. Pseudobulbs clustered, oblong-conical, 3–5 × 1–2 cm. Leaves solitary, elliptic-oblanceolate, 12–30 × 3–5 cm, on stout petiole. Inflorescence pendulous, 25–30 cm, with many imbricate persistent bracts. Flowers white to pink, arranged in two rows. Bracts large, concave, pinkish-brown. Sepals ovate; petals linear, falcate. Lip triangular, sac with three ridges; hypochile triangular; epichilebilobulate with median sinus. Column broad, winged in upper half.

Flowering: June–August.
Habitat: Rare, on moss-covered trunks in partial shade at elevations between 600–900 m elevations.
Distribution in Nainital: Bhujiaghat.
General Distribution: India (Andaman and Nicobar Islands, Andhra Pradesh, Arunachal Pradesh, Assam, Chhattisgarh, Goa, Jharkhand, Karnataka, Kerala, Madhya Pradesh, Maharashtra, Manipur, Meghalaya, Mizoram, Nagaland, Odisha, Sikkim, Tripura, Uttarakhand, W Bengal), Australia, Bangladesh, Bhutan, China, Indonesia, Malaysia, Myanmar, Nepal, New Guinea, Pacific Islands, Pakistan, Sri Lanka, Thailand.
Key References: Uniyal et al. (2007).

35. PLATANTHERA **Rich.**
From Greek *platys* = broad and *anthera* = anther, referring to the broad anthers. A genus of c. 200 species distributed in Europe, N Africa, temperate Asia, New Guinea, and North America. In India 15 species; six in Uttarakhand, four in Nainital.

35.1. ***Platanthera clavigera*** Lindl.
Terrestrial, 20–60 cm tall. Tubers ovoid, 1.2–2 cm. Stem stout, with basal sheaths. Leaves 4–5, narrowly ovate-oblong, 5–10 × 1.5–2.5 cm. Inflorescence 20–40 cm, many-flowered. Flowers yellowish green. Bracts longer than ovary, lanceolate, ciliate. Sepals elliptic-oblong, ciliolate, dorsal sepal concave, lateral reflexed. Petals

oblong, nearly as long as sepals. Lip linear, obtuse, with tooth-like callus at base. Spur clavate, pendulous, shorter than ovary. Column short, broad.

Flowering: July–August.
Habitat: Moist open grassy slopes at elevations between 2200–2500 m elevations.
Distribution in Nainital: Naina Peak, Rajbhawan, Snow View.
General Distribution: India (Arunachal Pradesh, Himachal Pradesh, Sikkim, Uttarakhand), Nepal.
Key References: Uniyal et al. (2007).

35.2. ***Platanthera japonica*** (Thunb.) Lindl.
Terrestrial, 20–30 cm. Stem erect, robust. Leaves oblong-lanceolate, 7–10 cm. Inflorescence a many-flowered spike. Flowers white, relatively large. Bracts linear-lanceolate, longer than ovary. Sepals: dorsal cymbiform, lateral deflexed, obtuse. Petals linear. Lip 2–3× longer than lateral sepals, tapering, puberulous, margins reflexed. Spur cylindric, slender, longer than ovary, incurved.

Flowering: June–July.
Habitat: Rare, in damp shaded places.
Distribution in Nainital: Reported by Duthie (based on Davidson coll.), but not recollected since.
General Distribution: India (Himachal Pradesh, Uttarakhand, Sikkim, Nagaland), Bhutan, Nepal, China, Japan, Korea.
Key References: Uniyal et al. (2007).

35.3. ***Platanthera edgeworthii*** (Hook.f. ex Collett)
Terrestrial, 10–40 cm. Tubers oblong to subglobose. Stem slender, leafy. Leaves 3–4, sheathing, 6–10 × 4.5 cm, ovate to lanceolate. Inflorescence spike 10–25 cm, many-flowered. Flowers green, crowded, deflexed in bud. Bracts lanceolate, shorter than ovary. Sepals green, ciliolate; dorsal concave, lateral larger, deflexed. Petals thick, triangular, forming hood with dorsal sepal. Lip yellow, strap-shaped, longer than sepals, base orbicular, concave. Spur twice length of ovary, incurved. Column 2–3 mm, short.

Flowering: July–August.
Habitat: Frequent in forest edges and grassy slopes at elevations between 1800–2600 m elevations.
Distribution in Nainital: Badapather, Hanumangari, Kilbury, Naina Peak, Patwadangar, Snow View, Tiffin Top.
General Distribution: India (Himachal Pradesh, Kashmir, Uttarakhand), Pakistan.
Key References: Collett (1902), Uniyal et al. (2007).

35.4. ***Platanthera latilabris*** Lindl.
Terrestrial, 20–40 cm. Tubers ellipsoid to ovoid, 1–2.5 cm. Leaves 3–5, scattered, elliptic-oblong, 5–10 cm. Inflorescence a lax spike, several flowered. Flowers

yellowish green, 7–10 mm. Bracts lanceolate, acuminate. Sepals ciliolate; dorsal broadly ovate, concave; laterals lanceolate from triangular base. Lip longer than petals, triangular at base, tapering to linear limb with inrolled margin. Spur clavate, equalling or slightly longer than ovary. Column short.

Flowering: July–August.
Habitat: Frequent in moist shaded and mossy places at elevations between 1500–2500 m elevations.
Distribution in Nainital: Naina Peak range, Snow View, Tiffin Top.
General Distribution: India (Arunachal Pradesh, JandK, Madhya Pradesh, Sikkim, Uttarakhand), Nepal, Tibet.
Key References: Uniyal et al. (2007).

36. RHYNCHOSTYLIS **Blume**
From Greek *rynchos* = beak, *stylos* = pillar, referring to the beaked column. About 3–4 species, distributed in Sri Lanka, India, SE Asia, China, Philippines, Malaysia, Indonesia. Two species in India; one in Uttarakhand and Nainital.

36.1. ***Rhynchostylis retusa*** (L.) Blume
Epiphytic, 20–50 cm. Stem woody, covered by leaf sheaths. Leaves dense, distichous, linear, deeply channelled, 15–40 × 2.5 cm, bilobed. Inflorescence axillary, 20–25 cm, drooping, many-flowered. Flowers white to pink, with purple spots, 2 cm across. Bracts ovate-lanceolate, shorter than ovary. Sepals subequal, orbicular; petals oblong, obtuse. Lip oblong, spatulate, conduplicate, magenta-tipped, 8–10 mm, continuous with foot. Column stout, short. Capsule obovoid-oblong, 1.5–2 cm.

Flowering: June–July.
Habitat: Locally common on deciduous and evergreen trees, up to 1200 m elevation.
Distribution in Nainital: Bajoon, Bhimtal, Dogaon, Jeolikote, Mangoli, Naukuchiatal, Ranibagh.
General Distribution: India (widespread, including Andamans, Himalaya, NE and peninsular states), Sri Lanka, Myanmar, Thailand, Malaysia, Indonesia, Philippines.
Key References: Linnaeus (1753), Uniyal et al. (2007).

37. SATYRIUM **Sw.**
Named after *Satyrion* (Satyr), a name used by Dioscorides for orchids. About 100 species, mostly in Southern Africa; c. 12 species in Asia. Two species in India; one in Uttarakhand and Nainital.

37.1. ***Satyrium nepalense*** D.Don
Terrestrial, 20–50 cm. Tubers 1–2, ovoid. Stem leafy at base. Leaves 2, subradical, oblong to ovate-lanceolate, 4–18 × 1–8 cm, fleshy. Inflorescence 10–30 cm, densely flowered, with large foliaceous bracts. Flowers pink or white, fragrant,

non-resupinate. Sepals pink, dorsal oblanceolate, lateral oblong. Petals twisted at mouth. Lip conspicuous, globose, hooded, keeled, with two long spurs exceeding ovary. Column curved, dilated at apex.

Flowering: August–October.
Habitat: Common on grassy slopes, pine forests, and open habitats at elevations between 1500–3000 m elevations.
Distribution in Nainital: Badapathar, Bajoon, Bhowali, Ghorakhal, Kailakhan, Khurpatal, Kilbury, Lariakanta, Naina Peak, Pangot, Snow View, Tiffin Top, Ratighat.
General Distribution: India (Himalaya, NE, peninsular states), Bhutan, China, Myanmar, Nepal, Sri Lanka, Thailand, Malaya.
Key References: Uniyal et al. (2007).

38. SMITINANDIA **Holttum**
Named after Tem Smitinand, Thai orchidologist and former Director of Royal Forest Department. Three species distributed from India, Bhutan, Myanmar to SE Asia eastwards to Borneo and Sulawesi. One species occurs in India, Uttarakhand, and Nainital.

38.1. ***Smitinandia micrantha*** Lindl. Holttum
Epiphytic, 15–20 cm tall. Stem stout, terete, slightly flattened. Leaves 5–8, oblong, fleshy, bifid at apex, 8–10 × 1.2–2 cm. Inflorescence leaf-opposed raceme, 10 cm, many-flowered. Flowers pinkish, 5 mm across; sepals broadly ovate, spreading; petals smaller, oblong; lip 3-lobed, fleshy, pink-purple, mid-lobe oblong and convex. Column short, stout, with protuberance.

Flowering: June–July.
Habitat: Frequent epiphyte on tree trunks in riverine forests, up to 1400 m elevation.
Distribution in Nainital: Bhujiaghat, Dogaon, Jeolikote.
General Distribution: India (Andaman Islands, Arunachal Pradesh, Madhya Pradesh, Manipur, Meghalaya, Nagaland, Sikkim, Uttarakhand), Cambodia, China, Laos, Malaya, Myanmar, Nepal, Thailand, Vietnam.
Key References: Uniyal et al. (2007).

39. SPIRANTHES **Rich.**
From Greek *spira* = spiral and *anthes* = flower, referring to the spirally arranged flowers. About 50 species, mainly North American, with some in Africa, Asia, Australia, and Europe. Two species in India and Uttarakhand, one in Nainital.

39.1. ***Spiranthes sinensis*** Pers.
Terrestrial, 10–40 cm tall. Tubers fleshy, cylindrical. Stem erect, leafy below. Leaves 3–10 cm, linear-lanceolate, basal, acuminate. Inflorescence spike flexuous, pubescent, 15 cm, flowers spirally arranged. Flowers small, rose-purple; sepals lanceolate, dorsal forming a tube with petals and lip; lip sessile, obovate, basal whitish, apical

part rose-purple, undulate-crispate margin. Column short, prolonged into a flat pointed process above stigma.

Flowering: May–November.
Habitat: Rare; moist, shaded grassy slopes at elevations between 2000–2400 m elevations.
Distribution in Nainital: Kailakhan, Rajbhawan, Tiffin Top.
General Distribution: India (Arunachal Pradesh, Assam, Madhya Pradesh, Manipur, Meghalaya, Nagaland, Peninsular India, Uttarakhand), Australia, China, Japan, Malaya, Mongolia, Myanmar, Nepal, New Zealand, Siberia, Sri Lanka.
Key References: Persoon (1807), Uniyal et al. (2007).

40. THUNIA **Rchb.f.**
Named in honour of Count von Thun Hohenstein of Bohemia. About six species distributed in SE Asia, Bhutan, China, India, Nepal. Three in India, two in Uttarakhand, one in Nainital.

40.1. ***Thunia alba*** (Lindl.) Rchb.f.
Epiphytic or lithophytic, 40–90 cm tall. Stem stout, pendulous, tufted, with many basal sheaths. Leaves numerous, narrowly elliptic-lanceolate, 10–20 × 2.5–5 cm, acuminate. Inflorescence drooping raceme, 4–6-flowered. Flowers large, white, c. 5 cm across; sepals and petals oblong, acute; bracts ovate-oblong, as long as ovary. Lip yellow, pandurate, lower half tubular, upper portion cucullate, shell-like, margins undulate. Spur short, slightly bifid. Column short, winged near apex, with very short foot.

Flowering: June–August.
Habitat: Frequent, on trees and rocks in forests at elevations between 500–1800 m elevations.
Distribution in Nainital: Bhimtal, Dogaon, Jeolikote, Naina Gaon, Bajoon, Naukuchiyatal, Patwadangar, Sattal.
General Distribution: India (Arunachal Pradesh, Sikkim, Uttarakhand), Bhutan, Nepal.
Key References: Uniyal et al. (2007).

41. VANDA **Jones**
About 40 species occur in tropical Asia to New Guinea and Australia; 11 species in India, five in Uttarakhand, three in Nainital.

41.1. ***Vanda cristata*** Lindl.
Epiphytic orchid up to 15 cm. Stem erect, stout, covered with old leaf sheaths. Leaves crowded, bifarious, 6–10 cm, narrowly oblong, bilobed. Inflorescence axillary raceme, 2–5-flowered, equal or shorter than leaves. Flowers yellowish or olive-green with purplish blotches, 4–5 cm across; lip 3-lobed, with horn-like fleshy beak.

Flowering: May–July.

Habitat: Common epiphyte between 1000–1800 m elevations, growing on trunks and branches of various trees.

Distribution in Nainital: Bajoon, Baldiyakhan, Bhimtal, Dogaon, Jeolikote, Khurpatal, Mangoli, Naukuchiyatal, Patwadangar, Ranibagh, Sattal, Ratighat.

General Distribution: India (Arunachal Pradesh, Assam, Meghalaya, Sikkim, Uttarakhand), Nepal, Bhutan, Bangladesh, Tibet, China (Yunnan).

References: Uniyal et al. (2007).

41.2. ***Vanda tessellata*** (Roxb.) Hook. ex G. Don

Epiphyte with climbing stem 15–30 cm, roots thick and fleshy. Leaves distichous, linear, keeled, auriculate at base, 10–20 × 1–1.5 cm. Inflorescence axillary raceme 3–10-flowered, zig-zag, longer than leaves. Flowers yellowish-green, bluish tessellated, fragrant, 3.5–5 cm. Lip white, 3-lobed, mid-lobe pink or purple-dotted, panduriform. Spur hairy within.

Flowering: June–July.

Habitat: Extremely rare epiphyte up to 600 m elevation, usually on *Mangifera indica* in open warm sites.

Distribution in Nainital: Haldwani.

General Distribution: India (Bengal, Bihar, Gujarat, Karnataka, Madhya Pradesh, Odisha, Uttarakhand), Bangladesh, China (Yunnan), Myanmar, Nepal, Sri Lanka.

References: Roxburgh (1795), Uniyal et al. (2007).

41.3. ***Vanda testacea*** Lindl. Rchb.f.

Epiphyte 10–25 cm. Stem erect, covered with sheaths. Leaves linear-oblong, 7–10 × 1–1.3 cm, bilobed at apex. Inflorescence erect raceme, many-flowered, as long as or longer than leaves. Flowers yellow, 1.5 cm. Lip spurred, 3-lobed, reflexed, lateral lobes oblong, mid-lobe ligulate, fleshy; spur half as long as lip.

Flowering: June–July.

Habitat: Rare, between 600–800 m elevations, trunk epiphyte in warmer valleys, often gregarious.

Distribution in Nainital: Kathgodam, Ranibagh.

General Distribution: India (widespread including Uttarakhand), Bhutan, Myanmar, Nepal, Sri Lanka, Thailand.

References: Uniyal et al. (2007).

42. ZEUXINE **Lindl.**

Genus name from Greek *zeuxis* ("yoking"), referring to partial union of column and lip. About 80 species in tropical Africa, Asia, New Guinea and Australia; 20 in India, three in Uttarakhand, one in Nainital.

42.1. ***Zeuxine flava*** Wall. ex Lindl.
Terrestrial herb up to 15 cm. Stem slender with sheathing base. Leaves ovate to ovate-lanceolate, 3–6 × 1–2 cm, often withered at flowering. Inflorescence terminal spike 2–4 cm, few-flowered. Flowers yellow; sepals oblong, lateral pair smaller; petals white, oblong-elliptic; lip yellow, cymbiform at base, anterior lobe bilobed, lobules divergent, T-shaped. Column short, stout, curved at apex (Figs. 4.3 and 4.4).

Flowering: April–May.
Habitat: Rare but locally frequent up to 1400 m elevation, forest floor in moist, shaded, litter-rich soil.
Distribution in Nainital: Bajoon, Patwadangar, Sattal.
General Distribution: India (Arunachal Pradesh, Assam, Meghalaya, Sikkim, Uttarakhand, West Bengal), Bhutan, Borneo, China, Java, Myanmar, Nepal, Sri Lanka, Thailand, Vietnam.
References: Uniyal et al. (2007).

4.2.8 Threats and Conservation Concerns

Many Nainital orchids face threats from habitat loss due to deforestation, urbanization, and agricultural expansion (Jyoti and Tewari 2020). Overharvesting for horticulture, particularly of attractive epiphytes like Vanda and Phalaenopsis, also threatens populations. Climate change impacts phenology, altitudinal ranges, and pollinator relationships (Anjum et al. 2025). Conservation efforts must include habitat protection, restoration, sustainable forestry, and propagation programs policies (Swarts and Dixon 2009; Mahapatra et al. 2021).

4.3 Conclusion

Nainital orchids are key for taxonomic, ecological, and evolutionary studies, displaying intraspecific morphological variability that indicates microhabitat adaptation (Jalal et al. 2009a). The diverse topography and microclimates support a variety of epiphytic, terrestrial, and lithophytic orchids, reflecting the mix of tropical and temperate floristic elements. Morphological adaptations such as tubers, spurs, and specialized lips demonstrate ecological strategies for survival, pollination, and reproduction across different habitats and elevations. Flowering periods from April to November ensure temporal niche differentiation, promoting reproductive success and pollinator interactions.

Despite their ecological importance, several species are rare and locally restricted, making them vulnerable to habitat loss, over-collection, and climate change. Conservation strategies, including in situ protection, habitat management, and ex situ propagation, are essential to preserve this unique orchid heritage. Future research should focus on pollination ecology, mycorrhizal relationships, genetics, and population dynamics for resilience and conservation planning (McCormick et al. 2004;

Fig. 4.3 Few species of orchids found in Nainital

Fig. 4.4 Few species of orchids found in Nainital

Mahapatra et al. 2021). The region's rich orchid diversity underscores its status as a biodiversity hotspot demanding systematic study and conservation (Arora 1982; Uniyal et al. 2007).

References

Agrawala DK, Chowdhery HJ, Chaudhury S (2010) New combinations in Indian Orchidaceae. Kew Bull 65(1):105–106

Anjum N, Khan S, Verma S, Gaira KS, Rawat B, Chettri N et al (2025) Plant-pollinator interactions along the altitudinal gradient in Berberis lycium royle: an endangered medicinal plant of the Himalayan region. PLoS One 20(5):e0310572

Arditti J (1992) Fundamentals of orchid biology. Wiley

Aryal S, Baniya MK, Danekhu K, Kunwar P, Gurung R, Koirala N (2019) Total phenolic content, flavonoid content and antioxidant potential of wild vegetables from Western Nepal. Plants 8(4): 96

Arora CM (1982) Orchids of North-Western Himalaya. Botanical Survey of India

Atkinson ET (1882) Gazetteer of Himalayan districts of the North West provinces and Oudh Government Press

Banday M, Bhardwaj DR, Pala NA (2017) Variation of stem density and vegetation carbon pool in subtropical forests of North-western Himalaya. J Sustain For 36:1–14

Blume CL (1825) Bijdragen tot de flora van Nederlandsch Indië (Vol 1). ter lands Drukkerij

Champion GC (1921) XXXVII.—Notes on various African and Asiatic species of Laius, Guérin, with an account of their accessory♂-characters [Coleoptera]. Ann Mag Nat Hist 7(40):322–343

Champion HG, Seth SK (1968) Forest types of India. Government of India Press

Chase MW, Cameron KM, Freudenstein JV, Pridgeon AM, Salazar G, Van den Berg C, Schuiteman A (2015) An updated classification of Orchidaceae. Bot J Linnean Soc 177(2): 151–174

Chauhan NS (1996) Medicinal orchids of the Himalaya. Scientific Publishers

Chowdhery HJ (2009) Orchid diversity in Arunachal Pradesh. Bishen Singh Mahendra Pal Singh

Chowdhery HJ, Giri GS, Pal GD (2010) Materials for the flora of Arunachal Pradesh (Vol. 3). Botanical Survey of India

Chowdhery HJ, Agrawala DK (2013) A century of West Himalayan Orchids. Bishen Singh Mahendra Pal Singh. New Connaught Place, Dehra Dun, India, pp 318

Colebrooke HT (1824) A discourse read at a meeting of the Asiatic society of Great Britain and Ireland, on the 15th of March, 1823. Trans R Asiatic Soc Great Br Irel 1(1):xvii-xxiii

Collett H (1902) Flora simlensis. Thacker Spink & Co. Calcutta and Shimla, Reprinted 1971. Bishen Singh Mahendra Pal Singh, Dehradun

Cramer J, Fritsch K (1888) Allgemeine Geologie (Vol 5). Engelhorn

Cribb P, Govaerts R (2005) Just how many orchids are there? Orchids 74(3):16–21

De LC, Vij SP (1992) Commercial orchids. Bishen Singh Mahendra Pal Singh, Dehradun

Deva S, Naithani HB (1986) The orchid flora of North-West Himalaya. Print and Media Associates

DFRS (2015) State of Nepal's Forests. Forest Resource Assessment (FRA) Nepal, Department of Forest Research and Survey (DFRS)

Don D (1825) Prodromus Florae Nepalensis, London

Dressler RL (1993) Phylogeny and classification of the orchid family. Cambridge University Press

Duthie JF (1902) The orchids of northwest Himalaya. Annals of the Royal Botanic Garden Calcutta 9(2):81–211

Duthie JF (1906a) The orchids of the North-western Himalaya. Ann R Bot Garden (Calcutta) 9(2): 1–211

Duthie JF (1906b) Catalogue of plants of Kumaun and of the adjacent portions of Garhwal and Tibet. Government Press, Calcutta

Gandhi KANCHI, Ormerod P (2012) A new combination in Dithrix (Orchidaceae). Phytoneuron 61:1–3
Givnish TJ, Spalink D, Ames M, Lyon SP, Hunter SJ, Zuluaga A, Doucette A et al (2016) Orchid historical biogeography, diversification, Antarctica and the paradox of orchid dispersal. J Biogeogr 43(10):1905–1916
Gupta RK (1968) Flora nainitalensis: a handbook of the flowering plants of Nainital. Navayug Traders, New Delhi. Hooker JD 1872–1897. The flora of British India Vol (1–7). London
Hooker JD (1890) Flora of British India. Reeve and Co., London, p 6
Hooker WJ (1823–1827). Exotic flora, containing figures and descriptions of new, rare, or otherwise interesting exotic plants (Vols. 1–3). W. Blackwood & T. Cadell
Hooker JD (1872–1897) The flora of British India (Vols. 1–7) L. Reeve & Co
Hooker JD (1906) A sketch of flora of British India. London
Jackson JK (1994) Manual of afforestation in Nepal. Forest Research and Survey Centre, Kathmandu
Jain SK (1956) On a botanical trip to Nainital. Indian For 82:22–38
Jalal JS, Jayanthi J (2012) Endemic orchids of India and their conservation. Rheedea 22(2):93–101
Jalal JS, Pangtey YPS (2010a) *Gastrodia falconeri*: a saprophytic orchid rediscovered after a century from Kumaun Himalaya. India Curr Sci 98:1038–1040
Jalal JS, Rawat GS, Pangtey YPS (2005) *Peristylusaffinis* (DDon) Seidenf (Orchidaceae): a new record from Kumaun Himalaya (Uttaranchal). J Econ Taxon Bot 30:401–402
Jalal JS, Rawat GS, Pangtey YPS (2007) A new distributional record of *Ornithochilusdifformis* (Wall ex Lindl) Schltr (Orchidaceae) from Nainital, Kumaun hills. Indian J For 30:523–524
Jalal JS, Lalit Tewari M, Pangtey YPS (2009a) Nervilliagammieana (Hookf) Pflitzer (Orchidaceae): a new record for Kumaun Himalaya. J Am Sci 5:91–94
Jalal JS, Kumar P, Rawat GS, Pangtey YPS (2009b) *Habenaria pubescens* Lindley: an interesting orchid from West Himalaya. Richardiana 9:76–84
Jalal JS, Pangtey YPS (2011) Gastrodia falconeri recollected from Kumaun, Western Himalayas, India. Richardiana 11(2): 53–60
Jalal JS, Kumar P, Rawat GS (2012) *Nerviliapangteyanaspnov*; a terrestrial orchid from Western Himalaya, India. Nord J Bot 30:407–411
Joshi GC, Tewari LM, Lohani N, Upreti K, Jalal JS, Tewari G (2009) Diversity of orchids in uttarakhand and their conservation strategy with special reference to their medicinal importance. Report and Opinion 1:47–52
King G, Pantling R (1898) The Orchids of the Sikkim-Himalaya Vol VIII
Kumar S, Manilal KS (1994) A catalogue of Indian orchids. Botanical Survey of India, Calcutta
Lal A, Pant M, Kumar A, Palni LMS, Singh A, Siddiqui MR, Husen A (2024) DNA barcoding of epiphytic orchids for species level identification. J Hortic Sci Biotechnol 100:1–11
Lindley J (1835) The genera and species of orchidaceous plants. Ridgways
Linnaeus C (1753) Genera plantarum, C. Wishoff, Amsterdam
Lodhiyal N, Lodhiyal LS (2012) Tree layer composition and carbon content of oak and pine in Lohaghat forests of Kumaun Himalaya. J Plant Dev Sci 4:55–62
Loureiro J de (1790) Flora cochinchinensis: sistens plantas in regno Cochinchina nascentes (Vols. 1–2). Typis et Expensis Academicis
Mahapatra SK, Schoenherr T, Jayaram, J (2021) An assessment of factors contributing to firms' carbon footprint reduction efforts. Int J Prod Econ 235:108073
McCormick MK, Whigham DF, O'neill J (2004) Mycorrhizal diversity in photosynthetic terrestrial orchids. New Phytol 163(2):425–438
Misra S (2007) Orchid flora of India. Bishen Singh Mahendra Pal Singh
Pangtey YPS, Kalakoti BS (1983) A note on the occurrence of *Cheirostylisgriffithii* Lindl (Orchidaceae) from north-west Himalaya. Indian J For 6:170
Pangtey YPS, Samant SS (1987) Some orchids new to the flora of Nainital. J Econ Taxon Bot 11:1–4
Pangtey YPS, Samant SS (1991) Orchids of Nainital. J Orchid Soc India 5:19–34

Pangtey YPS, Samant SS, Rawat GS (1991a) Orchids of Kumaun Himalaya. Bishen Singh Mahendra Pal Singh, Dehradun

Pangtey YPS, Samant SS, Pant S (1991b) Orchids of Kumaun Himalaya: a checklist. J Orchid Soc India 5:89–101

Pellissier L, Anne Bråthen K, Pottier J, Randin CF, Vittoz P, Dubuis A, Yoccoz NG, Alm T, Zimmermann NE, Guisan A (2010) Species distribution models reveal apparent competitive and facilitative effects of a dominant species on the distribution of tundra plants. Ecography 33(6): 1004–1014

Persoon CH (1807) Neottia sinensis. Synopsis Plantarum 2:511

Prahlad VC, Tewari JC, Singh SP (2018) Stand structure and growth pattern of deodar forests in Garhwal Himalaya. Int J Curr Microbiol Appl Sci 7:2319–7706

Radford AE (1986) Fundamentals of plant systematics. Harper and Row

Radford AE, Dickison WC, Massey JR, Bell CR (1974) Vascular plant systematics. Harper and Row

Rasmussen HN (1995) Terrestrial orchids: from seed to mycotrophic plant. Cambridge University Press

Rau MA, Arora CM (1973) On the occurrence of *Diplomerishirsuta* Lindl (Orchidaceae) in Western Himalaya. Bull Bot Surv India 15:138–139

Rawat GS (2005) Alpine meadows of Uttarakhand: ecology, landuse and conservation. Bishen Singh Mahendra Pal Singh

Rawat GS, Pangtey YPS (1987) Forest vegetation of Nanda Devi Biosphere Reserve. Int J Ecol Environ Sci 13:73–93

Rawat S, Bhatt ID, Rawal RS (2011) Total phenolic compounds and antioxidant potential of Hedychium spicatum Buch. Ham. ex D. Don in west Himalaya, India. J Food Compos Anal 24(4-5):574–579

Reichenbach H (1886) Studien zur Entwicklungsgeschichte des Flusskrebses (Vol 4). M. Diesterweg

Reichenbach K (1862) Odische Begebenheiten zu Berlin in den Jahren 1861 und 1862. Schroeder

Renz J (1978) New species of *Peristylus* and *Platanthera* (Orchidaceae) from India and New Guinea. J Orchid Soc India 1:23–28

Roxburgh W (1795) An account of a new species of Swietenia (Mahogany); and of experiments and observations on its Bark, made with a view to ascertain its powers, and to compare them with those of Peruvian Bark, for which it is proposed as a substitute: being an abstract of a paper on this subject, addressed to the Honourable Court of Directors of the United East-India Company. Med facts observations 6:127

Roxburgh W (1820) Flora indica, or, Descriptions of Indian plants (Vol 1). Mission Press

Royle JF (1833–1839) Illustrations of the botany and other branches of natural history of the Himalayan Mountains and of the Flora of Kashmir. W.H. Allen & Co., London. https://doi.org/10.5962/bhl.title.449

Savard JPL, Clergeau P, Mennechez G (2000) Biodiversity concepts and urban ecosystems. Landscape urban plann 48(3-4):131–142

Seidenfaden G, Arora CM (1982a) The orchids of north-western Himalaya. Nord J Bot 2:1–77

Sigdel SR, Wang Y, Camarero JJ, Zhu H, Liang E, Peñuelas J (2018) Moisture-mediated responsiveness of treeline shifts to global warming in the Himalayas. Glob change biol 24(11):5549–5559

Seidenfaden G, Arora CM (1982b) An enumeration of the orchids of North Western Himalaya. Nord J Bot 2:7–27

Simpson MG (2019) Plant systematics, 3rd edn. Academic Press

Singh JS, Singh SP (1987) Forest vegetation of the Himalaya. Bot Rev 53:80–192

Summerhayes VS (1955) A revision of the genus Brachycorythis. Kew Bull 10(2):221–264

Tiwari PC, Joshi B (2010) Land use changes and conservation of water resources in Himalayan Headwaters. In proceedings of the 2nd German-Indian conference on research for sustainability: energy and land use, science and technology-drivers for a common future

Swarts ND, Dixon KW (2009) Terrestrial orchid conservation in the age of extinction. Ann Bot 104: 543–556

Uniyal BP (2007) Checklist of orchids of Uttarakhand Himalayan. J Bot 1:56–72

Uniyal SK, Swami A, Uniyal BP (1997) Monocotyledonous plants of Uttar Pradesh—(excluding Cyperaceae and Poaceae): a checklist. Bishen Singh Mahendra Pal Singh

Uniyal BP, Sharma JR, Choudhery U, Singh DK (2007) Flowering plants of Uttarakhand (a checklist)

Vij SP, Gupta S, Jamwal PS (1983a) A checklist of orchids of Nainital and adjacent hills in Central Himalaya. J Econ Taxon Bot 4:765–784

Vij SP, Navdeep Shekhar SK, Kashyap S, Garg AK (1983b) Observations on the orchids of Nainital and adjacent hills in the Central Himalaya (Ecology and Distribution). Res Bull (Sci) Panjab Univ 34:63–76

Waud M, Busschaert P, Ruyters S, Jacquemyn H, Lievens B (2014) Impact of primer choice on characterization of orchid mycorrhizal communities using 454 pyrosequencing. Mol Ecol Resour 14(4):679–699

Willdenow KL (1805) The principles of botany, and of vegetable physiology. W. Blackwood

Yam TW, Arditti J (2009) History of orchid propagation: a mirror of the history of biotechnology. Plant Biotechnol Rep 3(1):1–56

Pollination Ecology and Conservation Perspectives on Orchid Species in the Western Himalayan Landscape

5

V. P. Uniyal, Manu Pant, Agni Chandra, and Vandana Mehrwar

5.1 Introduction

The Western Himalayan mountain system is one of the most ecologically complex and biogeographically important systems on earth and includes dramatic altitudinal gradients that generate spectacular heterogeneity of habitat and species abundance (Körner 2003; Myers et al. 2000). The Orchidaceae seems to stand as a flagship family within this woodland, displaying impressive evolutionary capacity in adaptation to the montane environment, where more than 240 species are recorded across elevational ranges of subtropical foothills (300 m) to alpine meadows (4500 m and above) (Misra 2004; Jalal and Jayanthi 2012). Orchids are one of the most diverse blooming plant families, with a variety of behaviours including terrestrial, epiphytic, lithophytic, and saprophytic. Approximately 70% of the species are epiphytic (Fischer et al. 2011; Marwah et al. 2021). The diversity of this orchid comprises not only the tropical epiphytic lineages adopting a temperate environment but also the old temperate-alpine terrestrial groups to form assemblages of great biogeographical interest (Hooker 1890; Duthie 1906). In India orchids are represented by 1129 species and 184 genera, and show maximum diversity in the eastern Himalaya, including the north-eastern region, Western Ghats, eastern Himalaya, and eastern part of Western Himalaya (Kumaon Himalaya) (Jalal et al. 2008; Lal et al. 2023).

The relationships between orchids and pollinators are key factors that have enabled orchids to evolve and contributed to the development of some of the most

V. P. Uniyal (✉) · M. Pant · V. Mehrwar
Department of Biosciences, Graphic Era (Deemed to be University), Dehradun, Uttarakhand, India
e-mail: vpuniyal.bt@geu.ac.in

A. Chandra
Snow Leopard Conservation Regenerative Project, Uttarakhand Forest Department, Dehradun, Uttarakhand, India

A. Husen, M. Pant (eds.), *Forest-based Orchids*,
https://doi.org/10.1007/978-981-92-0096-2_5

complex reproductive mechanisms in the universe (Faegri and van der Pijl 2013; Johnson and Steiner 2000). Simple nectar rewards attract generalist pollinators through complex mechanical structures and deceptive mechanisms that ensure that the right amount of pollen is delivered to the flowers (Nilsson 1988; Tremblay et al. 2005). The accuracy and stability of such systems of pollination are even more important to the existence and reproductive functionality of orchids in montane zones, such as the Western Himalaya, where growing seasons are brief and environmental factors are often hostile (Körner 2003; Williams et al. 2015).

Nevertheless, the pollination ecology of orchids in the Western Himalayas, regardless of the established global importance of both pollination and Western Himalayan biodiversity as a whole, remains fundamentally unexplored. Most published work is dedicated to taxonomic descriptions and surveys of the flora and conservation status without considering the ecological processes that sustain orchid populations (Ved et al. 2003; Jalal and Jayanthi 2012; Sharma et al. 2017). There is a conspicuous lack of scientific literature on direct observation of pollinator visitation, experimental research on pollination efficacy, and quantitative research of pollen transfer (King et al. 2013). Likewise, the phenological interactions between orchid flowering and pollinator activity that are central to the study of climate change effects are practically undocumented (Memmott et al. 2007; Shrestha et al. 2012). This knowledge gap is a significant challenge to effective conservation planning and ecosystem management in a region that experiences rapid environmental change. The climate warming in the Himalaya is occurring at faster rates than global averages with potentially dire consequences to the fine temporal coordination of plants and their pollinators (Shrestha et al. 2012). In addition, orchid populations are becoming more isolated due to increasing habitat fragmentation and tourism, which also affects pollinator diversity and distribution (Pandit et al. 2006). Overharvesting of economically important orchids continues to offset densities below critical thresholds for maintaining viable pollination services (Samant et al. 1998; Ved et al. 2003).

The present review collects all the peer-reviewed literature available on orchid diversity, distribution and ecology of the Western Himalayas and their insect pollinators. We review systematic knowledge about orchid-pollinator interactions and clearly indicate where empirical evidence is lacking. Our research raises fundamental questions: Which insect taxa are specialised pollinators of various orchid species? Can flowering phenology and pollinator activity coordinate with elevation gradients? What do habitat fragmentation and climate change do to pollination networks?

This review provides a scientific basis for the intended empirical research to gain knowledge and ensure the safety of one of the most extraordinary and yet little-researched plant-pollinator systems in the world by recording what we already know and what we still do not know or comprehend, as well as the gaps in our knowledge that require closure.

5.2 Orchid Diversity and Biogeographical Context

The orchids in the Western Himalaya represent a rich repertoire of species, due to the region's location at some of the most important biogeographical features, as well as its diverse environments. Naturalistic work by colonial-era explorers and systematic surveys has documented high taxonomic diversity within the mountainous terrain. In Western Himalaya, John Firminger Duthie was the first botanist who compile the Orchidaceae of the region based on his own collections and the collections made by the earlier explorers, especially Strachey and Winterbottom (Jalal et al. 2008). Hooker's Flora of British India provided a taxonomic foundation for the Himalayan orchids by describing many species with altitudinal zonation (Hooker 1890), and subsequent work by Duthie (1906) also provided an account of the orchid species, their distribution and habitat from the northwestern Himalaya with detailed descriptions of 85 new species. This has been supplemented by modern floristic surveys, and Jalal and Jayanthi have developed a comprehensive checklist of over 240 orchid species in Uttarakhand alone, which are representative of approximately 40 genera within the different ecological regions in that state (Jalal and Jayanthi 2012). These contemporary collections, along with local surveys of Himachal Pradesh and Jammu & Kashmir, indicate that the entire Western Himalaya contains 240–280 species of orchids, which is almost one-third of Indian orchid diversity (Misra 2004; Sharma et al. 2017). Such orchids vary in their adaptation, having different ecological strategies to adapt to varying altitudes and ecological zones. In the humid mid-elevation forests (1200–2500 m) there are widespread epiphytic species including those of the genera *Dendrobium*, *Coelogyne* and *Bulbophyllum*. They grow on the bark of oak, rhododendron, and coniferous trees as their growing substrate (Hooker 1890; Jalal and Jayanthi 2012). All of these epiphytes developed specialised aerial root systems and water storage reserves that enable them to survive these seasonal dry conditions typical of their arboreal environments (Benzing 1990). Orchids thrive in diverse environments, which is their most distinguishing quality. Orchids require a specialised microhabitat to grow and develop (Jalal and Rawat 2009). Microhabitat conditions vary between environments. The presence of particular mycorrhizal fungi in the microclimate may potentially influence the habitat of orchids (Hegde 1982).

A more extreme form of altitudinal specialisation is seen in terrestrial orchids; the various genera are not only grouped into separate elevation bands but also represent large climatic zones. *Dactylorhiza hatagirea*, *Orchis latifolia*, and *Platanthera clavigera*, among others, are confined to alpine meadows and wetlands at an altitude above 2800 m, where they have adapted to the severe seasonal temperature changes and short growing periods of the high-altitude environment (Cribb 2003; Sharma et al. 2017). The adaptations of such alpine specialists are commonly remarkable: they are compact, heavily pigmented, and have storage organs that allow them to withstand a long freezing period and snow cover (Körner 2003).

Dactylorhiza hatagirea, also referred to as Salampanja or Hatajari, is one of the most ecologically and economically vital species in the region, found in marshy meadows and wetlands ranging between 2800 and 4500 m in elevation (Ved et al.

2003). This species has gained special significance through its assumed medicinal value, and as a result, it has experienced high harvesting pressures amounting to significant population losses in most of its habitat (Samant et al. 1998). The species is exemplary of the conservation issues of many Western Himalayan orchids, where patterns of traditional use have been enhanced with commercial demand, leading to extraction rates that are higher than the natural rate of population replacement.

The genus *Cypripedium* (e.g., *C. cordigerum*, *C. elegans*) has a very particular ecology in the shade of the forest floor and the alpine highlands, between 2500 and 4500 m (Cribb 2003). Commonly known as lady slippers, they are identifiable by their inflated labellum and specialised pollination mechanism, but they have fallen victim to overcollection as ornamentals since they look nice and are perceived to be rare (Raskoti 2009). The orchid *Platanthera clavigera* (club-tailhabenaria or club-spur bog orchid), with nectar spurs more than 8 cm long (one of the longest recorded in the orchid family), is perhaps the most morphologically distinctive orchid in the region (Nilsson 1988). This exuberant floral arrangement is a strong indication of coevolution with long-tongued nocturnal pollinators, probably sphingid moths, but direct pollination has not been reported in the literature (Alexandersson and Johnson 2002). The ecological vulnerability of Western Himalayan orchids can be seen through their conservation status and through increasing human pressures to which they are exposed. Even though the IUCN Red List has not formally evaluated most of the regional species, available information suggests that many of them are endangered at different levels (Ved et al. 2003).

5.3 Pollinator Communities and Foraging Ecology

The insect fauna which may be involved in orchid pollination in the Western Himalaya includes species of several different orders, but Hymenoptera and Lepidoptera appear to be the most important in terms of their morphology and comparison with other similar ecosystems. Nevertheless, observational evidence that directly connects particular insect taxa to orchid pollination is extremely limited, and this evidence is one of the most severe deficits in our knowledge about these ecological systems (Johnson and Steiner 2000; Singh et al. 2022).

Among the Hymenoptera, bumblebees of the genus *Bombus* seem to have important roles in the pollination of high-altitude orchids, depending on their known occurrence and their known pollination interactions with orchids in similar temperate mountain environments (Williams et al. 2015). Most recent species distribution mapping by Singh et al. (2022) has found a number of *Bombus* species, such as *B. luteipes* or *B. breviceps*, in the Great Himalayan National Park and found suitable habitat between 3000–4500-m elevations. Due to special morphological and physiological characteristics enabling them to regulate their temperature and cover thick setae that keeps them active at temperatures down to zero or below which normally paralyse other insects make these fuzzy creatures able to forage in high altitude regions very easily (Heinrich 1979). Bumblebees are likely to be most effectively adapted to orchid pollination in the montane because of their foraging

Fig. 5.1 Honeybees found in the Western Himalayan region, India

ecology. Their size and relative strength allow them to cope with the complex pollination regimes characteristic of most orchid species, and the ability to buzz pollinate by sonication may enhance pollen removal by orchid blooms that have complex mechanical obstacles (King et al. 2013).

Moreover, bumblebees show flower constancy on foraging trips, making it more likely pollen taken off one flower of a particular species will be deposited on another flower of the same species instead of being discarded on another incompatible flower (Waser 1986). There are also records of solitary bees belonging to genera like *Andrena* and *Halictus* in the Western Himalaya, but their relationship with orchids has not been investigated yet (Michener 2007). These bees can be more specialised in their foraging behaviours than their social relatives and can have significant roles in pollinating those early-flowering orchid species that bloom soon after the snow melts when bumblebee colonies are still settling (Stone et al. 1996). They may be able to offer essential pollination services to orchids with small phenological profiles because of their patterns of emergence, which occur in synchrony with warming soil temperatures during the spring.

Honeybees (*Apis cerana* and *A. dorsata*) are found across the entire area of the lower elevation forests where epiphytic orchids thrive the most (Chandra and Mattu 2017) (Fig. 5.1).

These species are mainly generalist foragers and would be expected to visit orchid flowers as random foragers, but their pollination of orchids has not been reported. The morphological disparity between honeybee size, foraging behaviour, and the specialised pollination systems of most orchids indicates that honeybees could be ineffective pollinators of native bee species (Johnson and Steiner 2000).

The lepidopteran fauna that may be involved in the pollination of orchids comprises not only day-neutral butterflies but also nocturnal moths, and various groups may be differentially attractive to different orchid taxa, depending on floral structure and timing of floral display. A survey of moth diversity conducted by Singh et al. (2022) in the Tirthan Valley recorded 524 species, including many species of the Sphingidae family with long proboscides that might be used to reach deep nectar spurs on species like *Platanthera clavigera*. Circumstantial evidence indicating the possibility of a pollination relationship between orchids with long spurs and *Agrius convolvuli* is given by the presence of species with proboscis lengths longer than 10 cm, although no direct references have been made (Kitching and Cadiou 2000).

The butterfly communities of the area mainly belong to the families Papilionidae, Nymphalidae and Pieridae, and it is reported that many species visit flowers for nectar in other montane ecosystems (Chauhan et al. 2021). The connection between orchid pollination efficacy and butterfly foraging, however, has been highly speculative. The comparatively short proboscis of the majority of butterfly species restricts their capacity to reach nectar in long-spurred orchids, and their low mass and frail structure could restrain successful manipulation of complicated mechanical pollination structures (Faegri and van der Pijl 2013; Proctor et al. 1996). Temporal interactions complicate orchid pollination activities in the Himalayan ecosystem as they influence pollinator behaviour and limit their movement to a brief spell of good weather between sunny mornings and early afternoons (Kevan 1975). Many orchid species are adapted for diurnal pollination; the timing of nectar and volatile compound release is coordinated with the activity of the active pollinators, making it more suitable for pollinator interactions (Herrera 1990). It exposes nocturnal moths to a range of opportunities and challenges regarding orchid pollination since atmospheric conditions in mountains can become relatively stable and humid enough to allow increased activity of pollinators and dispersal of floral fragrances (Raguso 2008).

5.4 Pollination Mechanisms and Reproductive Strategies

Western Himalayan orchids are highly varied in the number of different pollination mechanisms that mirror the refinement of coevolution over millions of years, but we still have little insight on these mechanisms, as we lack observational studies. Floral morphology and presentation patterns of local orchids are strong evidence of a number of discrete pollination strategies, each evolved to take advantage of some facet of pollinator behaviour and ecology (Johnson and Steiner 2000; Nilsson 1988). Nectar-rewarding systems have also been found to typify much of the terrestrial orchids in the region, including orchids like *Dactylorhiza hatagirea* that are relatively open-flowered and whose nectaries are readily accessible (Baker and Baker 1983). The nectar produced by the plants has sugar concentrations ranging from 30 to 50 (typically a high-energy reward), which attract and reward pollinators (Petanidou 2005). This intense sugar level of mountain orchid nectar could reflect an evolutionary response to the energetic requirements of high-altitude pollinator traffic, as insects require a lot of energy to thermoregulate and fly in thin air and cold environments (Boose 1997).

Additional fine-tuning of pollinator attraction is provided by the temporal control of nectar-schedule in rewarding orchids. It seems that many species seem to synchronise nectar formation at certain times that coincide with maximum pollinator activity, but in-depth analysis of the nectar production cycle has not been described in the Western Himalaya literature (Zimmerman 1988). This temporal coordination between nectar availability and pollinator foraging could be particularly important in montane environments where favourable conditions for both plant and animal activity may occur within narrow windows (Totland 1994). A number of Western Himalayan orchid genera, such as *Orchis*, are suspected of having deceptive pollination mechanisms where they not only obtain pollination services without offering a reward but also have morphologically similar flowers to the rewarding plants in their communities (Johnson and Steiner 2000). These orchids take advantage of pollinator learning behaviour and search image formation and act on the fact that pollinators are at times forced to visit unrewarding flowers in order to find out new sources of resources (Jersáková et al. 2006). The success or failure of food deception is strongly contingent on the low densities of populations compared to rewarding flowers so that the pollinators are presented with enough rewards to sustain their foraging behaviour and occasional access to deceptive orchids (Ferdy et al. 1998). This frequency-dependence of deceptive pollination generates intriguing population-level interactions, with orchid reproductive success potentially decreasing when local populations in the community get too large in comparison to profitable flowers (Johnson et al. 2003). This association can potentially justify the largely patchy distributions of many species of orchids and can have significant consequences on conservation management projects that seek to expand the population by controlling the habitat or restoring it.

The most complex pollination mechanism recorded to date in the Western Himalaya is perhaps in the genus *Cypripedium*, where a ribbon-like labellum acts as a temporary trap to attract small insect pollinators (Bänziger et al. 2005). In this system, pollinators are attracted to the flower by visual and olfactory cues and enter the inflated labellum through the large frontal opening but find themselves unable to exit the same way due to the smooth, inward-curving walls and strategically positioned hairs that prevent backward movement (Li et al. 2006). The only escape route leads through narrow lateral openings positioned adjacent to the reproductive structures, ensuring that departing pollinators contact both the stigma and anthers in precise sequence. *Cypripedium* trap pollination is one of the most evolved forms of floral engineering in the plant kingdom, with its specific coordination of attractive signals, mechanical barriers and escape paths (Cribb 2003). None of the scientific literature, however, has explicitly examined the coevolution of orchids and pollinators with respect to specific pollinators being utilised, as well as the efficiency of pollen transfer among Western Himalayan *Cypripedium* populations.

Another specialised pollination mechanism that seems to be highly specific to particular pollinator taxa is the extreme nectar spurs present in species like *Platanthera clavigera*. These flowers need insects with sufficiently long proboscises in order to be effectively pollinated; the insects responsible in this case are probably sphingid moths, operating in the evening (Nilsson 1988). The development of such hashed floral morphology suggests high levels of convolutional pressure and implies highly derived pollinator interactions, which might be especially sensitive to environmental perturbation.

Orchid pollination systems are heavily dependent on chemical communication, wherein floral fragrances are used to attract pollinators both at distances of several kilometres and to give them species-specific recognition cues (Schiestl 2005). The orchids of the Western Himalayas have generated various arrays of volatile compounds, both sweet and vanilla-like odours that resemble the benzenoid compounds and more complex bouquets that can consist of terpenoid, fatty acid derivatives and nitrogen-based compounds (Johnson et al. 2003). Combinations of visual and mechanical signals and scent lead to highly specific communication systems, yet there are no empirical studies of floral volatiles and pollinator behaviour in such orchids.

5.5 Temporal Dynamics and Phenological Constraints

Due to the temporal constraints on reproductive behaviour at high-altitude regions, crucial pollinator-orchid phenological interactions are set up between both species for successful reproduction (Kudo and Ida 2013). In the Western Himalaya, there are only four to five months of good growth. It involves a relationship where flowering, the appearance of pollinators, and favourable weather conditions all occur within limited time periods (Körner 2003; Sharma et al. 2017). Early-flowering plants, such as certain *Cypripedium* taxa, begin to flower within weeks after the snows are melted, when soil temperatures are close to freezing, which makes them uncompetitive for pollinator attention but means they have to endure adverse conditions (Mosquin 1967).

Flowering takes place in July and August when the environment is good and pollinators are most active. However, in the course of this period, there is a great diversity of flowering vegetation competing intensely with each other (Primack 1985). By contrast, later flowering plants will tend to possess far more diverse ranges of flowering periods and will exploit the resulting reduction of competition achieved once other plants have finished their own reproductive process (Rathcke and Lacey 1985). There are further temporal differentiations in the patterns of the hourly activities of the pollinators. During favourable flying and foraging weather, most hymenopteran pollinators show their peak activity in the mid-morning to early afternoon (Stone et al. 1996).

A second key component is weather patterns causing stochastic variation in reproductive success (Totland 2001). The low temperatures, precipitation, wind and low light intensity reduce the numbers of pollination activities with the receptive flora as a significant proportion and may cause a lack of successful pollination (García Camacho and Totland 2009). Additionally, the onset of the monsoon is also considered to be a leading factor in flowering and pollination periods (Shrestha et al. 2012). Phenological associations are changing with climate change, which may have dire effects on specialised interactions (Both et al. 2006), causing temporal mismatch in flower and pollinator emergence, which is a threat to population viability (Miller Rushing and Primack 2008). The warming in the Western Himalaya has been very rapid, particularly at more elevated altitudes where the orchids are endemic (Shrestha et al. 2012). *Dendrobium hesperis* (Seidenf.) Schuit. & P.B. Adams (Orchidaceae: Epidendroideae: Dendrobieae), an interesting but lesser-known orchid, referred to in earlier literature as *Flickingeria hesperis*

Seidenf., was formerly believed to be endemic to the Gori Valley (Uttarakhand), the westernmost orchid hotspot in the Indian Himalayas. Originally described from Western Himalayas in 1982, this species was later discovered in Manipur in north-east India in 2014, thereafter reported from Laos in 2020 (Kumar et al. 2020).

5.6 Conservation Challenges and Anthropogenic Impacts

With the sheer amounts of local habitat destruction and the overall impacts of global climate change, orchid-pollinator interactions are becoming threatened at alarming rates (Ved et al. 2003; Pandit et al. 2006; Shrestha et al. 2012). These problems endanger the pollinator communities upon which these plants depend as well as the survival of the individual orchid species. Consequently, thousands-year-old ecological networks find themselves at serious risk of losing their stability.

Direct exploitation depicts short-term perils, and the species suffer from excessive depletion due to overharvesting in terms of medicine and horticulture (Samant et al. 1998; Ved et al. 2003). Therapeutically useful *Dactylorhiza hatagirea* have been harvested in an unsustainable manner throughout their range, leading to local extinctions and the fragmentation of populations. Equally, *Cypripedium* species are victims of excessive harvesting in the international horticultural trade despite CITES measures (Cribb 2003; Raskoti 2009). Continuous harvesting does not just destroy the orchid population; it also interrupts pollination networks. Selective harvesting reduces attractiveness to pollinators and has removed supporting flowering displays by selecting vigorous reproductive individuals (Ticktin 2004).

The common threats include habitat destruction and fragmentation, which impact the entire landscape (Pandit et al. 2006). Construction of roads, hydro-electric power plants and tourist sites has ruined thousands of hectares and torn up the portions that remain. Tourism exerts greater pressure on the environment because it tramples on vegetation and alters microhabitat (Pickering and Hill 2007). Fragmentation greatly interferes with orchid-pollinator interactions. Little patches of habitat can often not support sustainable populations of pollinators, and the distances between patches can often exceed the foraging ranges of the pollinators. This situation greatly limits gene flow and increases chances of inbreeding (Aguilar et al. 2006). Moreover, edge effects have an undoubted impact on the microclimate conditions and shift the pollination community composition (Murcia 1995).

Climate change compounds the relationships by various means (Memmott et al. 2007). The increase in temperature at high altitudes can cause shifts in both species, which may cause the phenological mismatch and decouple mutualism (Lenoir et al. 2008). Flowering and timing of pollinators are influenced by precipitation changes in unpredictable ways (Both et al. 2006). Interactive stressor effects can also be especially troublesome, as discontinuous populations are not resilient to added problems (Brook et al. 2008).

5.7 Research Priorities and Future Directions

Empirical research is strongly needed to fill in the gaps in our knowledge that demand a scientific foundation to be able to conserve effectively (King et al. 2013; Memmott et al. 2007). There remains a striking paucity of natural history observations concerning which pollinators visit orchids and how effective these visitors actually are—a gap that makes it harder to grasp the true workings and adaptations of mountain ecosystems. There is an urgent need to witness, firsthand, how insects interact with orchids in the wild, especially in the Western Himalaya, where reliable records are few and fragmented (Johnson and Steiner 2000). If we could embark on systematic studies—patiently watching representative orchids and mapping their habitats—so much so that remains guesswork could be brought to light, quickly providing a foundation for understanding the architecture of pollination networks. Experimental approaches add another dimension: hand-pollination, for example, allows us to measure just how much seed potential a plant holds, while exclusion experiments prove the essential role animal visitors play (Knight et al. 2005).

What stands out about the Valley of Flowers is the way it quietly holds some of the last truly intact habitats for wild lady's slipper orchids (*Cypripedium cordigerum*, *C. elegans*, *C. himalaicum*), the continued presence of all three signals that microhabitat conditions, population structure and gene pools here remain robust (Rai et al. 2010; Chandra et al. 2023). As science delves deeper, molecular tools offer a new lens: DNA metabarcoding of pollen now reveals which species pollinators truly visit, while seed genotyping can trace the intricate paths by which pollen travels from plant to plant (Bell et al. 2016). Thus, our understanding of pollination expands, weaving together field notes, genetic markers and phenological observations.

To truly anticipate and face the coming changes wrought by shifting climates, we should track the timing of blooms, the appearance of pollinators, and the rhythms of weather along the curves of elevation, such information could serve as an early warning system (Kudo and Ida 2013). Added to this is the promise of chemical ecology. Through gas chromatography, the fleeting language of volatile floral compounds becomes legible, opening new chapters in how flowers and insects communicate. Conservation genetics deepens the picture still further, illuminating population trends and the impact of fragmentation (Young et al. 1996; Schiestl 2005).

5.8 Conclusion

While taxonomic knowledge of orchids is extensively developed, empirical knowledge about plant-pollinator interactions is nearly non-existent in all subfields of pollination ecology. More than 240 species documented across the elevational gradients to date are a testament to evolutionary achievement, the result of millions of years of montane evolution and coevolution with pollinators. There is fascinating evidence for the evolution of specialised relationships from extreme nectar spurs to complex trap mechanisms. However, the lack of direct observational data linking orchids to pollinators is the single largest barrier to understanding system function. While there is some anecdotal evidence of bees and Lepidoptera playing important roles, their effectiveness is unknown. This knowledge gap is problematic as environmental changes ramp up and conservation decisions need to be made based on science.

Temporal dimensions seem to be especially at risk from climate disruption, but this has been addressed in phenological monitoring. Given shortened growing seasons and strict phenological constraints on reproductive timing, small fluctuations in the environment may have a profound effect on synchrony and reproductive success. Habitat alteration, degradation, and loss owing to the development of motor roads, structures, unmanaged forest fires, invasive alien species, and overexploitation are the main threats to epiphytic orchid species. The conservation issues faced are complex and severe and include exploitation, habitat destruction and fragmentation, and climate change. Yet, pollinator community features actually important for orchid persistence are currently largely overlooked. Research priorities are at once a way to advance science and also a way to address pressing conservation needs. The Western Himalaya offers a rare chance to further our knowledge of the relationship between plants and pollinators and help in preserving biodiversity of global significance. Nevertheless, it takes a huge amount of resources and institutional and stakeholder coordination in order to succeed. The orchids and the pollinators belong to the priceless biological heritage of millions of years of evolution. Only our ability to discern, save and help them adjust to accelerated environmental change will ensure their survival in the long term.

References

Aguilar R, Ashworth L, Galetto L, Aizen MA (2006) Plant reproductive susceptibility to habitat fragmentation: review and synthesis through a meta-analysis. Ecol Lett 9(8):968–980

Alexandersson R, Johnson SD (2002) Pollinator mediated selection on flower tube length in a hawkmoth pollinated *Gladiolus* (Iridaceae). Proc R Soc Lond 269(1491):631–636

Baker HG, Baker I (1983) Floral nectar sugar constituents in relation to pollinator type. In: Handbook of experimental pollination biology. Scientific and Academic Editions, pp 117–141

Bänziger H, Sun H, Luo YH (2005) Pollination of a slippery lady slipper orchid in Southwest China: *Cypripedium guttatum* (Orchidaceae). Bot J Linn Soc 148(3):251–264

Bell KL, Burgess KS, Botsch JC, Dobbs EK, Read TD, Brosi BJ (2016) Quantitative and qualitative assessment of pollen DNA metabarcoding using constructed species mixtures. Mol Ecol 25(20): 4962–4978

Benzing DH (1990) Vascular epiphytes: general biology and related biota. Cambridge University Press

Boose DL (1997) Sources of variation in floral nectar production rate in *Epilobium canum* (Onagraceae): implications for natural selection. Oecologia 110(4):493–500

Both C, Bouwhuis S, Lessells CM, Visser ME (2006) Climate change and population declines in a long distance migratory bird. Nature 441(7089):81–83

Brook BW, Sodhi NS, Bradshaw CJ (2008) Synergies among extinction drivers under global change. Trends Ecol Evol 23(8):453–460

Chandra A, Mattu VK (2017) Studies on major pests and predators of *Apis cerana* and *Apis mellifera* in the Chamba valley, Himachal Pradesh. J Entomol Zool Stud 5(6):728–731

Chandra N, Singh G, Rai ID, Mishra AP, Kazmi MY, Pandey A, Jalal JS, Costache R, Almohamad H, Al-Mutiry M, Abdo HG (2023) Predicting distribution and range dynamics of three threatened Cypripedium species under climate change scenario in Western Himalaya. Forests 14(3):633

Chauhan M, Uniyal VP, Chandra A, Thakur P, Mehrwar V (2021) Preliminary assessment and conservation of insect pollinators in Lahaul & Spiti, Himachal Pradesh. Curr Sci 120(5): 883–887

Cribb P (2003) The genus Cypripedium. Kew Publishing

Duthie JF (1906) The orchids of the north-western Himalaya. Ann R Bot Gard (Calcutta) 9(2): 81–211

Faegri K, Van Der Pijl L (2013) Principles of pollination ecology

Ferdy JB, Gouyon PH, Moret J, Godelle B (1998) Pollinator behaviour and deceptive pollination: learning process and floral evolution. Am Nat 152(5):696–705

Fischer A, Blaschke M, Baessler C (2011) Altitudinal gradients in biodiversity research: the state of the art and future perspectives under climate change aspects. AFSV 11:35–47

García Camacho R, Totland Ø (2009) Pollen limitation in the alpine: a meta-analysis. Arct Antarct Alp Res 41(1):103–111

Hegde SN (1982) Observations on the habitat-distribution of orchids of Arunachal Pradesh. J Bombay Nat Hist Soc 82:114–125

Heinrich B (1979) Bumblebee economics. Harvard University Press

Herrera CM (1990) Daily patterns of nectar secretion in bilabiate flowers. Oikos 58(1):73–79

Hooker JD (1890) Flora of British India, Orchideae, vol 6. L. Reeve & Co.

Jalal JS, Jayanthi J (2012) Annotated checklist of orchids of Uttarakhand, Western Himalaya. Taiwania 57(4):394–432

Jalal JS, Rawat GS (2009) Habitat studies for conservation of medicinal orchids of Uttarakhand, Western Himalaya. Afr J Plant Sci 3(9):200–204

Jalal JS, Rawat G, Kumar P, Pangtey Y (2008) Orchidaceae, Uttarakhand, Western Himalaya, India. Check List 4(3):304–320

Jersáková J, Johnson SD, Kindlmann P (2006) Mechanisms and evolution of deceptive pollination in orchids. Biol Rev 81(2):219–235

Johnson SD, Steiner KE (2000) Generalisation versus specialisation in plant pollination systems. Trends Ecol Evol 15(4):140–143

Johnson SD, Peter CI, Nilsson LA, Ågren J (2003) Pollination success in a deceptive orchid is enhanced by co-occurring rewarding magnet plants. Ecology 84(11):2919–2927

Kevan PG (1975) Sun-tracking solar furnaces in High Arctic flowers: significance for pollination and insects. Science 189(4204):723–726

King C, Ballantyne G, Willmer PG (2013) Why flower visitation is a poor proxy for pollination: measuring single visit pollen deposition, with implications for pollination networks and conservation. Methods Ecol Evol 4(9):811–818

Kitching IJ, Cadiou JM (2000) Hawkmoths of the world: an annotated and illustrated revisionary checklist (Lepidoptera: Sphingidae). Natural History Museum, London

Knight TM, Steets JA, Vamosi JC, Mazer SJ, Burd M, Campbell DR, Dudash MR, Johnston MO, Mitchell RJ, Ashman TL (2005) Pollen limitation of plant reproduction: pattern and process. Ann Rev Ecol Evol Syst 36:467–497

Körner C (2003) Alpine plant life: functional plant ecology of high mountain ecosystems. Springer, p 4

Kudo G, Ida TY (2013) Early onset of spring increases the phenological mismatch between plants and pollinators. Ecology 94(10):2311–2320

Kumar P, Sharma J, Isaac Williams M, Rawat GS (2020) 943. Dendrobium hesperis: Orchidaceae. Curtis's Bot Mag 37(2):180–190

Lal A, Pant M, Kumar A, Palni LMS, Singh A, Wabaidur SM, Husen A (2023) Clonal propagation and hardening of epiphytic orchids (*Aerides multiflora* and *Rhyncostylis retusa*) for sustainable conservation and utilization. S Afr J Bot 160(2023):402–411

Lenoir J, Gégout JC, Marquet PA, De Ruffray P, Brisse H (2008) A significant upward shift in plant species optimum elevation during the 20th century. Science 320(5884):1768–1771

Li P, Luo YB, Bernhardt P, Kou Y, Perner H (2006) Pollination of *Cypripedium plectrochilum* (Orchidaceae) by *Lasioglossum* spp. (Halictidae): the roles of generalist attractants versus restrictive floral architecture. Plant Biol 8(2):220–230

Marwah G, Bisht S, Rawat GS (2021) Patterns of distribution among epiphytic orchids and environmental factors in Gori valley, western Himalaya. Orchid Soc India 35:159–167

Memmott J, Craze PG, Waser NM, Price MV (2007) Global warming and the disruption of plant–pollinator interactions. Ecol Lett 10(8):710–717

Michener CD (2007) The bees of the world, 2nd edn. Johns Hopkins University Press

Miller Rushing AJ, Primack RB (2008) Global warming and flowering times in Thoreau's Concord: a community perspective. Ecology 89(2):332–341

Misra R (2004) Orchids of India: a glimpse. B.S. Vishal Publications

Mosquin T (1967) Evidence for autopolyploidy in *Epilobium angustifolium* (Onagraceae). Evolution 21(4):713–719

Murcia C (1995) Edge effects in fragmented forests: implications for conservation. Trends Ecol Evol 10(2):58–62

Myers N, Mittermeier RA, Mittermeier CG, Da Fonseca GA, Kent J (2000) Biodiversity hotspots for conservation priorities. Nature 403(6772):853–858

Nilsson LA (1988) Evolution of flowers with deep corolla tubes. Nature 334(6178):147–149

Pandit MK, Sodhi NS, Koh LP, Bhaskar A, Brook BW (2006) Unreported yet massive deforestation is driving the loss of endemic biodiversity in the Indian Himalaya. Biodivers Conserv 15(12):4097–4105

Petanidou T (2005) Sugars in Mediterranean floral nectars: an ecological and evolutionary approach. J Chem Ecol 31(5):1065–1088

Pickering CM, Hill W (2007) Impacts of recreation and tourism on plant biodiversity and vegetation in protected areas in Australia. J Environ Manag 85(4):791–800

Primack RB (1985) Longevity of individual flowers. Annu Rev Ecol Syst 16(1):15–37

Proctor M, Yeo P, Lack A (1996) The natural history of pollination. Timber Press

Raguso RA (2008) Wake up and smell the roses: the ecology and evolution of floral scent. Annu Rev Ecol Evol Syst 39:549–569

Rai ID, Adhikari BS, Rawat GS (2010) A unique patch of timberline ecotone with three species of Lady's slipper orchids in Garhwal Himalaya, India. J Threatened Taxa 2(3):766–769. https://doi.org/10.11609/JoTT.o2121.766-9

Raskoti BB (2009) The orchids of Nepal. Bhakta Bahadur Raskoti and Rita Ale

Rathcke B, Lacey EP (1985) Phenological patterns of terrestrial plants. Ann Rev Ecol Syst 16(1): 179–214

Samant SS, Dhar U, Palni LMS (1998) Medicinal plants of Indian Himalaya: diversity distribution potential values. Gyanodaya Prakashan

Schiestl FP (2005) On the success of a swindle: pollination by deception in orchids. Naturwissenschaften 92(6):255–264

Sharma H, Thakur S, Thakur M (2017) Diversity and distribution of orchids in the Indian Himalaya. J Mount Sci 14:768–780

Shrestha AB, Wake CP, Mayewski PA, Dibb JE (2012) Maximum temperature trends in the Himalaya and its vicinity: an analysis based on temperature records from Nepal for the period 1971-94. J Clim 25(6):1999–2018

Singh AP, Chandra A, De K, Uniyal VP, Sathyakumar S (2022) Decreasing potential suitable habitat of bumblebees in the Great Himalayan National Park Conservation area. Orient Insects 56(1):1–18

Stone GN, Willmer P, Rowe JA (1996) Partitioning of pollinators during flowering in an African *Acacia* community. Ecology 77(8):2808–2827
Ticktin T (2004) Ecological implications of harvesting non-timber forest products. J Appl Ecol 41(1):11–21
Totland Ø (1994) Influence of climate, time of day and season, and flower density on insect flower visitation in alpine Norway. Arc Alp Res 26(1):66–71
Totland Ø (2001) Environment dependent pollen limitation and selection on floral traits in an alpine species. Ecology 82(8):2233–2244
Tremblay RL, Ackerman JD, Zimmerman JK, Calvo RN (2005) Variation in sexual reproduction in orchids and its evolutionary consequences: a spasmodic journey to diversification. Biol J Linn Soc 84(1):1–54
Ved DK, Kinhal GA, Ravikumar K, Prabhakaran V, Ghate U, Sankar RV, Indresha JH (2003) Conservation assessment and management prioritization for the medicinal plants of Indian Himalaya. Foundation for Revitalization of Local Health Traditions
Waser NM (1986) Flower constancy: definition, cause, and measurement. Am Nat 127(5):593–603
Williams PH, Thorp RW, Richardson LL, Colla SR (2015) Bumble bees of North America: an identification guide. Princeton University Press
Young A, Boyle T, Brow T (1996) The population genetic consequences of habitat fragmentation for plants. Trends Ecol Evol 11(10):413–418
Zimmerman M (1988) Nectar production, flowering phenology, and strategies for pollination. In: Plant reproductive ecology. Oxford University Press, pp 157–178

Ex Situ and In Situ Conservation Strategies for Threatened and Endemic Orchids

6

Giriraj Singh Panwar and Bhavana Joshi

6.1 Introduction

Orchids are unique assemblage of highly advanced monocotyledonous plants belongs to the family Orchidaceae. The term 'orchid' originates from the Greek word 'orchis', meaning 'testicle', which refers to the shape of the plant's tubers. John Lindley (1845) has used the term 'orchid' in his work 'School Botany' to represent the plant group. This etymology reflects the unique morphology of the orchid's roots, which resemble testicles. Orchids are valued for their attractive long-lasting flowers available in diverse sizes, shapes, and colours. Orchids are known to mankind for their floricultural and medicinal values since time immemorial. These are characterized by distinct floral morphology, pollination mechanism, association with unique fungal partners (Mycorrhizae) and miniscule seeds. Globally, they are represented by 29,199 species of 850 genera, and orchidaceae is the second largest family of flowering plants in the world (Govaerts et al. 2017; De and Pathak 2018). Orchids grow in diverse habitats such as epiphyte, terrestrial, and lithophyte (Fig. 6.1) and dominated by epiphytes (75%) (Arditi 1992). Orchids are cosmopolitan plants and are found in almost every habitat on the Earth except glaciers. In India orchids are represented by 1331 species of 186 genera and range from sea to trans Himalaya. They are dominated in the Himalaya (815 species), North-East (714 species), Western Ghats (305 species), Peninsular India (191 species), and Andaman & Nicobar Islands (158 species) (Kumar et al. 2022; De 2020) (Table 6.1).

Orchids have an international business of around 8% in the global floriculture trade and have the potential to alter the economics of a country (Kalimuthu et al.

G. S. Panwar (✉)
Botanical Survey of India, Botanic Garden of Indian Republic, Noida, India

B. Joshi
Botanical Survey of India, Northern Regional Centre, Dehradun, India

A. Husen, M. Pant (eds.), *Forest-based Orchids*,
https://doi.org/10.1007/978-981-92-0096-2_6

Fig. 6.1 Habit of Epiphytic and terrestrial orchid species of India: (**a**) *Vanda cristata* Wall. ex Lindl., (**b**) *Rhynchostylis retusa* (L.) Blume, (**c**) *Calanthe tricarinata* Lindl., (**d**) *Dactylorhiza hatagirea* (D.Don) Soo, (**e**) *Goodyera* spp. (**f**) *Habenaria intermedia* D.Don, (**g**) *Malaxis muscifera* (Lindl.) Kuntze and (**h**) *Eulophia dabia* (D.Don) Hochr

2007). Orchids like *Cymbidium*, *Dendrobium*, *Paphiopedilum*, *Phalaenopsis*, *Oncidium* have become commercially important as potted plants and cut flowers (Lee 2011). Besides the horticultural dominance of orchids, they also contain various

Table 6.1 Distribution of orchids in different states of India (Singh et al. 2019)

State	No. of species	No. of genus	Endemic species
Arunachal Pradesh	612	123	65
Sikkim	560	118	36
West Bengal	479	103	31
Meghalaya	446	102	36
Manipur	422	83	17
Nagaland	376	88	11
Assam	293	75	15
Mizoram	287	77	3
Kerala	253	69	111
Uttarakhand	238	68	13
Tamil Nadu	215	62	92
Karnataka	198	60	79
Andaman and Nicobar	158	65	24
Odisha	141	48	15
Maharashtra	104	30	40
Himachal Pradesh	84	43	2
Andhra Pradesh	73	34	13
Jharkhand	69	23	6
Goa	67	29	31
Chhattisgarh	52	21	7
Tripura	51	29	0
Madhya Pradesh	48	17	7
Jammu & Kashmir	46	24	1
Bihar	36	15	2
Gujarat	29	14	11
Uttar Pradesh	17	8	0
Rajasthan	15	8	2
Punjab	12	6	0
Dadra & Nagar Haveli and Daman & Diu	9	9	2
Haryana	3	3	0
Delhi	1	1	0

phytochemical substances and thus imparting therapeutic significance for mankind. Orchid species like *Dactylorhiza hatagirea* (D. Doon) Soo, *Eulophia dabia* (D.Don) Hochr, *Rhynchostylis retusa* (L.) Blume, *Calanthe tricarinata* Lindl, *Coelogyne prolifera* Lindl, etc. are the prominent. Ayurvedic remedies like Chavyanprasa include a considerable amount of Ashtavarga Plant, also referred to as the 'herb of immortality'. Orchids are one of the important components of the Astavarga medicinal system and out of eight herbs, four are orchid species viz. *Herminium edgeworthii* (Eruthvi), *Habenaria intermedia* (Vruthvi), *Malaxis muscifera* (Rishabaka) and *Crepidium acuminatum* (Jeevak) (Singh and Duggal 2009).

6.2 Conservation Status and Threat

Orchids are the most threatened flowering plants species on the earth. Illicit harvesting, habitat destruction, loss of pollinators, climate change and shifting agriculture has further compounded the risk and consequently causing the decline of orchids (Swarts and Dixon 2009; Husen and Pant 2024; Svahnstrom et al. 2025). Worldwide, due to various anthropogenic activities significant number of orchid species are at the brink of extinction and are figured prominently in the Red Data book. IUCN Global Red List has assessed about 1641 orchid species out of which 174 orchid species have been included in the red list categories. Out of this 9 are critically endangered, 47 Endangered, 42 Vulnerable, 16 near threatened, 36 Least concern and 24 are Data Deficient (IUCN 2021). It has been reported that around 250 orchids species in India are threatened while others like *Anoectochilus rotundifolius*, *Aphyllorchis gollani*, *Coelogyne treutleri*, etc. have disappeared from their Indian haunts (Yadav et al. 2024; Medhi and De 2019) (Table 6.2).

The nine orchid species are listed in Appendix-I and remaining species are included in Appendix-II of Convention on International Trade in Endangered Species of Wild Fauna and Flora (CITES), where the international trade of such species is strictly controlled and monitored (CITES 2025) (Table 6.3). Similarly, the

Table 6.2 Threatened and endemic orchid species of India (Medhi and De 2019)

Habitat	Species	Endemic	Extinct	Threatened
North Eastern India	675	76	18	34
Eastern Himalayas	730	88	18	105
Western Himalayas	255	10	–	44
Peninsular India	267	13	5	25
Eastern India	130	6	–	5
Andaman And Nicobar Islands	117	15	2	2
Central India & Gangetic Plains	60	–	–	–

Table 6.3 Orchid species listed in Appendix-I and II of CITES (CITES 2025)

S. no.	Appendix-I	Appendix-II
1.	*Aerangis ellisii*	ORCHIDACEAE spp. P3 #4 (Except the species included in Appendix I)
2.	*Cattleya jongheana*	
3.	*Cattleya lobata*	
4.	*Dendrobium cruentum*	
5.	*Mexipedium xerophyticum*	
6.	*Paphiopedilum* spp.	
7.	*Peristeria elata*	
8.	*Phragmipedium* spp.	
9.	*Renanthera imschootiana*	

Table 6.4 Legally protected orchid species of India and their IUCN status

S. no.	Species	Habit	IUCN status[a]	Distribution
1.	*Paphiopedilum charlesworthii*	Epiphyte	EN	N-E India
2.	*Paphiopedilum druryi*	Terrestrial	CR	Southern India
3.	*Paphiopedilum fairrieanum*	Lithophyte	CR	N-E India
4.	*Paphiopedilum hirsutissimum*	Epiphyte	VU	N-E India
5.	*Paphiopedilum insigne*	Terrestrial	EN	N-E India
6.	*Paphiopedilum spicerianum*	Lithophytic	EN	N-E India
7.	*Paphiopedilum venustum*	Terrestrial	EN	N-E India
8.	*Paphiopedilum villosum*	Epiphyte	VU	N-E India
9.	*Paphiopedilum wardii*	Lithophyte	EN	N-E India
10.	*Vanda coerulea*	Epiphyte	CR	N-E India
11.	*Renanthera imschootiana*	Epiphyte	CR	N-E India

[a] IUCN (2025)

Government of India has legally protected 11 orchid species by placing them under Schedule-VI of Wild Life Protection Act (1972) which are categorised as threatened by the IUCN (Table 6.4) (De and Pathak 2018).

6.3 Conservation Measures

Orchids are among the most threatened plant species on the earth and needs imminent holistic action on their in situ and ex situ conservation. The conventional orchid propagation methods have several limitations including slow growth, number of seedlings and economic viability. Moreover, lack of endosperm in the orchid seeds and requirement of mycorrhizal association for the germination further diminishes the regeneration potential of orchids in the wild. Alternatively, establishment of in vitro seed germination protocol for orchids vary from species to species and depends on several factors such as capsule maturity, etc. (Lee 2011; Kauth et al. 2011). Some terrestrial orchids also propagated through perennating rhizomes, however only a limited number of propagules can be produced only in their favourable seasons (Joshi et al. 2022). In vitro techniques can be used for the conservation of endemic, and important species, and for the production of large number of plantlets in short period of time. It also facilitates to meet out the increasing floriculture demands and to reduce the unsustainable exploitation pressure on natural populations (Panwar et al. 2022). Thus, for the effective conservation of orchids, both in situ and ex situ conservation measures are needs to be mobilized to achieve the tangible goals of conservation and preservation of germplasm for the posterity of future generation.

6.3.1 In Situ Conservation

Orchids grow in a varied of habitat which ranges from sea level to trans Himalayan region and in situ conservation of orchids aims to protect orchid species in their natural habitat i.e. within their natural environment. To effectively implement the aim of in situ conservation of biodiversity, Ministry of Environment Forest and Climate Change, Govt of India has launched the Protected Area Network (PAN) Program. The PAN program was strategically designed to achieve long-term conservation of biodiversity and other ecological values such as ecosystem services, sustainable management and culture, etc. These networks include national parks, wild life sanctuaries, community reserves and conservation reserves (Table 6.5). Under the protected area network program govt has designated 1134 protected areas which are encompassing an area of 5.69% of countries total geographic land (www.wiienvis.nic.in).

A national park is an ecosystem with unique landscapes protected and maintained by the government to conserve the biodiversity and ecosystem services of the particular phytogeographic region. These are fully undisturbed areas with peculiar floral and faunal diversity and have a special scientific, educational, and recreational interest. Human activities in such protected areas are prohibited to conserve the biodiversity and other ecological significance of the areas. These are created under the provision of Chapter IV 'Protected Area' of the Wildlife (Protection) Act, 1972. Presently 106 National parks are present in India and actively engaged in the in situ conservation of the biodiversity of the respective phytogeographic region.

North-East India is the treasure of orchid biodiversity and fostering more than 714 species of the orchids. For the conservation of this unique treasure of nature a dedicated sanctuary known as Sessa Orchid Sanctuary (Arunachal Pradesh), is the only sanctuary in India devoted for the conservation of naturally growing orchids. Besides this, many other protected areas are also involved in the conservation of naturally growing orchid species such as Similipal National Park in Odisha (96 species), Buxa Tiger Reserve in West Bengal has (150 species, the Kaziranga Orchid National Park (600 species) is the largest orchid national park having approximately 600 types of wild orchids collected from North-east India (De and Pathak 2018; Kumar et al. 2024). Apart from North-East India, Western Himalaya and Eastern Ghats are also serving as natural habitats for the orchids and these areas are also harbouring a good number of orchid species. The protected areas designated in these areas are also effectively helping in the in situ conservation of orchid species of the

Table 6.5 Protected areas of India (as on February 2025)

S. no.	Protected area (PAs)	No.	Area (km^2)	Coverage % of country
1.	National Parks	106	44,402.94	1.35
2.	Wild life Sanctuaries	574	1,27,241.24	3.87
3	Community Reserves	309	8464.13	0.26
4.	Conservation Reserves	145	7054.60	0.21
	Total PAs	**1134**	**1,87,162.91**	**5.69**

particular phytogeographic region. These includes, Valley of Flower National Park (32 species) Nanda Devi National Park (20 species), Govind Pashu Vihar National Park, Gangotri National Park (10 Species), Raja Ji National Park, Jim Corbett national Park, Great Himalayan National Park and Silent Valley National Park (50), etc. (Samant and Joshi 2005; Pusalkar and Singh 2012; Bisht et al. 2018).

Besides these there are 18 biosphere reserves in India which were created under the 'Man and Biosphere' initiative, of United Nations educational, scientific, and cultural organization (UNESCO) in 1971 (De and Pathak 2018). Nilgiri Biosphere Reserve (NBR) is the first biosphere reserve of India and harbours about 240 species of orchids belonging to 72 genera with 15 endemic species (Anusuba et al. 2023). Similipal, the ninth biosphere reserve in Odisha, India, has 96 orchid species listed by UNESCO with some endemic species like *Eria meghasaniensis* and *Tainia hookeriana* (Kumar et al. 2021).

6.3.2 Ex Situ Conservation

Ex situ conservation is an approach of conserving species outside its natural habitat/environment. In this case plant propagules (seeds, live plants, plant parts) are collected from wild and relocated to different places for the propagation, preservation and research purpose. This approach of conservation is considered as the most feasible, economical and widely being practiced for the effective conservation of germplasm. In India, Botanical Survey of India (BSI) has established National Orchidarium at Yercaud, Tamil Nadu, Eastern Regional Centre (ERC), Shillong and AJC Bose Indian Botanic Garden (AJCBIBG), Howrah, where the representative live plant collections of orchids are maintained and conserved. Apart from BSI, many states have also established ex situ conservation sites such as the Orchid Research Centres (in Tipi, Itanagar, Dirrang, Jenging, and Sessa) in Arunachal Pradesh, three (i.e. Kodagu, Kudremukh & Dhandeli) in Karnataka and Jawaharlal Nehru Tropical Botanic Garden & Research Institute (TBGRI), Palode, Thiruvananthapuram, Kerala, carries out ex situ conservation by preserving a variety of orchid species (Rao and Sridhar 2007; De and Pathak 2018; Kumar et al. 2024). Similarly, the Indian Council of Agriculture Research (ICAR) has established dedicated orchid research centre, the National Research Centre for Orchids (NRCO) at Sikkim working on the conservation and sustainable use of orchid biodiversity.

In vitro propagation, cryopreservation, gene banks, and seed storage, and botanical gardens are some of the most popular ex situ conservation methods. Wyse Jackson proposed that 'A botanic garden is an institution holding documented collections of living plants for scientific research, conservation, display and education' (Jackson 1999). Over 3100 botanical gardens exist in the world, according to Botanic Gardens Conservation International (BGCI). There are nearly 8 million plant specimens in these gardens, which have locations in more than 170 nations (www.bgci.org). Many internationally acclaimed organizations like the New York Botanical Garden and Royal Botanic Garden (Kew) have conserved living orchid collections, and continuously working on the ex situ conservation and preservation of germplasm (Swarts and Dixon 2017). In India, currently there are 13 botanical gardens actively engaged in conserving orchid species. The Lloyd Botanical Garden in Darjeeling, West Bengal, has 43 collected orchid species (De and Pathak 2018). The Botanical Garden and National Orchidarium, Barapani, Shillong has 419 orchid species from 94 genera collected from various states of North-east India.

6.3.2.1 Seed Germination

The seed germination process is more intricate in terrestrial orchids than that of epiphytic orchids which is ascribed due to the presence of impermeable seed coat. The histochemical studies of terrestrial orchid seeds revealed that the embryo is enclosed tightly and compactly by inner seed coat (i.e. carapace) and various chemical substances such as cuticular substances, lignin, suberin and phenolic compounds (Lee et al. 2007) keep on depositing as the carapace and seed coat mature (Fig. 6.2) (Lee 2011). The deposition of these hydrophobic compounds plays a key role in the impermeability of the seed coat and consequently affect the germination rate. Moreover, high level of endogenous abscisic acid (ABA) was observed to accumulate in mature seeds which further retard the germination percentage in terrestrial orchid species (Lee et al. 2007, 2015). These findings reveals that the hard seed coat and increase ABA level might be the cause of low germination ratio in terrestrial orchids. Thus, severely affecting the natural regeneration potential of these species and causing population decline in the wild.

Through in vitro technology both immature and mature seeds can be germinated, but germination of mature seeds require some extra treatment to bleach the impermeable seed coat in comparison to immature seeds. Therefore, in case of seed

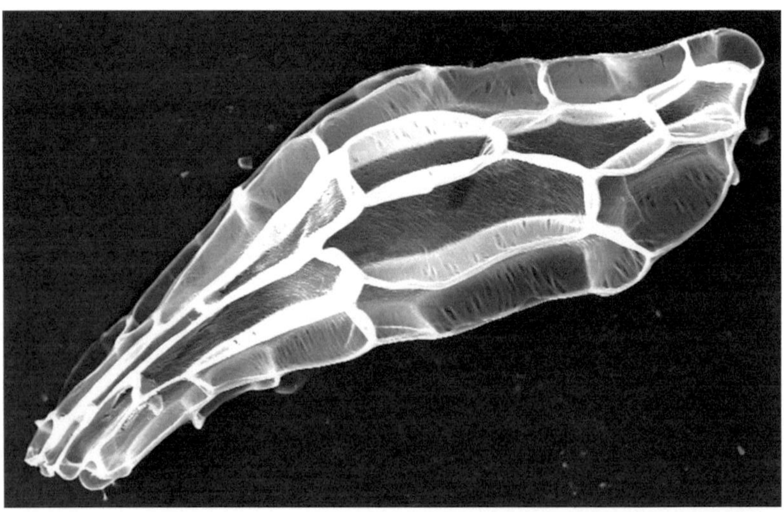

Fig. 6.2 The scanning electron microscope image showing the seed coat structure of *Eulophia dabia*, a terrestrial orchid

germination of terrestrial orchids, the seed selection stage (prior the development of hard seed coat) is very important for achieving maximum germination percentage (Lee 2011).

Further, lack of endosperm in orchid seeds is one of the prominent causes of reduced germination percentage. It is further aggravated by the variation in reserve food material in immature and mature seeds. It was observed that, during the process of seed maturation starch and other carbohydrates of immature seeds is converted into lipids in mature seeds. The lack of metabolic machinery in mature seeds is unable to utilize the reserve food material and thus diminishing the rate of seed germination (De et al. 2016). Dormancy of mature seeds also ascribed by the accumulation of germination inhibitors in the seed coat and it can be overcome by low temperature treatment (De and Pathak 2018) or by using sucrose (Weatherhead et al. 1986).

6.3.2.2 In Vitro Propagation

The in vitro technique is a biotechnological tool used for the clonal propagation of threatened, medicinal and ornamental plant species (Oseni et al. 2018). Different explants like shoots, roots, leaves, seeds, rhizomes, pseudobulbs, etc. are commonly used for the micropropagation of different orchid species (Lal and Singh 2020; Lal et al. 2020a, b). This technique ensures large scale production of true to type and disease-free plants. It is highly effective in conserving the gene pool and biodiversity. The in vitro culture is the only viable method to preserve and reintroduce threatened genetic material into the wild under the species recovery program. It is considered the best technique for conservation of threatened and endemic orchids with ornamental and medicinal significance (Bhowmik and Rahman 2017).

A large number of orchid species has been successfully propagated through tissue culture and proved to be one of the viable methods for the conservation of orchids species (Table 6.6). Further, it is also helpful in meeting the increasing commercial demands of the cut flower and medicinal orchids, thus play a pivotal role in the sustainable development and conservation of biodiversity.

Eulophia dabia, a critically endangered, endemic and medicinal orchid of North-West Himalaya. It is facing severe threat in wild due to various anthropogenic activities. Similarly, *Rhynchostylis retusa* an epiphytic and medicinal species, is also facing severe pressure due to various developmental activities and unsustainable exploitation. Both the species viz. *Eulophia dabia* and *Rhynchostylis retusa* were multiplied in vitro by using seed germination method and protocols were successfully standardized for them (Figs. 6.3 and 6.4) (Lal et al. 2020a, b; Panwar et al. 2022; Panwar and Joshi 2024).

6.3.2.3 Seed Bank and Cryopreservation

A seed bank is a repository used to preserve and protect the seed germplasm of plant species. The Millennium Seed Bank (MSB), Sussex is home to over 2.4 billion seeds, representing over 39,000 different plant species of the world's storable seeds. In MSB seeds are stored in Deep freeze chambers and stored at −20 °C, using international standards (www.kew.org).

Table 6.6 Some of the orchid species successfully conserved through in vitro propagation approach

Name of species	Explant	Nutrient medium and PGRs used	Reference
Aerides multiflora Roxb	Leaf	MPR + 2 mg/L BAP + 0.5 mg/L NAA	Vij and Aggarwal (2003)
Anoectochilus sikkimensis King & Prantl.	Nodes	Shoots developed in WPM and rooting in 2.7 μM NAA + 0.2% AC	Gangaprasad et al. (2000)
Cymbidium ebumeum Lindl.	Axenic shoot tip and leaf base	Mitra medium + 10% CW, 500 mg/L CH + 8.8 μM BA + 4.1 μM NAA	Seeni and Latha (2000)
Vanda spathulata (L.) Spreng	Stem node	Mitra medium + 44.4 μm BA + 17.1 or 28.5 μm IAA + 66.6 mM BA with 28.5 or 40.0 μm IAA	Decruse et al. (2003)
Cymbidium insigne Rolfe	Shoot	MS—1% quartz porphyry	Shimasaki et al. (2003a, b)
Cymbidium atropurpureum (Lindley) Rolfe	Shoot tip	MS + 2.5 mg/L BAP	Subramaniam and Taha (2003)
Doritaenopsis	Root tip from in vitro plantlets	Modified MS + 2.3 μM TDZ	Park et al. (2003)
Micropera pallid Lindl.	Leaf	½ MS + 2 mg/L NAA + 2 mg/L BAP	Bhadra and Hossain (2004)
Vanda teres (Roxb.) Lindl.	Protocorm	VW + 1 BAP 0.5 NAA	Sinha and Roy (2004)
Aerides crispum L.	Leaf	MS + 2.0 mM BAP	Sivakumar et al. (2005)
Vanilla planifolia Andr.	Shoot tip	MS with 1.0 mg/L BAP + 150 mL CW	Kalimuthu et al. (2006)
Phalaenopsis amabilis (L.) Blume	Thin sections of Inflorescence axis	NDM medium 0.5 mg/L NAA, 2% sucrose, 10% CW, 2 g/L peptone + 1 g/L AC	Sinha et al. (2007)
Dendrobium lituiflorum Lindl.	Axillary buds from 5-month-old in vitro grown plants	MS medium + 0.5 mg/L 2,4-D	Das et al. (2008)
Vanilla planifolia Andrews	Leaf from field growing plant	MS medium + 4.52 μM 2,4-D + 2.22 μM BAP	Janarthanam and Seshadri (2008)
Ponerorchis graminifolia Rchb. f.	Inflorescence explants	½ MS + 4.44 mM BAP + 0.54 mM NAA	Mitsukuri et al. (2009)
Coelogyne cristata Lindl.	Segment of PLBs	½ MS with 1.0 mg/L NAA and 0.5 mg/L BA with coconut powder	Naing et al. (2011)
Cymbidium giganteum Wall.	Pseudostem from seedling	MS with 0.909 μM TDZ	Roy et al. (2012)

(continued)

Table 6.6 (continued)

Name of species	Explant	Nutrient medium and PGRs used	Reference
Cymbidium aloifolium and *Cymbidium iridioides*	Axenic aerial roots (5–6) week old	MS + sucrose (3%, w/v) and Kn (3 μM) and formed PLBs. MS + sucrose (3%), AC (0.1%) and IAA (3 μM) and formed roots	Deb and Pongener (2012)
Eulophia nuda Lindl.	Tuber explants	MS + 44.4 μM BA and additives (283.8 μM ascorbic acid, 67.8 μM adenine sulphate, 143.5 μM arginine and 118.9 μM citric acid)	Panwar et al. (2012)
Eulophia cullenii (Wight) Bl.	Seeds	½ MS, ¼ MS, Knudson C, or Mitra liquid medium	Decruse et al. (2013)
Thunia marshalliana Rchb.f.	Pseudonodes from plants grown in glasshouse	MS with 22.0 μM BAP	Singh et al. (2014)
Vanda stangeanna Rchb. F.	Seed	Seed germination in ½MS and PLB formation in ½MS + NAA (2.7 μM) and KN (2.3 μM)	Pebam et al. (2016)
Vanda thwaitesii Hook. f.	Seed	Symbiotic germination and protocorm development	Decruse et al. (2018)
Cymbidium goeringii (Rchb. f.) Rchb.f.	Rhizome explants from seedlings	MS with 20 μM 2,4-D and 2 μM TDZ MS with 2 μM NAA	Park et al. (2018)
Paphiopedium villosum (Lindl.) Stein	Seed	Asymbiotic seed germination and PLBs formation in Nitsch + BAP (0.25) + NAA (0.50)	Raj et al. (2018)
Cattleya 'Almakee'	Epidermal (foliar) peel root explants from the greenhouse plants	Mitra + Kn (1 mg/L) + NAA (1 mg/L)	Sharma (2018)
Cymbidium ensifolium (L.) Swartz.	Seeds	Seed germination in MS medium protocorm development in MS medium + BAP (4.0 μM), rooting in 8 μM IBA	Ramesh et al. (2019)
Dendrobium sp.	Leaf tip	MS + 0.5 mg/L NAA + 0.5 mg/L BAP	Goswami et al. (2020)
Heminium lanceum (Thunb. Ex Sw.) Vujik	Seed	KC medium + IBA (1.0 mg/L) and Kn (1.0 mg/L)	Thakur and Pathak (2020)
Vanda cristata Wall. Ex Lindl.	Green pod culture	½ MS + Auxin + Cytokinin in PLB formation and MS medium + NAA (5.3 μM) in root development	Sunita and Mahan (2021)
Eulophia dabia (D.Don) Hochr	Seed	MS medium + 4.4 μM BAP and AC	Panwar et al. (2022)

(continued)

Table 6.6 (continued)

Name of species	Explant	Nutrient medium and PGRs used	Reference
Rhynchostylis retusa (L.) Blume	Seeds	MS medium + BAP (4.4 μM) and AC (2.0 g/L)	Panwar and Joshi (2024)
Cattleya schilleriana Rchb.f.	–	MS medium + 2% and 3% mannitol were the most effective for the growth and survival of the species	Andrade et al. (2025)

KC Knudson C, *MPR* Mitra-Prasad-Roychowdhury, *MS* Murashige and Skoog, *NDM* New Dogashima Medium, *VW* Vacin & Went, *WPM* Woody Plant Medium, *AC* activated charcoal, *TDZ* thidiazuron, *CW* coconut water, *CH* casein hydrolysate, *BA* benzyladenine, *NAA* naphthalene acetic acid, *IBA* indole-3-butyric acid, *Kn* kinetin, *BAP* 6-benzylaminopurine, *2,4-D* 2,4-dichlorophenoxyacetic acid, *PLBs* protocorm like bodies

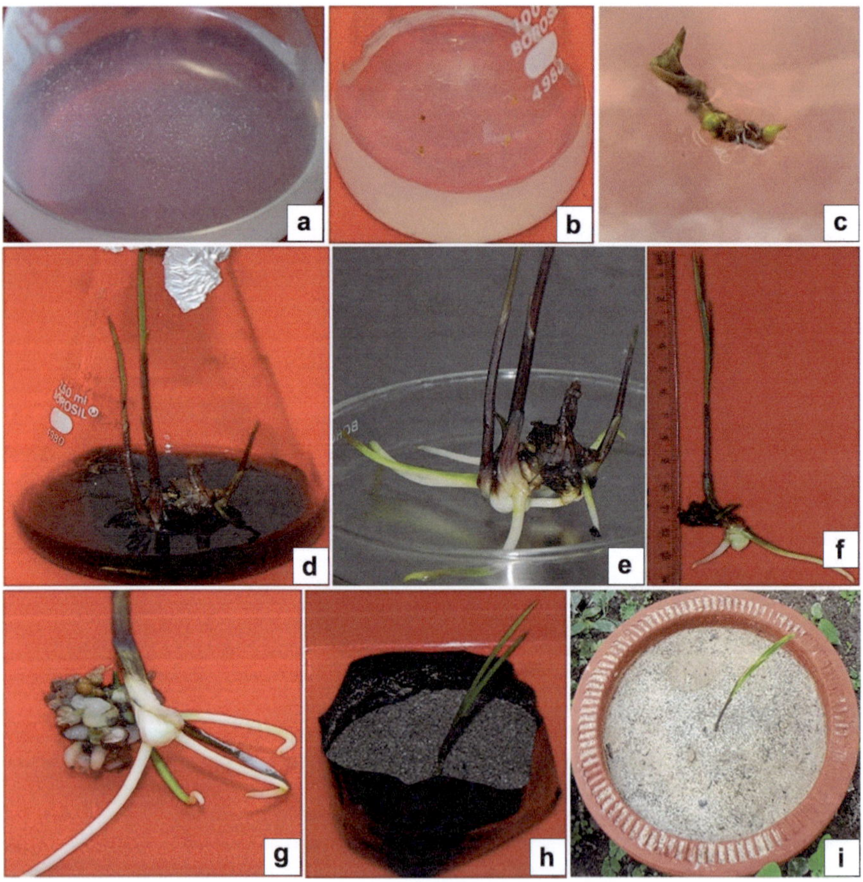

Fig. 6.3 In vitro propagation of *Eulophia dabia*, a terrestrial orchid: (**a**, **b**) inoculation and germination of seeds, (**c**) development of rhizome-like bodies (RLB), (**d**) development of shoots from rhizome, (**e–g**) fully developed roots and shoots, (**h**) fully developed plantlets maintained in the glasshouse, (**i**) plantlets transferred to the open environment

Fig. 6.4 In vitro propagation of *Rhynchostylis retusa*, an epiphytic species: (**a**) germination of seeds, (**b**) development of protocorm like bodies (PLBs), (**c**, **d**) development of shoots, (**e**, **f**) proliferation of shoots and development of plantlets, (**g**) fully developed plantlets, (**h**) plantlets transferred to the open environment

A single capsule of orchid produces large number of seeds but having typically low germination rate. Seeds are devoid of endosperm and require a specific orchid mycorrhizal fungus for the germination (Chakraborty et al. 2023). The dry seeds and pollen of many species have the potential to survive in storage for a long time particularly at low and cryogenic temperatures. The seeds of *Dactylorhiza*, *Dendrobium*, *Eulophia*, and *Paphiopedilum*) are preserved at −70 °C to maintain the high viability and excellent potential was observed for long-term storage at low temperatures (Seaton et al. 2018).

Cryopreservation is another technique used to preserve plant organs and reproductive parts by storing in liquid nitrogen at −196 °C. The tissues and explants of various orchid species have been successfully cryopreserved in liquid nitrogen by treating them with cryoprotectants and plant vitrification solutions (Nikishina et al. 2007, 2019; Imsomboon and Thammasiri 2020; Das et al. 2021; Kolomeitseva et al. 2022; Decruse et al. 2023). In India, Cryobank facilities have been established in India by the National Bureau of Plant Genetic Resources, with over 2.5 lakh conserved germplasm lines (De and Pathak 2018).

6.4 Conclusion

Orchids need specific microclimates and stable forest ecosystems with high humidity to survive. But their multifarious significance and other anthropogenic activities are rendering them towards the threat of extinction. Besides other factors, habitat loss has the devastating implications on orchid survival. The lack of extinction risk assessment information and the IUCN status of orchids is further limiting the conservation actions that need to be mobilised for the effective preservation of germplasm. For the effective conservation of orchids, one should have a clear understanding of their niche requirement and then prioritise the scaling up of protection of their natural habitats. The in situ conservation of orchids has many fold implications and simultaneously they also need to be complemented by the ex situ conservation approach such as seed banks, in vitro propagation and maintaining living collections in orchidarium and botanic gardens. Thus, multipronged, holistic conservation actions need to be implemented effectively by national and international agencies to safeguard this fascinating group of plants before they vanish from the planet. Intensive biotechnological interventions are required to increase the post-harvest shelf life of orchid flowers, development of disease-resistant varieties, etc., which can provide a robust base to the floriculture industry. Further, intensive studies are needed to untap the phytochemical potential of this wonder plant group to safeguard human health, well-being, provide sustainable livelihoods, and thus support sustainable development.

References

Andrade GVS, Nadal MC, Pasqual M et al (2025) Adaptive response to conservation *of Cattleya schilleriana* Rchb.f. under slow growth conditions. Plant Cell Tissue Organ Cult 162:29. https://doi.org/10.1007/s11240-025-03132-0

Anusuba V, Murugesan M, Sharief MU, Karthik B, Tharani R (2023) A checklist of the orchids in Nilgiri Biosphere Reserve, India. Biol Forum 15(5):1274–1292

Arditi J (1992) Fundamental of orchid biology. Wiley, New York

Bhadra SK, Hossain MM (2004) Induction of embryogenesis and direct organogenesis in *Micropera pallida* Lindl, an epiphytic orchid of Bangladesh. J Orchid Soc India 18:5–9

Bhowmik TK, Rahman MM (2017) Effect of different basal media and PGRs on *in vitro* seed germination and seedling development of medicinally important orchid *Cymbidium aloifolium* (L.) Sw. J Pharmacogn Phytochem 6:167–172

Bisht M, Chandra Sekar K, Kant R, Kumar A, Singh P, Arya D (2018) Floristic diversity in Valley of Flowers National Park, Indian Himalayas. Phytotaxa 379(1):1–26

Botanic Gardens Conservation International (2026). https://www.bgci.org/

Chakraborty O, Agrawala DK, Chakraborty AP (2023) Studies on orchidoid mycorrhizae and mycobionts, associated with orchid plants as plant growth promoters and stimulators in seed germination. In: Mathur P, Kapoor R, Roy S (eds) Microbial symbionts and plant health: trends and applications for changing climate. Rhizosphere biology. Springer, Singapore. https://doi.org/10.1007/978-981-99-0030-5_16

Convention on International Trade in Endangered Species (CITES) (n.d.) *Vanda coerulea*. https://bsienvis.nic.in/CITES/Vanda%20coerulea.pdf

Convention on International Trade in Endangered Species of Wild Fauna and Flora (CITES) (2025) Appendices I, II and III, p 73

Das MC, Kumaria S, Tandon P (2008) In vitro propagation and conservation of *Dendrobium lituiflorum* Lindl through protocorm-like bodies. J Plant Biochem Biotechnol 17:177–180

Das MC, Devi SD, Kumaria S, Reed BM (2021) Looking for a way forward for the cryopreservation of orchid diversity. Cryobiology 102:1–14. https://doi.org/10.1016/j.cryobiol.2021.05.004

De LC (2020) Conservation of rare and threatened orchids of India. Proc Mordovian State Nat Reserve 25:97–122

De LC, Pathak P (2018) Conservation, management, and utilization of orchid genetic resources. J Orchid Soc India 32:81–91

De LC, Rao AN, Singh DR (2016) Endangered orchids and their conservation in North East India. In: Bioprospecting of indigenous bioresources of North-East India. Springer, Germany, pp 61–75

Deb CR, Pongener A (2012) Studies on the in vitro regenerative competence of aerial roots of two horticultural important *Cymbidium* species. J Plant Biochem Biotechnol 21:235–241. https://doi.org/10.1007/s13562-011-0099-5

Decruse SW, Gangaprasad A, Seeni S, Menon VS (2003) Micropropagation and eco restoration of *Vanda spathulata*, an exquisite orchid. Plant Cell Tissue Organ Cult 72:199–202

Decruse SW, Reny N, Shylajakumari S, Krishnan PN (2013) *In vitro* propagation and field establishment of *Eulophia cullenii* (Wight) Bl., a critically endangered orchid of Western Ghats, India through culture of seeds and axenic seedling-derived rhizomes. In Vitro Cell Dev Biol Plant 49(5):520–528

Decruse SW, Neethu RS, Pradeep NS (2018) Seed germination and seedling growth promoted by a Ceratobasidiaceae clone in *Vanda thwaitesii* Hook.I., an endangered orchid species endemic to South Western Ghats, India and Sri Lanka. S Afr J Bot 116:222–229

Decruse SW, Ajeesh Kumar S, Vargheese S (2023) Pollinia cryopreservation of selected orchid species of Western Ghats, India. In: Rajasekharan P, Rohini M (eds) Pollen cryopreservation protocols, Springer protocols handbooks. Humana, New York. https://doi.org/10.1007/978-1-0716-2843-0_35

Gangaprasad A, Latha PG, Seeni S (2000) Micropropagation of terrestrialorchids, *Anoectochilus sikkimensis* and *Anoectochilus regalis*. Indian J Exp Biol 38:149–154

Goswami K, Yasmin S, Nasiruddin KM, Khatun F, Akte J (2020) In vitro regeneration of *Dendrobium* sp. of orchid using leaf tip as explant. J Environ Sci Nat Resour 8:75–78

Govaerts R, Bernet P, Kratochvil K, Gerlach G, Carr G, Alrich P, Pridgeon AM, Pfahl J, Campacci MA, Baptista DH, Tigges H, Shaw J, Cribb P, George A, Kreuz K, Wood JJ (2017) World checklist of orchidaceae. Royal Botanic Garden, Kew, London

Husen A, Pant M (2024) Exploring medicinal orchids. Taylor & Francis Group. https://doi.org/10.1201/b23375

Imsomboon T, Thammasiri K (2020) Cryopreservation of *Paphiopedilum* exul (Ridl.) Rolfe seeds using encapsulation-vitrification and encapsulation-dehydration methods. Acta Hortic 1298: 195–204. https://doi.org/10.17660/ActaHortic.2020.1298.28

IUCN (2021) The IUCN Red List of threatened species. Version 2021

IUCN (2025) The IUCN Red List of threatened species. Version 2025-1

Jackson PSW (1999) Experimentation on a large scale—an analysis of the holdings and resources of botanic gardens. Bot Gardens Conserv News 3(3):27–30

Janarthanam B, Seshadri S (2008) Plantlet regeneration from leaf derived callus of *Vanilla planifolia* Andr. In Vitro Cell Dev Biol Plant 44:84–89

Joshi B, Panwar GS, Singh SK (2022) *Eulophia dabia*—a lesser known critically endangered terrestrial orchid. Indian For 148(1):83–87

Kalimuthu K, Senthikumar R, Murugalatha N (2006) Regeneration and mass multiplication of *Vanilla planifolia* Andr.—a tropical orchid. Curr Sci 91:1401–1403

Kalimuthu K, Senthilikumar R, Vijayakumar S (2007) In vitro micropropagation of orchid, Oncidum sp. Afr J Biotechnol 6(10):1171–1174

Kauth P, Kane ME, Vendrame WA (2011) Comparative *in vitro* germination ecology of *Calopogon tuberosus* var *tuberosus* (Orchidaceae) across its geographic range. In Vitro Cell Dev Biol Plant. https://doi.org/10.1007/s11627-010-9316-5

Kolomeitseva GL, Nikishina TV, Babosha AV, Ryabchenko AS, Vysotskaya ON (2022) Morphophysiology and cryopreservation of seeds of *Dendrobium nobile* Lindl. (Orchidaceae) at different stages of development. Acta Physiol Plant 44:36. https://doi.org/10.1007/s11738-022-03372-z

Kumar S, Mishra S, Mishra AK (2021) Diversity of orchid species of Odisha state, India. With note on the medicinal and economic uses. Richardiana 5:1–26

Kumar S, Devi RS, Choudhury R, Mahapatra M, Biswal SK, Kaur N, Tudu J, Rath SK (2022) Orchid diversity, conservation, and sustainability in Northeastern India. In: Furze JN, Eslamian S, Raafat SM, Swing K (eds) Earth systems protection and sustainability. Springer, Cham. https://doi.org/10.1007/978-3-030-85829-2_5

Kumar J, Katoch D, Thakur A, Pathania A, Anand A, Choudhary K, Shelja S (2024) A comprehensive review on threats and conservation status of orchids. J Appl Biol Biotechnol 12(2): 43–47. https://doi.org/10.7324/JABB.2024.150084

Lal N, Singh M (2020) Prospects of plant tissue culture in orchid propagation: a review. Indian J Biol 7:103–110

Lal A, Pant M, Palni LMS, Kumar A (2020a) *In vitro* propagation of *Rhynchostylis retusa* (L.) Blume through immature seed culture. Ecol Environ Conserv 26:46–51

Lal A, Pant M, Palni LMS, Kumar A (2020b) Development of rapid micropropagation protocol for germplasm conservation of two orchid species—*Aerides multiflora* Roxb. and *Rhyncostylis retusa* (L.) Blume. Asian J Conserv Biol 9(2):341–347

Lal A, Pant M, Kumar A, Palni LMS, Singh A, Wabaidur SM, Husen A (2023) Clonal propagation and hardening of epiphytic orchids (*Aerides multiflora* and *Rhyncostylis retusa*) for sustainable conservation and utilization. S Afr J Bot 160(2023):402–411

Lee Y-I (2011) In vitro culture and germination of terrestrial Asian orchid seeds. Methods Mol Biol 710:53–62

Lee Y-I, Lu CF, Yeung EC, Chung MC, Lee N (2007) Developmental changes in endogenous abscisic acid concentrations and asymbiotic seed germination of a terrestrial orchid, *Calanthe tricarinata* Lindl. J Am Soc Hortic Sci 132:246–252

Lee YI, Chung MC, Yeung EC, Lee N (2015) Dynamic distribution and the role of abscisic acid during seed development of a lady's slipper orchid, *Cypripedium formosanum*. Ann Bot 116: 403–411

Medhi RP, De LC (2019) Orchid biodiversity in North East India. https://doi.org/10.13140/RG.2.2. 17566.69448

Mitsukuri K, Mori G, Johkan M, Shimada Y, Mishiba KI, Morikawa T, Oda M (2009) Effects of explant position and dark treatment on bud formation in floret culture of *Ponerorchis graminiflolia* Rchb. f. Sci Hortic 121:243–247

Naing AH, Chung JD, Park IS, Lim KB (2011) Efficient plant regeneration of the endangered medicinal orchid, *Coelogyne cristata* using protocorm like bodies. Acta Physiol Plant 33:659–666. https://doi.org/10.1007/s11738-010-0586-7

National Wildlife Database (2025). https://www.wiienvis.nic.in

Nikishina TV, Popova EV, Vakhrameeva MG, Varlygina TL, Kolomeitseva GL, Burov AV, Popovich EA, Shirokov AI, Shumilov VY, Popo AS (2007) Cryopreservation of seeds and protocorms of rare temperate orchids. Russ J Plant Physiol 54(1):121–127

Nikishina TV, Kozlova ON, Levitskaya GE, Vysotskaya ON (2019) Study of *Dactylorhiza* Seeds (*D. baltica* and *D. maculata*) from the Orchid Collection of the Cryobank at Timiryazev Institute of Plant Physiology, Russian Academy of Sciences. Biol Bull Russ Acad Sci 46:242–250. https://doi.org/10.1134/S1062359019030063

Oseni OM, Pande V, Nailwal TK (2018) A review on plant tissue culture, a technique for propagation and conservation of endangered plant species. Int J Curr Microbiol Appl Sci 7: 3778–3786

Panwar GS, Joshi B (2024) In vitro propagation and clonal fidelity assessment of *Rhynchostylis retusa* (L.) Blume: a threatened ornamental and medicinal orchid. Vegetos. https://doi.org/10. 1007/s42535-024-00831-5

Panwar D, Ram K, Harish SNS (2012) In vitro propagation of Eulophia nuda Lindl., an endangered orchid. Sci Hortic 139:46–52

Panwar GS, Joshi B, Joshi R (2022) Axenic rhizome culture and genetic fidelity assessment of *Eulophia dabia* (D. Don) Hochr: an endangered terrestrial orchid species. In Vitro Cell Dev Biol Plant. https://doi.org/10.1007/s11627-022-10258-9

Park SY, Murthy HN, Paek KY (2003) Protocorm like body induction and subsequent plant regeneration from root tip cultures of *Doritaenopsis*. Plant Sci 164:919–923

Park HY, Kang KW, Kim DH, Sivanesan I (2018) In vitro propagation of *Cymbidium goeringii* Reichenbach fil. through direct adventitious shoot regeneration. Physiol Mol Biol Plants 24: 307–313. https://doi.org/10.1007/s12298-017-0503-2

Pebam B, Kishor R, Bai VN (2016) *In vitro* immature embryo germination and propagation of *Vanda stangeana* Rchb.f., an orchid endemic to India. Hortic Environ Biotechnol 57(6): 615–624

Pusalkar PK, Singh DK (2012) Flora of Gangotri National Park, Western Himalaya, India. Botanical Survey of India, Kolkata, pp 575–581

Raj K, Mridul C, Ngursanzuala S, Bhutia TC, Singh DR (2018) Asymbiotic seed germination and *in vitro* seedling development of *Paphiopedilum villosum* (Lindi.) Stein, avaluable and vulnerable lady's slipper orchid from India. Curr Sci 114(2):266–269

Ramesh T, Renganathan P, Prabhakaran M (2019) Tissue culture studies on *Cymbidium ensifolium* (L.) Swartz. J Appl Adv Res 4(4). https://doi.org/10.21839/jaar.2019.v4i4.70

Rao TA, Sridhar S (2007) Wild orchids of Karnataka—a pictorial compendium. INCERT, Bangalore

Roy AR, Sajeev S, Pattanayak A, Deka BC (2012) TDZ induced micropropagation in *Cymbidium giganteum* Wall. Ex Lindl. and assessment of genetic variation in the regenerated plants. Plant Growth Regul 68:435–445

Royal Botanic Garden, Kew (n.d.) Millennium seed bank: research facilities. https://www.kew.org/science/collections-and-resources/research-facilities/millennium-seed-bank

Samant SS, Joshi HC (2005) Plant diversity and conservation status of Nanda Devi National Park and comparison with highland National Parks of the Indian Himalayan Region. Int J Biodiver Sci Manag 11:1–9. https://doi.org/10.1080/17451590509618081

Seaton PT, Hosomi ST, Custódio CC, Marks TR, Machado-Neto NB, Pritchard HW (2018) Orchid seed and pollen: a toolkit for long-term storage, viability assessment and conservation. In: Lee YI, Yeung ET (eds) Orchid propagation: from laboratories to greenhouses-methods and protocols, Springer protocols handbooks. Humana Press, New York. https://doi.org/10.1007/978-1-4939-7771-0_4

Seeni S, Latha PG (2000) *In vitro* multiplication and eco rehabilitation of the endangerd *Blue Vanda*. Plant Cell Tissue Organ Cult 61:1–8

Sharma V (2018) Regenerative competence of thin cell layer (epidermal peel) for in vitro propagation of *Cattleya* 'Almakee'. Asian J Microbiol Biotechnol Environ Sci 20:S65–S69

Shimasaki K, Ishikawa K, Fukumoto Y (2003a) Effect of ultrasonicated water treated with quartz porphyry "bakuhan-seki" on organogenesis in protocorm-like body (PLB) of *Cymbidium* and *Phalaenopsis* species cultured *in vitro*. Environ Control Biol 41:295–299

Shimasaki K, Shiraga T, Fukumoto Y (2003b) Effect of methyl jasmonate on organogenesis in shoot cultures of epiphytic and terrestrial *Cymbidium* species. Environ Control Biol 41:179–182

Singh A, Duggal S (2009) Medical orchids: an overview. Ethnobot Leaflets 13:351–363

Singh M, Kumaria S, Tandon P (2014) In vitro multiplication of *Thunia marshalliana* Rcbh. f. through mature seeds and pseudonode culture. Indian J Biotechnol 13:268–273

Singh SK, Agarwal DK, Jalal JS, Dash SS, Mao AA, Singh P (2019) Orchids of India: a pictorial guide. Botanical Survey of India, Kolkata

Sinha P, Roy SK (2004) Regeneration of an indigenous orchid, *Vanda teres* (Roxb.) Lindl. through *in vitro* culture. Plant Tissue Cult 14:55–61

Sinha P, Hakim M, Alam M (2007) Efficient micropropagation of *Phalaenopsis amabilis* (L.) BL. cv. 'Cool Breeze' using inflorescence axis thin sections as explants. Prop Ornament Plants 7:9–15

Sivakumar G, Yu KW, Hahn EJ, Paek KY (2005) Optimization of organic nutrients for ginseng hairy roots production in large-scale bioreactors. Curr Sci 89:641–649

Subramaniam G, Taha RM (2003) Morphogenesis of cymbidium atropurpureum in-vitro. Malays J Sci 22(1):1–5

Sunita PP, Mahan KC (2021) Green pod culture of an endangered and medicinally important orchid, *Vanda cristata* Wall. ex Lindi. from Himachal Pradesh. J Orchid Soc India 35:25–33

Svahnstrom VJ, Nic Lughadha E, Forest F, Leao TCC (2025) Geographic range size and rarity of epiphytic flowering plants. Nat Plants 11:1380–1389. https://doi.org/10.1038/s41477-025-02022-9

Swarts ND, Dixon KW (2009) Terrestrial orchid conservation in the age of extinction. Ann Bot 104: 543–556

Swarts ND, Dixon KW (2017) Conservation methods for terrestrial orchids. J Ross Publishing, Fort Lauderdale, p 240

Thakur B, Pathak P (2020) In vitro propagation of *Herminium lanceum* (Thunb. Ex Sw.) Vuijk (Orchidaceae), through as asymbiotic seed germination: a therapeutically important and endangered orchid from North Western Himalayas. J Orchid Soc India 34:61–67

Vij SP, Aggarwal S (2003) Regenerative competence of foliar explants: *Vanda coerulea* Griff. J Orchid Soc India 17:73–78

Weatherhead MA, Zee SY, Barretto G (1986) Some observations on the early stages of development of *Eulophia yushuiana*. Memoirs Hong Kong Nat Hist Soc 17:85–90

Yadav N, Kumar A, Sawariya M, Gulia H, Kumar N, Arya SS (2024) Orchid: research, medicinal uses, threats and conservation. Ecol Environ Conserv 30:226–234

Tissue Culture Approaches for Conservation of Floriculturally Important Forest-Based Orchid Species

7

Bijaya Pant, Sushma Pandey, Krissna Chand, and Bir bahadur Thapa

7.1 Introduction

Orchids commands attention in global flora through their remarkable beauty and extraordinary diversity. Their ecological and economic significance makes conservation critical, especially in biodiversity hotspots like Nepal. This chapter explores the crucial role of tissue culture in conserving floriculturally important forest-based orchid species in Nepal, addressing the challenges these remarkable plants face. Orchids require highly specific environmental and habitat conditions to survive. Beyond their restriction to particular ecological niches, they depend on precise microclimatic elements like temperature, humidity, light intensity and mycorrhizal relationships (Arditti, 1990). Consequently, most orchids proved extremely difficult to reproduce and cultivate under conventional horticultural conditions once removed from their native habitats. The Orchidaceae ranks among the most taxonomically diverse angiosperm families, with nearly 28,000 known species spanning virtually every terrestrial habitat, from arid environments to alpine meadows and equatorial rainforests (Govaerts et al. 2021). Their intricate interactions with pollinators and mycorrhizal fungi make them integral components of their ecosystems (McCormick et al. 2016; Pant et al. 2017; Shah et al. 2025). However, habitat destruction, climate

B. Pant (✉) · K. Chand
Central Department of Botany, Tribhuvan University, Kathmandu, Nepal
e-mail: bijaya.pant@cdb.tu.edu.np

S. Pandey
Annapurna Research Center, Maitighar, Kathmandu, Nepal

B. b. Thapa
Molecular Biotechnology Laboratory, School of Biotechnology, Suranaree University of Technology, Nakhon Ratchasima, Thailand

A. Husen, M. Pant (eds.), *Forest-based Orchids*,
https://doi.org/10.1007/978-981-92-0096-2_7

change and over-collection for the horticultural trade pose significant threats to orchid populations globally leading to alarming declines in many species. Conservation of orchids to maintain biodiversity and ecosystem health is very important. In Nepal, people use orchids extensively for culinary, medicinal and decorative purposes. Yet collectors harvest most orchids from the wild using destructive and unsustainable methods removing entire plants including roots, rhizomes, tubers, fruits and seeds. For generations, orchids have captivated people with their intricate floral structures and vibrant colors. Widespread use in floriculture has spawned a thriving international trade with substantial economic returns. This demand, however, encourages irresponsible harvesting practices that further deplete wild populations (Hinsley et al. 2018; Phelps and Webb 2015; Wraith and Pickering 2018). Sustainable cultivation methods, particularly tissue culture offer essential pathways to meet market demands while reducing pressure on natural populations (Pant 2013; Pant et al. 2018a; Paul and Kumaria 2024; Raskoti and Ale 2025). With over 501 taxa identified, it represents a significant center of orchid endemism within the Himalayan range (Shrestha et al. 2022). Preserving Nepal's distinctive orchid legacy requires urgent and comprehensive conservation efforts and targeted protective measures (Subedi et al. 2013).

7.2 Forest-Based Orchids of Floricultural Importance in Nepal

Nepal is home to an amazing variety of orchid species, many of which are found in its varied forest environments. Nepal's varied forest environments harbor remarkable orchid diversity, reflecting the country's dramatic landscapes and rich biodiversity. These forest-based orchids hold substantial cultural and commercial significance, particularly for floriculture, while playing essential roles in maintaining ecological equilibrium. Nepal's diverse terrain and climatic gradients from the subtropical lowlands to alpine heights of the Himalayas, create a mosaic of habitats supporting an extraordinary orchid diversity (Grytnes and Vetaas 2002; Tiruwa et al. 2024). The country's forests, from the lush tropical and subtropical Terai forests to temperate and subalpine forests at higher elevations, provide sanctuary for a vast array of orchid species (Acharya et al. 2011; Pant et al. 2020). Apart from their therapeutic uses, a number of orchid species are also eaten and fed to animals. Many high-value species have seen a rapid drop in population due to the ongoing harvesting of medicinal orchids from the wild and the substantial loss of habitat in recent years. Consequently, a number of these orchids are now included in Appendix II of CITES and classified as rare or critically endangered. Research on the medicinal properties and phytochemical composition of Nepal's orchids remains limited, though studies confirm their medicinal value (Paudel et al. 2019; Manandhar 2002; Joshi et al. 2024). Some of the floriculturally important orchids of Nepal which are not only used in floriculture but also has medicinal uses, is presented in (Fig. 7.1).

Fig. 7.1 Pictures of Floriculturally important wild orchids of Nepal (**A**) *Aerides odorata* (**B**) *Coelogyne cristata* (**C**) *Cymbidium aloifolium* (**D**) *Cymbidium devonianum* (**E**) *Dendrobium densiflorum* (**F**) *Dendrobium nobile* (**G**) *Phaius tankervilleae* (**H**) *Rhynchostylis retusa* (**I**) *Trudelia cristata*

The sustainable supply and demand for decorative and medicinal orchids are significantly out of balance in Nepal. Ex situ conservation, which aims at large-scale propagation, and in situ conservation, which preserves and restores natural ecosystems, are therefore critically needed. Both traditional propagation methods and cutting-edge techniques like tissue culture, which enable the mass multiplication of endangered orchid species without disturbing wild populations, can be successfully used in ex situ efforts. Many orchid species that are nearing extinction urgently require propagation and conservation efforts and new species identification are constantly being discovered in the area of orchids research (Shrestha et al. 2022; Tiruwa et al. 2024; Raskoti and Ale 2023; Kafle et al. 2024; Raskoti and Ale 2025). As a reflection of the various ecosystems they live in, these orchids display an amazing variety of growth patterns, floral morphologies and ecological adaptations. Different orchids possess exceptional floricultural value. *Coelogyne* species, known for its graceful, fragrant white or cream-colored flowers, is highly prized in the horticultural trade. *Cymbidium* species, with its long-lasting, showy flowers in a variety of colors, is another popular genus among orchid enthusiasts. *Dendrobium* species, a large and diverse genus, offers a wide range of flower shapes, sizes, and colours, making it a favourite for cut flowers and potted plants. *Paphiopedilum* species, distinguished by their characteristic pouch-shaped labella and intricate color patterns, commonly known as slipper orchids are particularly prized by specialty collectors. These genera, along with others like Rhynchostylis, Dendrobium and Vanda, contribute significantly to Nepal's floricultural potential. Climate change, with its associated shifts in temperature and precipitation patterns, adds another layer of complexity to orchid conservation (Willis 2017). Certain species are especially vulnerable, especially those with limited ecological tolerances or restricted distributions. To protect these priceless plants, precise evaluations of species-specific conservation status and focused conservation initiatives are essential. In Nepal, orchids which are floriculturally important are also culturally significant of which, several species are utilized in religious rituals and traditional medicine (Rajbhandari 2015). Numerous orchid species are used to treat a wide range of illnesses and are thought to have medicinal properties (Teoh 2016; Vaidya 2019; Pant and Raskoti 2023; Poudeyal et al. 2025). Some orchids are used in flower arrangements for festivals and celebrations, while some are prized for their visual attractiveness. This cultural link emphasizes how crucial it is to incorporate community involvement and local knowledge into orchid conservation initiatives (Kunwar et al. 2013). The long term survival of these amazing plants depends on encouraging sustainable harvesting methods and educating local populations about the value of orchid conservation.

7.3 Tissue Culture in Orchids

Biotechnology, especially plant tissue culture, provides powerful tools for orchid conservation (Pant 2013; Tiruwa et al. 2024; Sarmah et al. 2024). Tissue culture techniques enable rapid propagation of orchids in sterile lab conditions, generating large numbers of plants from minimal source material. This approach facilitates reintroduction of endangered species to native habitats, establishment of ex situ conservation, collections and promotion of sustainable commercial production, while reducing pressure on wild populations (Fay 2018; Gale et al. 2018). Fundamental concepts of orchid tissue culture, focusing on techniques successfully applied to Nepal's native species is being highlighted. Successful in vitro propagation studies demonstrate how species-specific protocol modifications and optimization strategies enhance plantlet survival and regeneration efficiency. We discussed current challenges in orchid tissue culture, contamination, somaclonal variation and acclimatization difficulties, alongside future directions for improving these biotechnological approaches. This chapter emphasizes the value of orchid tissue culture as a powerful tool for sustainable use, ex situ conservation, and the growth of Nepal's floriculture sector. The concepts of plant cell totipotency, aseptic manipulation and careful control of environmental and nutritional conditions are the foundation of orchid tissue culture, which together allow the regeneration of entire plants from tiny tissue pieces or even single cells. Despite being one of the largest and most diversified families of angiosperms, the Orchidaceae family natural propagation is sluggish and complicated due to its dust-like seeds that lack endosperm (Fig. 7.2) and its requirement for mycorrhizal fungi for germination and early growth.

Fig. 7.2 Seeds of different orchids (**A**) *Bulbophyllum leopardinum* (**B**) *Coelogyne stricta* (**C**) *Dendrobium bicameratum* (**D**) *Acampe rigida* (**E**) *Dendrobium aphyllum*

Consequently, in vitro propagation has become indispensable for commercial production, conservation and research. Since Knudson's (1922) pioneering work on asymbiotic seed germination, advances in basal media composition, growth regulators, and organic supplements have transformed orchid micropropagation. Standard nutrient formulations such as Murashige and Skoog (MS), Knudson C (KC), Vacin and Went (VW) and Mitra media are commonly used, but their strength and supplementation are often modified to suit species-specific requirements. Organic additives such as coconut water, banana and apple homogenates have been shown to significantly enhance protocorm development and seedling vigor in endangered orchids such as *Gastrochilus fuscopunctatus* and *Dendrobium moniliforme* (Hwang et al. 2024, 2025; Pant et al. 2018a, b). Growth regulators are central to morphogenesis, cytokinins like 6-benzylaminopurine (BAP) and kinetin stimulate shoot formation, whereas auxins including indole-3-acetic acid (IAA), indole-3-butyric acid (IBA) and naphthaleneacetic acid (NAA) induce root and callus development, with their ratio determining organogenic pathways. Recent studies have shown that reducing the basal salt concentration (e.g., half-strength

MS), incorporating organic additives improve seed germination, protocorm formation and plantlet conversion (Hwang et al. 2025; Pradhan et al. 2016). The use of temporary immersion systems (TIS) has gained prominence in orchid propagation because of superior nutrient uptake, aeration, and growth rates compared to conventional semi-solid media, while minimizing hyperhydricity (Nongdam et al. 2023). Protocorm based regeneration is a key method for clonal propagation and for obtaining pathogen-free plants, as demonstrated in *Cymbidium* sp., where meristem-derived protocorm like bodies (PLB) subcultures produced genetically stable and *Odontoglossum* ringspot virus-free plants verified by inter simple sequence repeat (ISSR) and quantitative real time polymerase chain reaction (qRT-PCR) analyses (Campol et al. 2024). Maintenance of genetic stability is a critical aspect of orchid tissue culture, as somaclonal variation may arise from prolonged culture or callus phases; thus molecular markers such as random amplified polymorphic DNA (RAPD), ISSR, and simple sequence repeat (SSR) are widely employed to confirm clonal fidelity (Sarmah et al. 2024, Pradhan et al. 2023; Oliya et al. 2021; Joshi et al. 2022; Pandey et al. 2020). Orchid tissue culture techniques are therefore invaluable for conserving endangered species, rescuing hybrid embryos, producing disease-free elite plants, and ensuring sustainable propagation of high-value commercial cultivars. Although challenges persist, such as species-specific responses, high infrastructural costs, acclimatization losses, the integration of omics tools, molecular diagnostics, and automated culture systems is revolutionizing orchid biotechnology. Collectively, these advancements reaffirm that orchid tissue culture is not merely a propagation technique but a cornerstone for genetic improvement, conservation, and sustainable commercialization of orchids worldwide.

7.4 Techniques of In Vitro Propagation in Orchids

Orchids are propagated in vitro using a variety of methods intended to produce entire plants in a sterile and regulated environment. These techniques, which have been modified to account for species-specific reactions, have been adopted from more general orchid tissue culture techniques for orchids. Asymbiotic seed germination, protocorm, vegetative explant regeneration, including, encapsulation (artificial seeds) and use of mycorrhiza on orchid growth and development are some of the methodological techniques emphasized in this chapter (Pant et al. 2018b; 2022; Pradhan et al. 2013, 2014; Shah et al. 2025; Thapa et al. 2020). Numerous in vitro culture methods have been used for germplasm conservation and large-scale propagation. Success have been achieved in selected orchids such as *Aerides odorata*, *Aerides multiflora*, *Coelogyne cristata*, *Coelogyne fuscescens*, *Coelogyne ovalis*, *Cymbidium aloifolium*, *Cymbidium devonianum*, *Cymbidium elegans*, *Cymbidium iridioides*, *Dendrobium* species, *Phaius tankervilleae*, *Pholidota articulata*, *Thunia alba*, *Spiranthes sinensis*, *Vanda* species, *Rhynchostylis retusa*, *Vanda tessellata,* through immature seed culture, protocorm culture, using different explants and through synthetic seed production (Pant and Shrestha 2011; Pant and Thapa 2012; Paudel and Pant 2012; Pant 2018b).

7.4.1 Asymbiotic Seed Germination

Orchid seeds are minute and lack endosperm, relying on fungal symbionts for nutrient acquisition in natural habitats. Asymbiotic seed germination circumvents this dependency by supplying nutrients and growth regulators directly through defined culture media. In Nepal, asymbiotic seed culture has been applied to genera such as *Cymbidium* sp., *Dendrobium* sp., *Coelogyne* sp., *Bulbophyllum* sp., *Phaius* sp., *Pleione* sp., *Vanda* sp., etc (Pradhan et al. 2016; Pant et al. 2011; Pant et al. 2018b, 2022; Thapa et al. 2024; Maharjan et al. 2020). Studies have demonstrated that immature green pods often yield higher germination percentages than mature, desiccated capsules, reflecting the reduced contamination and higher physiological viability of immature embryos. The success of asymbiotic seed germination is species specific and influenced by seed maturity, sterilization efficiency, light regime, and the composition of the medium. In our study it has been reported that the seed germination rate occurs within few weeks to several months after sowing, with protocorm formation followed by differentiation into shoots and roots over subsequent months. While asymbiotic seed germination is straightforward and cost-effective, scaling up production requires careful synchronization of seed availability and growth culture facilities (Parmar and Pant 2016). Some of the important seed germination research is highlighted in this chapter (Fig. 7.3, Table 7.1).

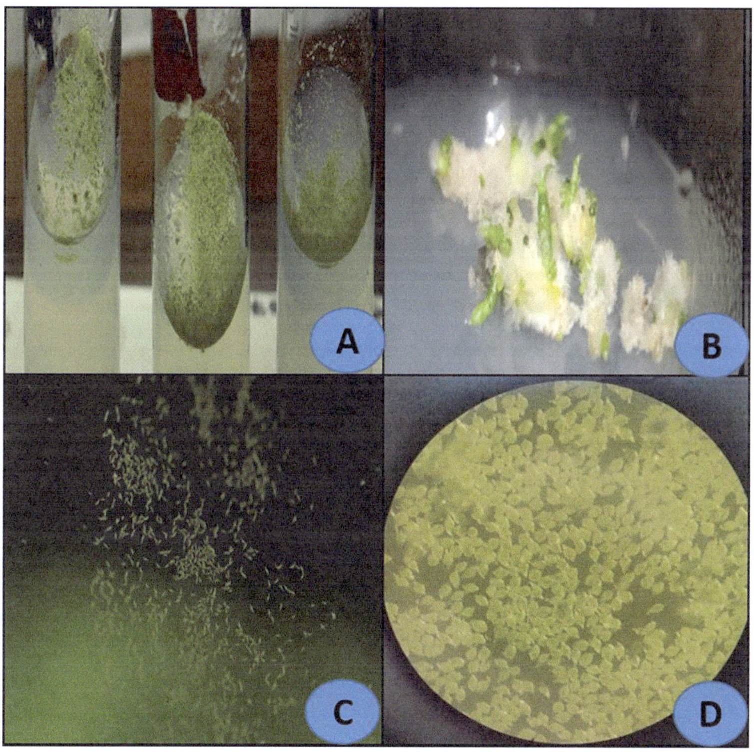

Fig. 7.3 (**A, B, C, D**): Seed germination of different orchid species (**A**) *Bulbophyllum leopardinum* (**B**) *Phaius tankervilleae* (**C**) *Dendrobium densiflorum* (**D**) *Dendrobium transparens*

7.4.2 Protocorm Culture

Protocorms are intermediate structures that develop from germinating orchid seeds before the formation of leaves and roots. In vitro, protocorms can be induced not only from seeds but also from vegetative tissues and yielding structures known as protocorm like bodies (PLB). These can be multiplied rapidly and converted into plantlets under suitable conditions. PLB induction has been achieved from seed-derived protocorms, shoot tips, and leaf segments. There are different media combinations with plant growth regulators like cytokinins, especially thidiazuron (TDZ) and benzyl amino purine (BAP), in combination with auxins such as naphthalene acetic acid (NAA), are frequently reported to enhance PLB formation and proliferation. Multiplication rates vary by species and explant type, with some reports indicating a several fold increase in protocorm number within 4–6 weeks of subculture as depicted in our research work (Figure 7.4, Table 7.2) PLB culture is advantageous for mass propagation as it allows exponential multiplication without repeated seed collection, reducing pressure on wild populations. We have reported mass propagation of different orchid genera by using seed derived protocorm culture. However, prolonged subculture can increase the risk of somaclonal variation, underscoring the importance of periodic genetic fidelity checks (Pant 2013; Paudel et al. 2019; Maharjan et al. 2019; Dhungana et al. 2022). Some of the in vitro propagation research carried out during our study is depicted in *Phaius tankervilleae*, *Bulbophyllum affine*, and *Rhynchostylis retusa* (Fig. 7.5, Table 7.1).

Table 7.1 In vitro culture of different orchid species reported in Nepal by seed culture

Sn	Orchid species	Explant used for in vitro culture	References
1.	*Bulbophyllum leopardinum*	Seed	Thapa et al. (2024)
2.	*Cymbidium aloifolium*	Seed	Regmi et al. (2017)
3.	*Cymbidium aloifolium*	Seed	Pradhan et al. (2013)
4.	*Cymbidium elegans*	Seed	Pradhan et al. (2009)
5.	*Cymbidium devonianum*	Seed	Parmar (2014)
6.	*Cymbidium iridioides*	Seed	Pant and Swar (2011)
7.	*Coelogyne flaccida*	Seed	Parmar and Pant (2015)
8.	*Coelogyne fuscescens*	Seed	Koirala et al. (2013)
9.	*Coelogyne ovalis*	Seed	Sitaula (2010)
10.	*Coelogyne nitida*	Seed	Ranabhat (2023)
11.	*Coelogyne stricta*	Seed	Thapa et al. (2020)
12.	*Dendrobium amoenum*	Seed	Rajkarnikar (2014)
13.	*Dendrobium chryseum*	Seed	Maharjan et al. (2020)
14.	*Dendrobium crepidatum*	Seed	Dhungana et al. (2022)
15.	*Dendrobium crepidatum*	Seed	Gurung et al. (2020)
16.	*Dendrobium densiflorum*	Seed	Pradhan et al. (2013)
17.	*Dendrobium densiflorum*	Seed	Pant et al. (2022)
18.	*Dendrobium fimbriatum*	Seed	Parmar and Acharya (2016)

(continued)

Table 7.1 (continued)

Sn	Orchid species	Explant used for in vitro culture	References
19.	*Dendrobium longicornu*	Seed	Rajbahak and Rajkarnikar (2017)
20.	*Dendrobium moniliforme*	Seed	Ghimire et al. (2025)
21.	*Dendrobium primulinum*	Seed	Pant and Thapa (2012)
22.	*Dendrobium transparens*	Seed	Joshi et al. (2023)
23.	*Esmeralda clarkei*	Seed	Paudel et al. (2012)
24.	*Otochilus albus*	Seed	Adhikari and Pant (2019)
25.	*Papilionanthe teres*	Seed	Pandey et al. (2020)
26.	*Phaius tankervilleae*	Seed	Shrestha (2008)
27.	*Satyrium nepalense*	Seed	Thapa et al. (2023)
28.	*Rhynchostylis retusa*	Seed	Oliya et al. (2021)
29.	*Vanda tessellata*	Seed	Malla et al. (2020)
30.	*Vanda pumila*	Seed	Maharjan et al. (2019)

Fig. 7.4 (**A, B, C, D**): Development and proliferation of protocorms of different orchid species

Table 7.2 Development of protocorm culture for mass propagation

Sn	Orchid protocorms	Growth medium	Formed structure	References
1.	*Coelogyne fuscescens* Lindl.	MS + 1 mg/L BAP + 0.5 mg/L NAA	Shoot formation	Koirala et al. (2013)
2.	*Cymbidium aloifolium* (L.) Sw.	MS (0.25, 0.5, 1) alone and 1.0 MS + 0.5 mg/L BAP + 0.5 mg/L NAA	Shoot, leaf, and plantlet formation	Pradhan et al. (2014), Regmi et al. (2017)
3.	*Cymbidium iridioides*	MS + BAP (0.5 mg/L), MS + 1 mg/L IBA	Shoot proliferation	Pant and Pradhan (2010), Pant and Swar (2011)
4.	*Dendrobium chryseum* Rolfe	½ MS + 2 mg/L Kinetin + 10% CW, ½ MS + 1 mg/L GA3 + 10% CW. ½ MS + 1.5 mg/L IAA + 10% CW	Root and shoot proliferation	Maharjan et al. (2020)
5.	*Dendrobium densiflorum*	MS + 2 mg/L BAP + 0.5 mg/L NAA, MS + 1.5 mg/L IBA	Plantlet formation	Pradhan et al. (2013)
6.	*Dendrobium longicornu*	MS + 2 mg/L BAP + 1 mg/L NAA + 10% CW	Plantlet formation	Rajbahak and Rajkarnikar (2017)
7.	*Otochilus albus*	MS + 0.5 mg/L BAP with 0.5 mg/ L NAA	Plantlet formation	Adhikari and Pant (2019)
8.	*Vanda pumila*	½ MS + 1.0 mg/L Kinetin + 10% CW and ½ MS + 0.5 mg/L IAA	Plantlet formation	Maharjan et al. (2019)
9.	*Dendrobium primulinum*	MS + 1.5 mg/L BAP followed by MS + 1.5 mg/L BAP + 0.5 mg/L NAA	Shoot and root formation	Pant and Thapa (2012)
10.	*Cymbidium longifolium*	MS + 10% CW + 1 mg/L NAA + adenine sulfate	Meristem culture to PLB induction	Shrestha and Rajbhandari (1988)
11.	*Aerides odorata*	MS media	Protocorms developments	Pant and Gurung (2005)
12.	*Coelogyne cristata*	MS + 1 mg/L BAP + 1 mg/L NAA + 10% CW	Shoot root formation	Thapa et al. (2020)
13.	*Dendrobium densiflorum*	MS + 1 mg/L BAP + 1 mg/L NAA + 10% CW + BA	Plantlet formation	Pradhan et al. (2013)

7.4.3 Vegetative Explant Regeneration Callus-Mediated Pathways

Vegetative explants provide an alternate starting material for some species, particularly those that produce few seeds or have resistant germination. Orchid cultures have been started using shoot tips, rhizome sections, pseudo bulb segments, and even leaf bases and roots (Islam 2015). In one experiment, in vitro germinated *Bletilla striata* seeds were used to create pseudobulb enhancing plant growth utilizing a Temporary Immersion System (Zhang et al. 2018). In one investigation,

Fig. 7.5 (**ABC**): Seed derived in vitro micropropagation of *Phaius tankervilleae* (**DEF**): Seed derived in vitro propagation of *Rhynchostylis retusa* (**GHI**): Seed derived in vitro propagation of *Bulbophyllum affine*

employing a leaf as an explant, direct somatic embryogenesis was seen on media enriched with Murashige and Skoog in *Vanilla planifolia* (Manokari et al. 2022). Using shoot tip explants, one study found that ½ MS media combined with BA generated the most shoots in *Dendrobium* Sonia "Earsakul." (Priyakumari et al. 2013). Research on the application of various explants in orchid regeneration was published by (Nongdam et al. 2023; Lal and Singh et al. 2020, 2023; Sonia et al. 2022). There are reports in orchids, applying to callus mediated culture research. Regmi et al. 2017, reported the formation of callus from protocorm in *Cymbidium aloifolium* (L.) SW. The callus was proliferated from a protocorm of *Cymbidium elegans* on the MS medium supplemented with 1.5 mg L-1 BAP (Pant and Pradhan, 2010). In the combination of BAP and NAA, yellow-greenish callus was observed after 16 weeks of culture. However, gibberellic acid (GA3) supplemented medium was found to be effective for the elongation of regenerated shoots in *Dendrobium crepidatum* (Dhungana et al. 2022). In organogenesis approaches, shoots emerge directly from the explant without an intervening callus phase, generally preserving genetic uniformity. In contrast, indirect regeneration via callus formation can increase propagation rates but carries a higher risk of genetic variation. Callus induction is typically promoted by higher auxin-to-cytokinin ratios, with subsequent shoot differentiation triggered by adjusting this balance (Nasiruddin et al. 2003).

7.4.4 Artificial Seed Technology

Encapsulation techniques involve enclosing small propagules, such as protocorms, within a gel matrix (commonly sodium alginate) to produce synthetic seeds that can be stored, transported, and sown like conventional seeds (Kumaria and Tandon 2001). *Dendrobium nobile*, an endangered medicinal orchid from northeastern India, was shown to have encapsulated protocorm-like structures utilizing artificial seed technology (Mohanty et al. 2013). A technique utilizing artificial seed technology in *Cymbidium devonianum* protocorm-like bodies (PLBs) encased in calcium alginate beads supplemented with MS basal media containing sucrose is applied for the storage and high frequency conversion of PLBs (Das et al. 2011). There are reports using artificial seeds technology in *Geodorum densiflorum* (Lam) Schltr (Datta et al. 1999). Some reports suggested the storage of produced artificial seeds using vitrification methods in *Cymbidium aloifolium* (Hirano et al. 2011). Orchids, encapsulation has been explored as a means of facilitating germplasm exchange among laboratories and distributing propagules for ex situ conservation or commercial planting (Pradhan et al. 2014) (Fig. 7.6). According to Pradhan et al. 2014, the artificial seeds were developed in vitro by encapsulating protocorms in 4% sodium alginate and 0.2 mol/L calcium chloride solution. Their germination and plantlet

Fig. 7.6 (**A, B, C, D, E, F, G, H, I**): Plant regeneration from artificial seed technology of *Cymbidium aloifolium* (Pradhan et al. 2014)

regeneration were evaluated on Murashige and Skoog (MS) liquid media of varying strengths (0.25, 0.5, and 1.0), as well as on MS liquid medium supplemented with 0.5 mg/L benzylaminopurine (BAP) and 0.5 mg/L naphthaleneacetic acid (NAA). Encapsulated protocorms have been reported to retain viability for several weeks to months under controlled storage conditions, with successful regeneration into plantlets upon sowing on suitable media. Factors such as alginate concentration, complexing agent type, and storage temperature influence germination rates and plantlet vigor. While promising, artificial seed technology in orchids from Nepal remains in the early stages of development and requires further optimization for large-scale use.

7.4.5 Use of Mycorrhiza on In Vitro Orchid Growth and Development

Physiological and biochemical studies reveal that factors like light quality, carbohydrate concentration and growth regulator balance influence chlorophyll content, antioxidant enzyme activity, and protocorm differentiation, while mycorrhizal symbiosis remains fundamental for nutrient exchange and early seedling development (Shah et al. 2018; Chand et al. 2020; Pant et al. 2017). Recent molecular insights have elucidated that orchids acquire carbon from symbiotic fungi via hydrolysis of fungal trehalose and subsequent sugar transport through SWEET and H^+-hexose symporters, providing deeper understanding of the nutritional and signaling interactions underlying symbiotic germination (Shah et al. 2025; Chamara et al. 2024). Root-associated endophytic fungi isolated from orchids in Nepal, including *Dendrobium moniliforme* and *Vanda cristata*, were shown to enhance in vitro protocorm and seedling growth, demonstrating significant plant growth-promoting potential (Chand et al. 2020). Comparison of growth promoting effect of fungal elicitor as a biotic plant growth regulators in orchid *Cymbidium aloifolium* has been successfully developed. (Fig. 7.7).

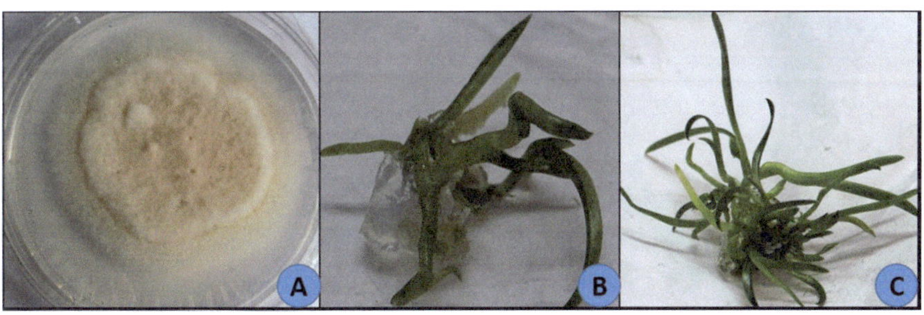

Fig. 7.7 In vitro shoot growth and development of *Cymbidium aloifolium*: Mycorrhizal association in *Cymbidium aloifolium* (**A**), Fungal colony *Paraconiothyrium hawaii* (Cvl1) (**B**), Control (MS media) (**C**), Plantlets treated with 10% fungal elicitor showing shoot multiplication

7.4.6 Acclimatization and Hardening of Tissue Culture Orchid

Tissue-cultured raised orchids, including *Cymbidium aloifolium* and *Dendrobium* species, require careful acclimatization and hardening before transfer to greenhouse or field conditions. Gradual exposure to substrates such as cocopeat moss, coconut husk, coir dust, and charcoal allows plantlets to adapt to fluctuating light, temperature, and humidity, resulting in 70–90% survival under ex vitro conditions (Regmi et al. 2017; Lakshanthi and Seran 2019; Hwang et al. 2024). Stepwise acclimatization ensures the development of functional roots and leaves, enhancing plantlet vigor and establishment in conservation and commercial propagation programs. Tissue culture propagation of orchids frequently succeeds in vitro, but the transition from sterile culture to ex vitro conditions, including acclimatization and hardening, remains the most challenging stage, with many plantlets losing viability due to stress from lower humidity, higher light intensity, poor root aeration and weak cuticle or stomatal function (Nowak and Pruski 2004). In fact, traditional semi solid culture systems are criticized for high plant losses during acclimatization because of hyperhydricity, poorly developed roots and physiological disorders arising when plantlets are transferred directly to greenhouse or field conditions. Different orchid genera have been successfully acclimatized under ex vitro conditions (Fig. 7.8).

Fig. 7.8 (**A, B, C, D, E, F**): Acclimatization of in vitro grown different orchid species (Source: Annapurna Research Center, Kathmandu data unpublished)

7.5 Reintroduction of In Vitro Grown Orchids in Nature

One promising approach to the conservation of wild orchids in the future is reintroduction. It has been used to reduce collecting pressure on endangered orchids and restore wild populations or commercial cultivation of various species (Shao et al. 2024; Zhang et al. 2025). By relocating seedlings into their natural habitats, a number of terrestrial orchid species have been successfully reinstated (Aggarwal and Zettler 2010; Zeng et al. 2012; Endres Junior et al. 2022; Reiter and Menz 2022). Both the public and corporate sectors should take orchid conservation seriously, working with local producers, research institutes, and non-governmental organizations. Tissue culture methods are one of the best solutions for the restoration and mass replication of endangered species because orchid seed germination is unpredictable (Pant 2013). One of the most pressing issues is cultivation for sustainable use. Although local communities and private nurseries in Kathmandu have started domesticating orchids on a modest scale, efficient propagation methods for growing high-potential species are still being developed. The Central Department of Botany, Tribhuvan University, Annapurna Research Center (ARC), Kathmandu, Nepal and Department of Plant Resources (DPR), Government of Nepal, has begun ex situ conservation and germplasm storage for a few commercially significant and endangered medicinal orchids (Fig. 7.9).

Additionally, through tissue culture propagation and acclimatization, the Annapurna Research Center has been instrumental in orchid conservation. ARC has successfully reintroduced species including *Aerides* species, *Cymbidium* species, *Rhynchostylis retusa* and *Dendrobium* species to different parts of Nepal, in collaboration with numerous organizations and trusts. In order to ensure long-term genetic preservation, the Annapurna Research Center also keeps a varied collection of germplasm (Fig. 7.10). Community education and awareness for habitat protection, ex situ conservation and commercial potential of orchids in some community forests in participation with the Community Forest Users (CFUs) of Daman and Chitlang of Makawanpur district, Nepal was studied (Pant et al. 2020).

Fig. 7.9 (A, B, C, D): Reintroduction of in vitro tissue cultured orchids in nature (Source: Annapurna Research Center, Kathmandu, Nepal, Central Department of Botany)

Fig. 7.10 (A, B, C): In vitro germplasms of different orchids species (Source: Annapurna Research Center, Kathmandu, Nepal)

7.6 Applications of Tissue Culture for Conservation

Tissue culture is a fundamental tool for plant conservation because it enables rapid mass propagation, germplasm rescue, and ex situ restoration of taxa that are otherwise difficult to conserve by conventional means, and this is particularly true for orchids whose dust-like seeds require specific fungal partners for germination (Pant et al. 2017; Cardoso et al. 2020; Shah et al. 2022). Protocorm-based micropropagation of *Dendrobium chryseum*, which uses immature seeds and modified ½ MS media with growth regulators and organic additives to produce high protocorm and shoot yields suitable for ex situ conservation and is one of the orchid-focused protocols developed and used for conservation and reintroduction in Nepal (Maharjan et al. 2020). Practical methods for creating planting stock for restoration and for valuing medicinal germplasm have been made possible by micropropagation, phytochemical/in vitro studies on the endangered medicinal orchid *Vanda cristata*, and related regeneration competence studies using leaf explants and whole-leaf techniques (Pathak et al. 2022). Because they enable rapid multiplication rates, generally uniform regenerants, and adaptability to scale-up approaches, a research thoroughly examined in recent literature, protocorms continue to be a key component of orchid clonal propagation (Cardoso et al. 2020). For mass-producing orchid plantlets, temporary immersion systems (TIS) and other bioreactor-based techniques have emerged as scalable solutions that lower production costs, improve nutrient uptake, and reduce hyperhydricity. These prospects are especially pertinent for scale-up options (Nongdam et al. 2023). Complementary conservation strategies include embryo rescue and in vitro hybridization to recover otherwise abortive hybrids or interspecific crosses, synthetic seed approaches for easier germplasm exchange, and cryopreservation/slow-growth storage for long-term safeguarding of recalcitrant germplasm (Engelmann 2018; Ahmed et al. 2023; Zhang et al. 2025). Molecular marker-based screening (RAPD, ISSR, SSR) is used to confirm genetic fidelity of in vitro regenerants and reduce the risk of somaclonal variation before reintroduction (Pradhan et al. 2023; Sarmah et al. 2024). Integration of tissue culture outputs with habitat restoration, where acclimatized, tissue-cultured orchids are used in reintroduction trials, has been shown to aid local biodiversity recovery while providing material for community-based conservation and sustainable use programs (Pant et al. 2018b; Timilsina et al. 2025).

7.7 Challenges and Future Directions

Orchid tissue culture has been considered as one of the important technology for conservation of orchid germplasm, but still in many developing countries, where the diversity of orchid is very high, it's very difficult to preserve this species viability due to weak infrastructure of lab. Establishing different micropropagation technology within the intra species is also a big challenge because of its species specific responses to culture media and different plant growth regulators (Pant and Raskoti 2013; Pathak 2022). Production of callus or protocorm which produces somaclonal variation, is also an important criteria for genetic homogenity, an important technology for conservation, but due to contamination from endogenous microbes and endophytes in explants from the wild, it's very difficult to conserve this species. (Sarmah et al. 2024). Different native orchids in Nepal which includes *Dendrobium densiflorum*, *Dendrobium chryseum* and *Coelogyne cristata* is reported to have genotype dependent morphogenic potential. In vitro development of protocorm and plantlet regeneration is possible because of the correct auxin and cytokinin ration, and different organic and inorganic sources within the culture media (Maharjan et al. 2019) However, hardening and acclimatization of in vitro grown plants survival rate is poor due to lack of mycorrhizal associations to obtain the essential nutrients from nature, which is a major concern in the field of orchid tissue culture, since it lacks endosperm (Nongdam et al. 2023). Therefore, future research must focus on multi-institutional collaboration among universities, botanical gardens, and conservation organizations in order to establish national orchid germplasm archives supported by modern biotechnological technology. Modern methods like as cryo-omics techniques, temporary immersion bioreactors, and synthetic seed technologies have the potential to increase the stability and efficiency of germplasm over the long term (Zhang et al. 2025). Additionally, combining genomic and transcriptomic techniques with conventional tissue culture could discover the identification of critical genes governing morphogenesis and stress tolerance in orchids, enabling adaptive reintroduction strategies under climate change (Wang et al. 2024). Mycorrhizal co-culture systems may also revolutionize in vitro orchid propagation by improving repeatability and post-transfer survival (Sah et al. 2019; Chand et al. 2020). Orchid conservation success depends on striking a balance between in vitro techniques, ecological restoration, and community involvement in order to reassure that tissue-cultured orchids can be successfully returned into their native environments (Pant and Raskoti 2013).

7.8 Conclusion

Orchid tissue culture is an important technology for the conservation and long-term usage of floriculturally important forest-based orchids. The development of species-specific techniques for asymbiotic seed germination, protocorm culture, and vegetative propagation has made it possible to create large amounts of healthy plantlets suitable for ex situ conservation and reintroduction back to the nature. These advancements have significantly reduced the dependency on wild populations and generated data for research and commercial floriculture. However, persistent issues such as genetic instability, contamination, and poor acclimatization highlight the need for continuous culture system improvement. Combining mycorrhizal co-culture, molecular and omics techniques will increase efficiency, genetic integrity and large-scale application. Research institutions, conservation organizations and local communities need to collaborate more closely to ensure that in vitro propagation leads to successful in situ conservation outcomes. All things considered, orchid tissue culture promotes economic opportunities through sustainable orchid production, and establishes Nepal as a regional hub for orchid biotechnology and conservation.

References

Acharya KP, Vetaas OR, Birks HJB (2011) Orchid species richness along Himalayan elevational gradients. J Biogeogr 38(9):1821–1833

Adhikari H, Pant B (2019) Effect of 6-benzylaminopurine and-naphthalene acetic acid hormonal supplements on the in vitro seed germination and seedling development of orchid Otochilus albus Lindl. Afr J Biotechnol 18(22):472–477

Aggarwal S, Zettler LW (2010) Reintroduction of an endangered terrestrial orchid, Dactylorhiza hatagirea (D. Don) Soo, assisted by symbiotic seed germination: first report from the Indian subcontinent. Nat Sci 8(10):139–145

Ahmed M, Razzaque S, Sultana R, Anisuzzaman M (2023) Synthetic seed technology for long-term conservation and clonal propagation of endangered medicinal plants: a review. Plants 12(16): 2911

Arditti J (1990) *Lewis Knudson*, 1884–1958: his science, his times and his legacy. American Orchid Society

Campol JR, Naing AH, Aung HM, Cho SB, Kang H, Chung MY, Kim CK (2024) Production of genetically stable and Odontoglossum ringspot virus-free Cymbidium orchid 'New True' plants via meristem-derived protocorm-like body (PLB) subcultures. Plant Methods 20(1):145

Cardoso JC, Zanello CA, Chen JT (2020) An overview of orchid protocorm-like bodies: mass propagation, biotechnology, molecular aspects, and breeding. Int J Mol Sci 21(3):985

Chamara RR, Rammitsu K, Minobe M, Kinoshita A, Kotaka N, Yukawa T, Ogura-Tsujita Y (2024) Mycorrhizal fungi of Phalaenopsis japonica (Orchidaceae) and their role in seed germination and seedling development. Diversity 16(4):218

Chand K, Shah S, Sharma J, Paudel MR, Pant B (2020) Isolation, characterization, and plant growth-promoting activities of endophytic fungi from a wild orchid Vanda cristata. Plant Signal Behav 15(5):1744294

Chettri Das M, Kumaria S, Tandon P (2011) Storage and high conversion frequency of encapsulated protocorm-like bodies of Cymbidium devonianum (orchid). J Hortic Sci Biotechnol 86(6):611–615. https://doi.org/10.1080/14620316.2011.11512811

Datta KB, Kanjilal B, Sarker DD (1999) Artificial seed technology: development of a protocol in Geodorum densiflorum (Lam) Schltr.–an endangered orchid. Curr Sci 1142–1145

de Oliveira SLA, Silva KLF, de Oliveira RJ, de Souza MJ, de Oliveira JRG, de Melo Ferreira W (2021) In vitro germination, initial development and acclimatization of Cattleya nobilior Rchb. f.(Orchidaceae): an approach to curb the eventual endangerment of this exuberant, near-threatened Cerrado species. Diversitas J 6(2):2167–2191

Dhungana S, Pradhan S, Paudel MR, Pant B (2022) In vitro propagation and genetic homogeneity assessment of Dendrobium crepidatum Lindley and Paxton. Plant Tissue Cult Biotechnol 32(1): 1–11

Priya KI, Sabina GT, Rajmohan K (2013) Influence of plant growth regulators on in vitro clonal propagation of Dendrobium Sonia 'Earsakul'

Endres Junior D, Sasamori MH, Müller A, Schmitt JL, Droste A (2022) Light drives vegetative phenology of Cattleya intermedia (Orchidaceae): a key factor for the establishment of plants translocated into a fragment of subtropical Atlantic Forest. Aust J Bot 70(6):409–420

Engelmann F (2018) Cryopreservation of clonal germplasm—progress and prospects. Acta Hortic 1209:17–26

Fay MF (2018) Orchid conservation: how can we meet the challenges in the twenty-first century? Bot Stud 59(1):16

Gale SW, Fischer GA, Cribb PJ, Fay MF (2018) Orchid conservation: bridging the gap between science and practice. Bot J Linn Soc 186(4):425–434

Ghimire M, Thapa M, Joshi PR, Paudel MR, Pant KK, Gurung P, Pant B (2025) Micropropagation and assessment of genetic homogeneity in *Dendrobium moniliforme* (L.) Sw. using molecular markers. BioRxiv. https://doi.org/10.1101/2025.08.31.673335

Govaerts R, Nic Lughadha E, Black N, Turner R, Paton A (2021) The world checklist of vascular plants, a continuously updated resource for exploring global plant diversity. Sci Data 8(1):215

Grytnes JA, Vetaas OR (2002) Species richness and altitude: a comparison between null models and interpolated plant species richness along the Himalayan altitudinal gradient, Nepal. Am Nat 159(3):294–304

Gurung PR, Shrestha A, Adhikari S, Limbu S, Pant B (2020) Asymbiotic seed germination and seedling development of a medicinally important epiphytic orchid, Dendrobium crepidatum Lindl. and Paxton. J Plant Res 18:226–234

Hinsley A, De Boer HJ, Fay MF, Gale SW, Gardiner LM, Gunasekara RS, Phelps J (2018) A review of the trade in orchids and its implications for conservation. Bot J Linn Soc 186(4):435–455

Hirano T, Yukawa T, Miyoshi K, Mii M (2011) Wide applicability of cryopreservation with vitrification method for seeds of some Cymbidium species. Plant Biotechnol 28(1):99–102

Hwang JE, Park HB, Jeon DY, Park HJ, Kim S, Lee CW, Yoon YJ (2024) Effect of different basal media and organic supplements on in vitro seedling development of the endangered orchid species *Dendrobium moniliforme* (L.) Swartz. Plants 13(19):2721

Hwang JE, Park HB, Tho JH, Kim M, Park HJ, Kim S, Kim YJ (2025) Influence of basal medium and organic additives on in vitro germination and plant growth of endangered orchid *Gastrochilus fuscopunctatus*. Plants 14(20):3133

Islam SM, Islam T, Bhattacharjee B, Mondal TK, Subramaniam S (2015) In vitro pseudobulb based micropropagation for mass development of *Cymbidium finlaysonianum* Lindl. Emir J Food Agric 27(6):469

Joshi PR, Pandey S, Maharjan L, Pant B (2022) Micropropagation and assessment of genetic stability of Dendrobium transparens wall. Ex Lindl. using RAPD and ISSR markers. Front Conserv Sci 3:1083933

Joshi PR, Pandey S, Maharjan L, Pant B (2023) Micropropagation and assessment of genetic stability of Dendrobium transparens wall. Ex Lindl. using RAPD and ISSR markers. Front Conserv Sci 3:1083933

Joshi PR, Thapa BB, Chand K, Maharjan L, Paudel MR, Pant B, Pant B (2024) Biological activities of protocorms and stems extracts of Dendrobium transparens. Trends Phytochem Res 8(2): 122–130

Kafle R, Magar MST, Parmar G (2024) Calanthe longipes Hook f, a new record of Orchidaceae for the flora of Nepal. Banko Janakari 34(1):51–56

Knudson L (1922) Nonsymbiotic germination of orchid seeds. Botanical gazette 73(1),:1–25

Koirala D, Pradhan S, Pant B (2013) Asymbiotic seed germination and plantlet development of Coelogyne fuscescens Lindl., a medicinal orchid of Nepal. Sci World 11(11):97–100

Kumaria S, Tandon P (2001) Orchids, the world's most wondrous plants. Orchids: science and commerce. Bishen Singh, Mahendrapal Singh, New Connaught Place, Dehradun, India, 17–28

Kunwar RM, Mahat L, Acharya RP, Bussmann RW (2013) Medicinal plants, traditional medicine, markets and management in far-west Nepal. J Ethnobiol Ethnomed 9(1):24

Lakshanthi JMT, Seran TH (2019) Survival rate and growth performance of in vitro raised plantlets of orchid (Dendrobium sp.) in different hardening substrates. Int J Adv Res Rev 4(3):01–09

Lal A, Pant M, Palni LMS, Kumar A (2020) Development of rapid micropropagation protocol for germplasm conservation of two orchid species-*Aerides multiflora* Roxb. and *Rhyncostylis retusa* (L.) Blume. Asian J Conserv Biol 9(2):341–347

Lal A, Pant M, Pant G, Palni LMS, Datta A, Kumar G (2023) ISSR marker assisted genetic diversity assessment in natural populations of two endemic orchids Aerides multiflora and Rhynchostylis retusa from Uttarakhand, India. S Afr J Bot 157:151–158

Maharjan S, Pradhan S, Thapa BB, Pant B (2019) In vitro propagation of endangered orchid, Vanda pumila Hook. f. through protocorms culture. Am J Plant Sci 10(7):1220–1232

Maharjan S, Thakuri LS, Thapa BB, Pradhan S, Pant KK, Joshi GP, Pant B (2020) In vitro propagation of the endangered orchid *Dendrobium chryseum* Rolfe from protocorms culture. Nepal J Sci Technol 19(1):39–47

Malla BB, Paudel MR, Pant B (2020) In vitro propagation of *Vanda tessellata* (Roxb.) Hook. ex G. Don from seed-derived protocorms. Botanica Orientalis J Plant Sci 14:14–20

Manandhar NP (2002) Plants and people of Nepal, p 599

Manokari M, Priyadharshini S, Cokulraj M, Dey A, Faisal M, Alatar AA, Shekhawat MS (2022) Exogenous implications of silver nitrate on direct and indirect somatic embryogenesis and germination of cold stored synseeds of Vanilla planifolia Jacks. ex Andrews. S Afr J Bot 150: 129–138

McCormick MK, Taylor DL, Whigham DF, Burnett RK Jr (2016) Germination patterns in three terrestrial orchids relate to abundance of mycorrhizal fungi. J Ecol 104(3):744–754

Mohanty P, Nongkling P, Das MC, Kumaria S, Tandon P (2013) Short-term storage of alginate-encapsulated protocorm-like bodies of Dendrobium nobile Lindl.: an endangered medicinal orchid from North-east India. 3 Biotech 3(3):235–239

Nasiruddin KM, Begum R, Yasmin S (2003) Protocorm like bodies and plantlet regeneration from Dendrobium formosum leaf callus. Asian J Plant Sci 2(13):955–957

Nongdam P, Beleski DG, Tikendra L, Dey A, Varte V, El Merzougui S, Vendrame WA (2023) Orchid micropropagation using conventional semi-solid and temporary immersion systems: a review. Plants 12(5):1136

Nowak J, Pruski K (2004) Priming tissue cultured propagules. Low cost options tissue cult technol developing countries 69

Oliya BK, Chand K, Thakuri LS, Baniya MK, Sah AK, Pant B (2021) Assessment of genetic stability of micropropagated plants of *Rhynchostylis retusa* (L.) using RAPD markers. Sci Hortic 281:110008

Pandey B, Neupane P, Tripathi S, Pant B (2020) Micropropagation of Papilionanthe teres (Roxb.) Schltr. by seed and shoot tip culture. Res J Biotechnol 15:1–8

Pandey S, Sundararajan S, Ramalingam S, Baniya MK, Pant B (2020) Rapid clonal propagation and valepotriates accumulation in cultures of Valeriana jatamansi Jones, a high-value medicinal plant. J Appl Bot Food Qual 93:177–185

Pant B (2013) Medicinal orchids and their uses: tissue culture a potential alternative for conservation. Afr J Plant Sci 7(10):448–467

Pant B, Gurung R (2005) In vitro seed germination and seedling development in *Aerides odorata* Lour. J Orchid Soc India 19(1 and 2):51–55

Pant B, Pradhan S (2010) Micropropagation of Cymbidium elegans Lindl. through protocorm and shoot tip culture. Plant Tissue Cult Biotechnol, pp 3–5

Pant B, Shrestha S (2011) In vitro mass propagation of a ground orchid-Phaius tancarvilleae (L'Her.) Blume through shoot tip culture. Plant Tissue Cult Biotechnol 21(2):181–188

Pant B, Raskoti BB (2013) Medicinal orchids of Nepal. Himalayan Map House, Pvt. Ltd.

Pant B, Swar S (2011) Micropropagation of Cymbidium iridioides. Nepal J Sci Technol 12:91–96

Pant B, Thapa D (2012) In vitro mass propagation of an epiphytic orchid, Dendrobium primulinum Lindl. through shoot tip culture. Afr J Biotechnol 11(42):9970–9974

Pant B, Shah S, Shrestha R, Pandey S, Joshi PR (2017) An overview on orchid endophytes. In: Mycorrhiza-nutrient uptake, biocontrol, ecorestoration. Springer, pp 503–524

Pant B, Paudel MR, Chand MB, Pradhan S, Malla BB, Raskoti BB (2018a) Orchid diversity in two community forests of Makawanpur district, central Nepal. J Threat Taxa 10(11):12523–12530

Pant B, Pradhan S, Paudel MR, Shah S, Pandey S, Joshi PR (2018b) Various culture techniques for the mass propagation of medicinal orchids from Nepal. In: IIIrd International orchid symposium, vol 1262, pp 109–124

Pant B, Paudel MR, Chand MB, Wagner SH (2020) A treasure trove of orchids in Central Nepal. Tribhuvan University, Central Department of Botany

Pant B, Chand K, Paudel MR, Joshi PR, Thapa BB, Park SY, Rajbhandari P (2022) Micropropagation, antioxidant and anticancer activity of pineapple orchid: Dendrobium densiflorum Lindl. J Plant Biochem Biotechnol 31(2):399–409

Parmar G (2014) In vitro seed germination and seedling development of *Cymbidium devonianum* Paxton (Orchidaceae). Bull Dept Plant Res 36:61–64

Parmar G, Acharya R (2016) In vitro seed germination of endangered Nepalese orchid species: Dendrobium fimbriatum Hook. Adv J Seed Sci Technol 3:72–74

Parmar G, Pant B (2015) In vitro seed germination and seedling development of Coelogyne flaccida Lindl. (Orchidaceae). Adv For Sci 2:85

Parmar G, Pant B (2016) In vitro seed germination and seedling development of the orchid Coelogyne stricta (D. Don) Schltr. Afr J Biotechnol 15(5):105–109

Pathak P, Kumari A, Thakur B (2022) Regeneration competence of an endangered orchid, Vanda cristata Wall. ex Lindl. using leaf explants: a study in vitro. S Afr J Bot 151:1018–1024

Paudel MR, Pant B (2012) In vitro plant regeneration of Esmeralda clarkei Rchb. f. via protocorm explant. Afr J Biotechnol 11(54):11704–11708

Paudel MR, Chand MB, Pant B, Pant B (2018) Antioxidant and cytotoxic activities of Dendrobium moniliforme extracts and the detection of related compounds by GC-MS. BMC Complement Altern Med 18(1):134

Paudel MR, Chand MB, Pant B, Pant B (2019) Assessment of antioxidant and cytotoxic activities of extracts of *Dendrobium crepidatum*. Biomolecules 9(9):478

Paudel MR, Sharma S, Joshi PR, Pant B, Wagner SH, Gurung P, Pant B (2025) Antioxidant and cytotoxic properties of protocorm-derived phenol-rich fractions of Dendrobium amoenum. BMC Complement Med Ther 25(1):61

Paul P, Kumaria S (2024) Conservation of medicinal orchids through plant tissue culture. In: Husen A, Pant M (eds) Exploring medicinal orchids. Taylor & Francis Group, pp 17–23. https://doi.org/10.1201/b23375

Phelps J, Webb EL (2015) "Invisible" wildlife trades: Southeast Asia's undocumented illegal trade in wild ornamental plants. Biol Conserv 186:296–305

Poudeyal MR, Saud K, Paudel MR, Singh UK (2025) The significance and utilization of Pholidota orchids in medical preparations. In: Exploring medicinal orchids. CRC Press, pp 131–140

Prabha A, Kushwaha V, Kumari A, Pathak P (2024) Cypripedium orchids and their utilization in traditional and modern medicine. In exploring medicinal orchids (pp 77–85). CRC Press

Pradhan S, Pant B (2009) In vitro seed germination in Cymbidium elegans Lindl. and Dendrobium densiflorum Lindl. ex Wall. (Orchidaceae). Botanica Orientalis J Plant Sci 6:100–102

Pradhan S, Paudel YP, Pant B (2013) Efficient regeneration of plants from shoot tip explants of *Dendrobium densiflorum* Lindl., a medicinal orchid. Afr J Biotechnol 12(12):1378–1383

Pradhan S, Tiruwa B, Subedee BR, Pant B (2014) In vitro germination and propagation of a threatened medicinal orchid, Cymbidium aloifolium (L.) Sw. through artificial seed. Asian Pac J Trop Biomed 4(12):971–976

Pradhan S, Regmi T, Ranjit M, Pant B (2016) Production of virus-free orchid Cymbidium aloifolium (L.) Sw. by various tissue culture techniques. Heliyon 2(10):e00176

Pradhan S, Paudel YP, Qin W, Pant B (2023) Genetic fidelity assessment of wild and tissue cultured regenerants of a threatened orchid, *Cymbidium aloifolium* using molecular markers. Plant Gene 34:100418

Rajbahak S, Rajkarnikar KM (2017) In vitro multiplication and protocorms development of Dendrobium longicornu Wall. ex Lindley. J Plant Res 15(1):100–105

Rajbhandari KR (2015) A handbook of the orchids of Nepal. Government of Nepal, Ministry of Forests and Soil Conservation, Department of Plant Resources

Rajkarnikar KM (2014) In vitro propagation of Dendrobium amoenum Wall. ex Lindl. from shoot-tip culture. Bull Dept Plant Res 36:52–55

Ranabhat S (2023) Biological activities and micropropagation of *Coelogyne nitida* (Wall. ex D. Don) Lindl. Doctoral dissertation, Department of Botany

Raskoti BB, Ale R (2025) An assessment of the perceptions, knowledge, and socioeconomic factors associated with illegal orchid trade in Nepal: implications for conservation. Ambio 55:176–188

Regmi T, Pradhan S, Pant B (2017) In vitro mass propagation of an epiphytic orchid, Cymbidium aloifolium (L.) Sw., through protocorm culture. Biotechnol J Int 19(1):1–6

Reiter N, Menz MH (2022) Optimising conservation translocations of threatened Caladenia (Orchidaceae) by identifying adult microsite and germination niche. Aust J Bot 70(3):231–247

Rokaya MB, Raskoti BB, Timsina B, Münzbergová Z (2013) An annotated checklist of the orchids of Nepal. Nord J Bot 31(5):511–550

Raskoti BB, Ale R (2023) A new variety of Habenaria (Orchidaceae; Orchideae; Habenariinae) and new records of orchids from Nepal. Taiwania 68(1)

Sarmah D, Mohapatra PP, Seleiman MF, Mandal T, Mandal N, Pramanik K, Wasonga DO (2024) Efficient regeneration of in vitro derived plants and genetic fidelity assessment of Phalaenopsis orchid. Front Sustain Food Syst 8:1359486

Shah S, Shrestha R, Maharjan S, Selosse MA, Pant B (2018) Isolation and characterization of plant growth-promoting endophytic fungi from the roots of *Dendrobium moniliforme*. Plants 8(1):5

Shah S, Thapa BB, Chand K, Pradhan S, Singh A, Varma A, Pant B (2019) Piriformospora indica promotes the growth of the in-vitro-raised Cymbidium aloifolium plantlet and their acclimatization. Plant Signal Behav 14(6):1596716

Shah S, Shah B, Sharma R, Rekadwad B, Shouche YS, Sharma J, Pant B (2022) Colonization with non-mycorrhizal culturable endophytic fungi enhances orchid growth and indole acetic acid production. BMC Microbiol 22(1):101

Shah S, Paudel MR, Thapa BB, Sharma H, Kashyap AK, Rekadwad BN, Pant B (2025) Extract from endophytic Fusarium isolates stimulates seed germination of the host and protocorm development of non-host orchids. Commun Integr Biol 18(1):2439798

Shao SC, Jacquemyn H, Selosse MA (2024) Improved use of mycorrhizal fungi for upscaling of orchid cultivation and population restoration of endangered orchids in China. Symbiosis 92(2): 149–158

Sharma S, Kumar V, Seth CA, Sourirajan A, El-Shazly M, Dev K (2024) A comprehensive review on the phytochemistry, pharmacological properties, and in vitro propagation of an endemic medicinal orchid, *Dactylorhiza hatagirea*. Naunyn-Schmiedebergs Arch Pharmacol 397(5): 2621–2635

Shrestha S (2008) Micropropagation of terrestrial orchid Phaius tancarvilleae (L'Her.) Blume. Doctoral dissertation, Central Department of Botany

Shrestha M, Rajbhandari SB (1988) Clonal propagation of *Cymbidium longiflorum* D. Don. by shoot tip culture. In: IInd National botanical conference on science and technology, 8–11 Jun 1994. RONAST, Kathmandu

Shrestha KK, Bhandari P, Bhattarai S (2022) Plants of Nepal: gymnosperms and angiosperms. Heritage Publishers and Distributors Pvt. Limited

Sitaula S (2010) Comparative study on in vitro growth rate of two orchid species viz. Coelogyne ovalis Lindley and Dendrobium densiflorum Lindley. Doctoral dissertation, Department of Botany

Sonia P, Apana N, Tikendra L, Dey A, Jamir I, Nongdam P (2022) Molecular clonal fidelity assessment of micropropagated orchids using DNA markers. In: Biotechnology and crop improvement (pp 143–179). CRC Press

Subedi A, Kunwar B, Choi Y, Dai Y, van Andel T, Chaudhary RP, Gravendeel B (2013) Collection and trade of wild-harvested orchids in Nepal. J Ethnobiol Ethnomed 9(1):64

Teoh ES (2016) Medicinal orchids of Asia, vol 16. Springer, Cham

Thapa BB, Thakuri LS, Joshi PR, Chand K, Rajbahak S, Sah AK, Pant B (2020) Ex-situ conservation and cytotoxic activity assessment of native medicinal orchid: *Coelogyne stricta*. J Plant Biotechnol 47(4):330–336

Thapa B, Baskey S, Pradhan S, Sharma P (2023) Satyrium nepalense D. Don.: a multi-faceted threatened medicinal orchid. Environ Ecol 41:103

Thapa BB, Chand K, Thakuri LS, Baniya MK, Pant B (2024) Ex-situ conservation of Bulbophyllum leopardinum, a threatened medicinal orchid of Nepal. J Nepal Biotechnol Assoc 5(1):1–7

Timilsina S, Poudel A, Ghimire N, Neupane S, Poudel PR (2025) Micropropagation of ornamental plants: a review on Nepalese protocols. Cogent Food Agric 11(1):2512165

Tiruwa BL, Neupane BD, Kadariya R, Pokheral CP, Pant B (2024) Diversity of orchids in terms of their distribution, uses and conservation in Annapurna Conservation Area of Nepal. Am J Plant Sci 15(6):422–440

Vaidya BN (2019) Nepal: a global hotspot for medicinal orchids. In: Medicinal plants: from farm to pharmacy. Springer International Publishing, Cham, pp 35–80

Wang J, Lee H, Chen Y (2024) Embryo rescue and hybridization in *Dendrobium orchids* using in vitro culture techniques. Sci Hortic 330:112073

Willis K (2017) State of the world's plants 2017. Royal Botanics Gardens Kew

Wraith J, Pickering C (2018) Quantifying anthropogenic threats to orchids using the IUCN Red List. Ambio 47(3):307–317

Zeng S, Wu K, da Silva JAT, Zhang J, Chen Z, Xia N, Duan J (2012) Asymbiotic seed germination, seedling development and reintroduction of *Paphiopedilum wardii* Sumerh., an endangered terrestrial orchid. Sci Hortic 138:198–209

Zhang B, Sarsaiya S, Pan X, Jin L, Xu D, Zhang B, Chen J (2018) Optimization of nutritional conditions using a temporary immersion bioreactor system for the growth of Bletilla striata pseudobulbs and accumulation of polysaccharides. Sci Hortic 240:155–161

Zhang W, Qin J, Liao QC, Zhou Y, Fu XW, Hu ZH, Zhang SB (2025) Reintroduction of an epiphytic orchid: Plant size matters. Glob Ecol Conserv 58:03496

Fungal Mycorrhizae and Their Role in Orchid Conservation and Propagation

8

Anuprabha and Vikas Kushwaha

8.1 Introduction

Orchidaceae, the largest and most evolved plant family representing around 8–10% of flowering plants, is more threatened than any other plant family. Orchids, the doyen among ornamentals, are credited with outstanding royalty as a potted floricultural plant and cut flower. Development of new hybrids is comparatively easier in orchids because of their feeble reproductive barriers. In the meantime, more than 100,000 gorgeous hybrid orchids have been created, and more than 1000 new hybrids are added every year (Hossain 2022). A German botanist, first coined the term mycorrhiza (mykorhiza), meaning "fungus-root" and described the phenomenon as "The entire body is neither tree root nor fungus alone, but like the thallus of lichens, a unique morphological organ which can be referred to as fungus-root (mycorrhiza) (Frank 1985)". According to Pandey et al. (2013) mycorrhiza is a type of biotrophic, endophytic, mutualistic symbiosis prevalent in more than 90% of higher plants in which both associates benefited with regard to nutrients exchange, fitness, and evolution (Pandey et al. 2013). Over time, research on mycorrhiza has greatly improved our understanding of symbiotic associations between higher plants and specific rhizosphere fungi (Bellgard and Williams 2011).

The biology of mycorrhiza includes the functions of different fungal species in relation to a wide range of plant species and the diverse habitats in which they coexist. This partnership may be primarily mutualistic or symbiotic, in which both partners gain from each other by providing the host with better nutrition, protection from the infecting agents (chemical and physical), and maintenance of the host's

Anuprabha (✉)
Department of Botany, Goswami Ganesh Dutta Sanatan Dharma College, Chandigarh, India

V. Kushwaha
Department of Biotechnology, Goswami Ganesh Dutta Sanatan Dharma College, Chandigarh, India

A. Husen, M. Pant (eds.), *Forest-based Orchids*,
https://doi.org/10.1007/978-981-92-0096-2_8

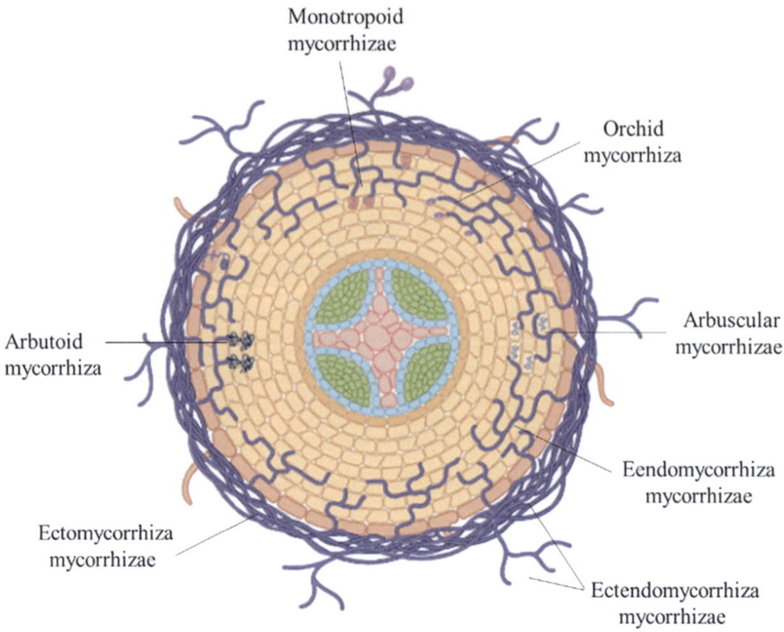

Fig. 8.1 Types of mycorrhizae association

hormonal balance (Lee et al. 2013). In a mycorrhizal interaction, the fungal partner colonizes the plant's root system and establishes specialized structures called mycorrhizae, which artfully function to facilitate the exchange of nutrients between the fungus and the plant. Mycorrhiza is a type of biotrophic, endophytic, mutualistic symbiosis prevalent in more than 90% of higher plants in which both associates benefited with regard to nutrients exchange, fitness, and evolution. Through this relationship plants gain enhanced access to critical water and nutrients especially phosphorus, nitrogen, and micronutrients delivered via an extensive fungal network that far exceeds the reach of roots alone, while fungi receive vital sugars from the plant, boosting both partners' fitness, growth, and evolutionary success (Correia et al. 2019).

On basis of location of fungal association in roots, mycorrhizae were categorized into three major groups: (1) Ectomycorrhiza (outside the root), (2) Eendomycorrhiza (inside the root), and (3) Ectendomycorrhiza (both in- and-outside the root). There are six types of mycorrhizae association that has been reported on basis of the distinctive morphological features of root-fungus association: (1) Arbuscular mycorrhiza (AM), (2) Ectomycorrhiza (EM), (3) Arbutoid mycorrhiza, (4) Monotropoid mycorrhiza, (5) Ericoid mycorrhiza, and (6) Orchid mycorrhiza (Roth-Bejerano et al. 2014) (Fig. 8.1).

Orchid mycorrhiza represents a remarkable and highly specialized form of plant–fungus mutualism and belongs to the Basidiomycota families Tulasnellaceae, Ceratobasidiaceae, and Serendipitaceae. Fungi from these families invade orchid roots or protocorms, colonize an orchid seed, and form a specialized coiled hyphal structure called pelotons within cortical cells (Li et al. 2021). These pelotons serve as dynamic interfaces for nutrient exchanger where fungus delivers essential carbon, nitrogen, phosphorus, and water to the orchid for seed germination and early development while the orchid may reciprocate with developing green leaves for photosynthesis. A majority of orchids are autotrophic and capable of photosynthesis; however, there are over 100 species that are completely myco-heterotrophic, without any of chlorophyll and completely depends on fungi for their nutritional needs throughout their lifespan. Certain orchids exhibit mixotrophy, integrating photosynthesis with carbon obtained from fungi. These mixotrophic orchids, while capable of photosynthesis as adults, continue to depend on fungi for a portion of their carbon supply and are regarded as evolutionary intermediates between fully autotrophic and achlorophyllous orchids (Yeh et al. 2019). Orchids exhibit a fascinating ability to regulate fungal invasion in mycorrhizal interactions, ensuring a balanced symbiosis rather than uncontrolled infection. When fungi colonize orchid tissues, the host initiates the response that enhances the activity of crucial defensive enzymes such as peroxidase, ascorbic acid oxidase, polyphenol oxidase, and catalase, which together facilitate the digestion of fungal pelotons. This regulation is further influenced by phytoalexins, natural plant defense compounds, that control the fungal overgrowth. The first phytoalexin discovered in *orchid* protocorms was orchinol which inhibits fungal growth. Mature pelotons inside host cells are eventually digested through the combined action of phytoalexins and orchinol enzymes, releasing vital nutrients for the orchid (Li et al. 2021; Leng et al. 2024). Digestion of pelotons in orchid cells released some bioactive substances such as peptides, proteins, polysaccharides, chitins, fatty acids, glycolipids, glycoproteins, and specific secondary metabolites, act as fungal elicitors. These fungal elicitors not only encourage further fungal colonization but also stimulate vigorous growth and development of the orchid's protocorms and seedlings (Zhao and Liu 2008; Gong et al. 2018). For instance, substances released by endophytic fungi such as *Fusarium oxysporum* can significantly enhance the synthesis of taxol and particular alkaloids in host cells (Shah et al. 2025). The antioxidant compounds may activate the orchid's intrinsic defense mechanisms by upregulating enzymes such as phenylalanine ammonia-lyase (PAL), thereby enhancing the young plant's disease resistance (Gómez-Vásquez et al. 2004). A study reported that co-cultivation of the medicinal orchid *Anoectochilus formosanus* with the fungus *Mycena* sp. F-23, remarkable improvements after 10 weeks: shoot height increased by 16.6%, leaf number by 22.5%, and shoot weight by 31.3%. Even more striking was the surge in compounds kinsenosides (by 85.5%), isorhamnetin-3-*O*-β-D-rutinoside (226.1%), and isorhamnetin-3-*O*-β-D-glucopyranoside (196.0%), respectively (Zhang et al. 2013).

8.2 Isolation of Orchid Mycorrhizal Fungi

The presence of mycorrhizal associations differs among orchids. The roots of epiphytic or lithophytic orchids that are attached to the substratum exhibited a higher prevalence of mycorrhizal associations, whereas terrestrial orchids with older roots showed a greater occurrence of these associations compared to their epiphytic counterparts. The mycorrhizal zone of orchid roots typically exhibits a yellowish or brownish color, while the aerial portions and root tips lack any mycorrhizal presence (Hossain 2019). Isolation of mycorrhizal fungi from orchids roots involves following steps:

8.2.1 Collection of Orchid from Field

Collection of orchids for fungal isolation generally occurred from actively growing roots of developed plants with natural seedlings. Although seedlings are the best source, mature orchids (also called as harbor fungi) can also contain fungi that support seed germination in both terrestrial and epiphytic species. After collection, roots will be detached with the help of a sterile scalpel to reduce the contamination (Yokoya et al. 2021).

8.2.2 Peloton-Based Fungi Isolation from Roots

For isolating orchid mycorrhizal fungi, fresh root segments of 1–4 cm with visible velamen and healthy cortex contain pelotons are collected in cold sterile bags and processed within 24 h. Further, roots are then surface-sterilized using 90% ethanol or bleach followed by rinse with sterile water to eliminate surface microbes without harming cortical tissues contain pelotons (Zhu et al. 2008).

Later on, peloton-rich cortical cells are carefully teased out from the split root and placed in sterile water or saline, often overlaid with molten low-nutrient agar such as modified Melin–Norkrans (MMN) or fungal isolation medium (FIM), which allows hyphae to grow out from the embedded pelotons (Zettler and Piskin 2011). Different culture media such as potato dextrose agar (PDA), cornmeal agar, oatmeal agar, malt extract agar, bran agar, and water Agar were used for growth of fungal endophytes. Various antibiotics like streptomycin, tetracycline, penicillin, and chloramphenicol were used to control the bacterial contamination during the fungal growth at various concentrations (20–150 mg/mL) (Hossain et al. 2013). FIM with antibiotics (e.g., streptomycin) and fungal hyphae are incubated at 20–25 °C in darkness, and rhizoctonia-like fungi are identified by their distinctive right-angle branching, constrictions, and monilioid cells. Repeated hyphal-tip transfers yield purified single-morphotype cultures (Zettler and Piskin 2011; Kendon et al. 2020). Many fungal species were reported from *Eulophia flava*, *Habenaria dentata*, *Goodyera procera*, and *Spiranthes hongkongensis* orchids after culturing individual root sections on PDA containing 50 mg/L streptomycin and 3 mg/L rose Bengal in dark condition at 25 °C (Shan et al. 2002). OMF were isolated using PDA medium enriched with 50 mg/ml each of penicillin and streptomycin from tropical orchids such as *Campylocentrum filiforme*, *C. fasciola*, *Erythrodes plantaginea*, *Ionopsis utricularioides*, *I. satyrioides*, *Oncidium altissimum*, *Oeceoclades maculata*, *Psychilis monensis*, and *Tolumnia variegate* (Tupac Otero et al. 2002). Hossain et al. (2013) reported that *Ceratobacidium AM* and *Ceratobasidium RR* fungus was isolated from fresh roots of naturally grown *Aerides multiflorum* and *Rhynchostylis retusa* using PDA medium with Streptomycin sulfate at 0.75 mg/mL. A novel mycorrhizal fungus dubbed Kyoto-Matsubara-Ishii ("KMI") was isolated from *Paphiopedilum thailandense*, using culture media PDA with antibiotic chloramines-T, streptomycin, and chloramphenicol (Matsubara et al. 2012). Another study on *Ophrys bertolonii* (Mediterranean orchid) shown that their roots are associated with fungi, belonging to families Basidiomycota and Ascomycota (Pecoraro et al. 2015). In the similar study, roots of *Bletilla striata* have the fungal isolate *Schizothecium fimbriatum*, those are extracted using various culture media, including potato dextrose agar (PDA), synthetic PDA, malt extract agar (MEA), and Czapek medium, supplemented with antibiotics (100 g/mL of streptomycin sulfate) (Xi et al. 2020). In a study, researchers reported that nine fungal isolates belonging to the genera *Cladosporium*, *Colletotrichum*, *Cylindrocarpon*, *Fusarium*, *Hypoxylon*, *Leptosphaerulina*, and *Trichoderma* were recovered from the roots of *Dendrobium moniliforme* using potato dextrose agar (PDA) media supplemented with 50 μg/mL chloramphenicol (Shah et al. 2019). Twelve different endophytic fungi were isolated from roots, stems, and leaves of *V. cristata*, belonging to Ascomycota (*Agaricus bisporus*, *Alternaria tenuissima*, *Fusarium tricinctum*, *Alternaria* sp., *Pseudochaetosphaeronema* sp., *Fusarium oxysporium*, *Fusarium solani*, *Fusarium* sp., *Mycoleptodiscus* sp., and *Paraconiothyrium hawaiiense* using PDA and Czapek's media (Chand et al. 2020). Recently, *Epulorhiza* sp. FQXY019 and *Tulasnella calospora* FQXY017 (beneficial mycorrhizal fungi) were extracted from the roots of wild *Paphiopedilum barbigerum* using PDA media, which proved to be highly effective in enhancing seedling development (Tian et al. 2022).

8.2.3 In Situ and Ex Situ Seed or Seedling Baiting

Higher yield of germination-enhancing orchid mycorrhizal fungi (OMF) seed and seedling baiting is used either in the field or lab to collect the compatible fungi. In situ seed baiting utilizes the placement of sterile seeds in mesh bags where orchids grow, after weeks or months, pelotons are collected from germinating seeds/ protocorms by dissecting, and fungi are collected. Ex situ seed baiting uses soil, bark, or moss with sterile seeds in controlled containers for testing multiple substrates with reduced contamination risks (Brundrett et al. 2003). Alternatively, seedling baiting with young orchids can attract fungi that support both germination and early growth. These methods are highly effective for isolating OMF such as *Tulasnella*, supporting orchid conservation and reintroduction efforts (Chen et al. 2021) (Fig. 8.2).

Fig. 8.2 (**a**) *Eulophia dabia orchid* with flower, (**b**) Transverse section of an *Eulophia dabia* root colonized with intracellular hyphal coils (pelotons), and (**c**) Fungal pelotons in cortex cells

8.3 Characterization of Orchid Mycorrhizal Fungi

Characterization of orchid mycorrhizal fungi involves the following approaches:

8.3.1 Microscopy and Morphological Traits

Pelotons are the core anatomical hallmark of orchid mycorrhiza in coiled formed inside cortical cells of orchid roots or seedling protocorms (Perotto and Balestrini 2024). Mycorrhizal fungi, such as Rhizoctonia-like basidiomycetes, including *Tulasnella*, *Ceratobasidium*, *Sebacina*, and *Russula*, form pelotons within the cortical cells of orchid roots, which can be visualized and identified through microscopy. Besides pelotons, the color of the young and mature colony, the growth of the colony, affinities for hyphal anastomosis, and the production of monilioid cells are distinctive morphological features for identifying the orchid mycorrhiza fungus. Besides that, certain microscopic features like size and shape of monilioid cells, formation of sclerotia, the ultrastructure of the septal pore, right angle branching, a constriction near branching point, a dolipore septum at the base of branch, cytomorphology of the vegetative and reproductive hyphae, number of nuclei in

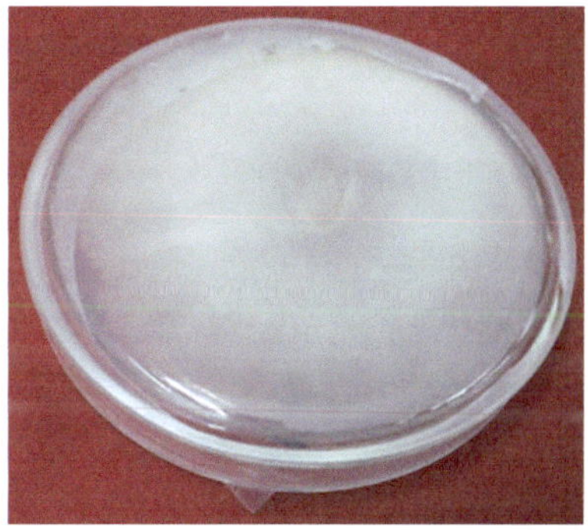

Fig. 8.3 Growth of fungal hyphae in petri dish

both vegetative and monilioid cells, teleomorphs, and enzyme activity are useful identifying characters for sterile mycelia of fungal isolates (Jiang et al. 2015; Hossain 2019). Researchers employed a range of staining techniques to explore the microscopic features of orchid mycorrhizal fungi (OMF) using common stains like trypan blue, aniline blue, lactophenol cotton blue, DAPI (4′,6-diamidino-2-phenylindole), orcein, Giemsa, and acridine orange. Lactophenol cotton blue, safranin (0.5% w/v), or a mixture of 3% KOH and trypan blue are typically used to visualized monilioid cells and measure hyphal diameter.

Fungal growth is assessed by inoculating a small fragment of live mycelium at the center of a Petri plate and recording the radial expansion of colonies over set time intervals (Bhuiyan et al. 2021). To study nuclear details of fungal cells, a tiny piece of fungal mat is first fixed in 2% formaldehyde for a few minutes, rinsed with distilled water, and then stained with DAPI. Further, the sample is mounted in a 50% glycerol solution and examined under a fluorescence microscope. This shows how many nuclei are in the hyphae and monilioid cells (Hossain 2019) (Fig. 8.3).

8.3.2 Molecular Identification (ITS Sequencing and Phylogenetics)

Molecular typing of orchid roots reveals complex microbial diversity associated with 19,896 different fungi. Methods based on DNA have been used for the identification of orchid fungus. In combination with the awareness that the OMF were confined to a single peloton, amplification of particular DNA segments with specific primers from this peloton using PCR directly has been deemed a reliable approach for characterization of fungi and gave much broader and more insightful information related to the nature of orchid-OMF partnerships. Molecular markers such as random amplified polymorphic DNA (RAPD) and restriction fragment length polymorphism (RFLP) are technically simple, inexpensive, and PCR-dependent. They are designed to amplify DNA fragments displaying polymorphisms that may be utilized to detect genetic variation (Bairu et al. 2011). Amplified fragment length polymorphism (AFLP) is a robust and reliable method used to fingerprint DNA which has been widely applied in other genetic studies as DNA profiling, genomics of organisms from all kingdoms (Agarwal et al. 2008). The ITS region of rDNA is currently the optimal method to offer a taxon with precision. Specialized ITS primer sets (ITS1, ITS2, ITS3, ITS4, ITS5) have proven particularly effective at amplifying key orchid mycorrhizal lineages. These important characteristic features of ITS regions help in the accurate identification of OMF through molecular techniques. (1) due to the high copy number ITS regions are practically easy to amplify, (2) comparatively a few primers are required to amplify the highly conserved large subunit (LSU) and small subunit (SSU) flanking regions, (3) differs slightly within the species but vastly between the species, and (4) well-represented DNA sequences in the Gene Bank data than any other genes of the fungal genome (Sun and Guo 2012). Another high-throughput sequencing (HTS) method uses ITS2-targeted primer pairs like ITS3/ITS4OF or ITS86F/ITS4, which help detect a wider range of fungal species and provide much more evolutionary information and phylogenetic relationship (Li et al. 2021).

8.3.3 Biochemical Profiling

The biochemical profiling of OMF provides essential information related to their adaptability, functional diversity, and metabolic activity throughout the orchid–fungus relationship. These fungal species released various extracellular enzymes like cellulases, ligninases, and phosphatases, which facilitate the mobilization and mineralization of nutrients rich in carbon, nitrogen, and phosphorus, which are required for the seed germination, protocorm development, and seedling establishment. Different biochemical markers, such as fatty acid profiles, lipid-based metabolites, sterols, and sulfur-containing compounds, are used for the identification and characterization of OMF, information about their taxa, and for the evaluation of symbiotic potential between orchids and fungal species. For example, fatty acid profile markers and sulfur metabolism markers are utilized for characterizing fungal species and for assessing active nutrient exchange during symbiosis, respectively. Nowadays, gas chromatography-mass spectrometry (GC-MS) and liquid chromatography-mass spectrometry (LC-MS) are utilized to characterize OMF and to study their metabolic interactions with orchids. These biochemical methods not only enhance fungal identification and their taxonomic classification but also provide strong tools for discovering the mechanism behind orchid–fungus synergy, symbiotic efficiency, and ecological adaptability (Pereira et al. 2011).

8.3.3.1 Fatty Acid Profiling

The analysis of fatty acid composition acts as an effective method for characterizing fungi isolates from orchids. This approach, validated by the MIDI method, has demonstrated sensitivity to variations among fungi within the same species as well as across different species and genera of Rhizoctonia-like fungi. This method distinguishes species and genera through unique fatty acid profiles and demonstrates a significant alignment with genetic clustering in the ITS region (Pereira et al. 2011).

8.3.3.2 Lipid-Based Metabolites

Lipids were the most prominent metabolites and major structural components of cell membranes, providing critical biological functions as energy and carbon storage, in signaling, stress response, and plant–microbe interactions (Siebers et al. 2016). Lipids have recently become an important topic in mycorrhizal research because a substantial increase in the amount of lipids was discovered in the hyphae of arbuscular mycorrhizal fungi during symbiosis. Structural lipids (glycerophospholipids, sphingolipids, fatty acid, and chitin precursors) reflect the fungus's architectural and proliferative readiness in symbiosis, whereas signaling lipids (phosphoinositides, oxylipins) represent active communication and regulatory responses. These lipidomic changes serve as powerful biochemical markers for monitoring and progression of orchid–fungus symbiosis (Ghirardo et al. 2020).

8.3.3.3 Sulfur Metabolism Indicators

Metabolomics is a powerful approach to understand the metabolic and biochemical processes occurred during the orchid-OMF synergy. This approach explores the identification of essential pathways and metabolic fluxes during the symbiotic relationship. For example, studies shown that using stable isotope ^{34}S has yielded strong evidence for the active transfer of sulfur from the orchid mycorrhizal fungus *Tulasnella calospora* to developing orchid protocorms. This study confirms the bidirectional metabolic interaction and also highlights the critical role of fungal-derived nutrients in early development of orchids. Further studies showed the enhancement of fungal sulfur assimilation pathways with increased expression of plant sulfotransferase (STRF) genes, indicating a synchronized regulatory mechanism between the host and its symbiotic partner. These findings suggest that sulfur-containing metabolites could serve as biochemical signature markers of OMF activity, presenting important indicators of functional symbiosis (De Rose et al. 2023).

8.3.3.4 Nitrogen Utilization Patterns and Transporters

In the orchid–fungal relationship, orchid seeds are nutrient-deficient, so the collection and transfer of nitrogen from the fungus is essential for the growth of young orchid seeds. OMF showed remarkable metabolic activity and has a strong tendency to assimilate inorganic nitrogen in the form of ammonium (NH_4^+), nitrate (NO_3^-), and nitrogen-containing amino acids such as arginine and glutamine. At the molecular level, it was evidenced that the role of specialized nitrogen transporters like fungal amino acid permeases (e.g., SvAAP1 and SvAAP2) has a significant role in the transport of organic nitrogen across the symbiotic interface. These permeases act as transport systems that mediate the uptake, transport, and delivery of amino acid-derived nitrogen to the orchid plant from its symbiont partner (Novotná et al. 2023).

8.4 Mycorrhiza and Conservation of Orchids

Orchids represent one of the largest and most diverse flowering plant families, comprising over 28,000 species across the ecosystem. These plants are recognized for their morphological diversity, limited availability, and complex ecological relationships. Most of the species exhibit high levels of specialization, depend on specific pollinators, and have close relationships with orchid mycorrhizal fungi (OMF) for successful seed germination, seedling establishment, and survival. However, the growth and survival of orchids in the natural environment face lots of challenges due to habitat degradation by deforestation, urban development, and agricultural practices. Moreover, climate change has further complicated these problems, disrupting temperature and rainfall patterns and affecting the ecological synchrony among orchids, their pollinators, and their fungal associates. Orchids are recognized as bioindicators of environmental health due to their extreme sensitivity to ecological disturbances (Table 8.1).

8.5 Strategies for Orchid Conservation

The conservation of orchids required collaborative strategies, which included both in situ and ex situ methods to protect their diversity and ecological significance. The in situ conservation method utilizes the protection of orchids in their natural habitats, including forests, wetlands, and grasslands. This method ensures that orchids grow in their ecological context and maintain their interactions with pollinators, orchid mycorrhizal fungi (OMF), and other biotic and abiotic factors. While ex situ conservation focuses on the safety of orchid species, it is those facing threats in the wild environment. Various ex situ strategies, such as seed banking, in vitro

Table 8.1 Different approaches for characterization of orchid mycorrhizal fungi

Methodology	Purpose	Key insights/examples
Microscopy (pelotons)	Confirm colonization morphologically	Standard in protocols involving root staining
Culture-based isolation	Recover live fungi; test symbiosis	Used with PDA, MEA, etc.
ITS sequencing and phylogenetics	Taxonomic identification	Matches to *Serendipita* or *Schizothecium* in specific orchids
Biochemical profiling	Their adaptability in nutrient acquisition and metabolic responses	Fatty acid composition, lipid-based metabolites, and sulfur metabolism indicators
Culture-independent NGS	Capture unculturable community diversity	Reveals broader fungal associations and host specificity
Ecological specificity analysis	Understand developmental/ecological patterns	Mycorrhizal preference varies with orchid species and life stage

propagation, cryopreservation, and living collections in botanical gardens, protect the genetic diversity of orchids beyond their natural habitats. Both these strategies create a comprehensive framework for orchid conservation, wherein in situ methods protect natural evolutionary processes, while ex situ approaches provide long-term genetic security and facilitate the recovery of declining populations.

8.5.1 In Situ Conservation

8.5.1.1 Protected Areas and Microhabitats

For the conservation of endangered orchid species, protected areas like forests, wetlands, and grasslands have been developed to protect them from various environmental threats like deforestation, land conversion, urbanization, etc. They provide a suitable environment and habitat for the growth of the orchids. In protected areas, some environmental elements such as deadwood, leaf litter, and forest canopy maintain the temperature and provide organic nutrients, which support the establishment and proliferation of orchids' symbiotic partner, orchid mycorrhizal fungi (OMF) (Zhou et al. 2021). Studies have shown that multiple fungi species like *Tulasnella calospora*, *T. asymmetrica*, and *T. bifrons* associated with the orchids' roots (*Cymbidium mannii*, *E. radicans*, and *C. goeringii*) are required for their seed germination and seedling formation (Yang et al. 2021).

8.5.1.2 Community Involvement

The formation of sacred groves and Community Reserves is an effective strategy for orchid conservation. Sacred groves have older trees, deadwood, thick leaf litter, a shaded canopy, and moisture in the soil, providing stable temperature, retaining moisture, recycling nutrients, and supporting mycorrhizal fungal communities. A case study reported that the Nakuleshwar sacred grove in the Himalayas protects and creates an environment with humid conditions and humus-rich soil for the growth of orchids such as *Goodyera hemsleyana*. Conserved areas, managed by local communities under legal frameworks, empower the local communities to participate in monitoring, decision-making, awareness-raising, and integrating conservation with livelihoods for protecting the endangered orchid species. The Kerala government has started a project "Protection and Conservation of Sacred Groves" that works in association with local communities to make a list of sacred groves, their management, and protection of endemic species (Sharma et al. 2025).

8.5.1.3 Habitat Management and Reintroduction

Orchid conservation involves the restoration of plants into their natural environments. The integration of seed-germination-promoting OMF (sgOMF) during restoration or reintroduction significantly enhances the success rates of endangered orchid species. OMF fungi facilitate the germination and the establishment of seedlings of orchids in their habitats (Zhao et al. 2021). Orchid species depend on seed-germination-promoting orchid mycorrhizal fungi (sgOMF) for germination (protocorm stage) and the transition to a seedling stage. Mycorrizal fungi derived from protocorms or germinating seedlings (sOMF) supply nutrients for growth because orchid seeds lack endosperm and cannot germinate without fungal support. Co-sowing orchid seeds with sgOMF in a suitable environment with ecological specificity, which are isolated and tested in both laboratory and field, have enhanced the seed germination and early seedling survival under natural habitat conditions. Co-sowing of orchids with sgOMF was successfully tested in several orchid species like *Dendrobium devonianum*, *D. aphyllum, and D. nobile*, *which led to the* establishment and growth of seedling populations in endemic areas.

8.5.2 Ex Situ Conservation

8.5.2.1 Seed Bank and Cryopreservation

Many orchid species produce recalcitrant seeds that cannot grow easily. In such cases, cryopreservation at very low temperatures helps in the preservation of genetic diversity over long periods. Cryopreservation of orchid germplasm, which includes mature or immature seeds, protocorm-like bodies, and their mycorrhizal fungi partner, is beneficial for preserving genetic diversity and later germination or growth. Survival rate of such recalcitrant seed has been enhanced by using advanced methods like encapsulation-dehydration and vitrification. Improve survival rates by carefully reducing water content and preventing ice crystal damage. The encapsulation-dehydration method utilizes the embedding of plant material in alginate beads, pre-treating with sucrose or other osmotic agents, partially drying, and then storing in liquid nitrogen, which enhances the germination survival rate of recalcitrant seed. *For* example, encapsulation-dehydration enhanced the germination rate of *Paphiopedilum javanicum* seeds by ~50%. Treatment of seed tissue with cryoprotectants (*Vitrification*) enhanced the seed germination and seedling. The study shown that ~91% higher germination rate in *Bletilla formosana* seeds after treatment through vitrification (Shah et al. 2025).

8.5.2.2 Botanic Gardens and Living Collections

There is an essential role of botanical gardens in orchid conservation by maintaining their living collections. These collections are important for ex situ conservation because they protect the orchid species from extinction. They help in preserving genetic diversity and facilitate public education about the importance of orchids in ecosystems. Botanic gardens play a big role in protecting and learning about orchid species through these efforts. However, endemic areas of orchids and tropical biodiversity hot spots lack sufficient representation in these collections, highlighting problems in global conservation initiatives (Zhao et al. 2021).

8.5.2.3 Tissue Culture and Micropropagation

Micropropagation is one of the effective biotechnological tools for the large-scale production of economically and ecologically important orchid species. Using highly sophisticated in vitro techniques like shoot tip culture, protocorm-like body (PLB) induction, and somatic embryogenesis facilitate the rapid clonal production of rare genotypes while maintaining genetic integrity among the regenerants. Micropropagation presents a highly efficient system compared to the traditional propagation methods that reduce plant production. Additionally, asymbiotic germination of orchid seeds in sterile environments dispels their natural dependence on mycorrhizal fungi, thus facilitating the production of large quantities of healthy, disease-free, and morphologically identical plantlets. This combination of tissue culture and seed germination methods not only aids commercial orchid cultivation but also delivers a platform for the ex situ conservation of endangered orchid species (Hemanta et al. 2021).

8.5.2.4 Symbiotic Propagation and OMF Utilization

Orchid seeds are too small and lack the nutrient reserve required for seed germination and seedling development. Therefore, incorporation of OMF is essential for orchid propagation and conservation. These fungal partners provide carbon, nitrogen, and essential minerals during the early stage of development. The symbiotic relation of orchids and OMF robustly supports the seedling compared to the asymbiotic methods. This orchid–fungal synergy increases the nutrient absorption, promotes growth, and increases tolerance to both biotic and abiotic stress in different orchid species. Another practice called biological hardening serves an important role in orchid propagation, where in vitro-raised seedlings are introduced to a specific fungal strain for their acclimatization to the surrounding environment. This method enhanced the root systems, photosynthetic efficiency, and resistance to environmental stress, ultimately resulting in higher survival rates and greater success in natural habitats and restoration initiatives (Zhao et al. 2021). This preliminary step enhanced the architecture of the root system, increased water and nutrient absorption capacity, and promoted the photosynthetic system. As a result, these inoculated seedlings demonstrate improved resistance against environmental changes, improved stress tolerance, and significantly higher survival rates. These strategies optimized the orchid reintroduction and restoration, also highlighting the importance of such orchid-OMF symbiotic synergy for the long-term sustainability of orchid populations.

8.6 Role of Orchid Mycorrhiza Fungus in Orchids Conservation

Orchids, the threatened plant species, are renowned for their elegance and diversity, but they rely on the orchid mycorrhizal fungi (OMF) for development. The fungi form symbiotic associations with the cortical cells of the orchid root and create specialized coiled structures like pelotons that enable the exchange of nutrients between the fungus and the orchid (Kaur and Deb 2025). The fungus provides essential nutrients like carbon, nitrogen, phosphorus, and water to the developing orchid during seed germination and seedling development when the orchid seeds do not have food reserves independently. This plant–fungal relationship supports the early development of orchid seeds in the natural environment (Cameron et al. 2007). This unique relationship between orchids and their fungal partners within their environments is for effective conservation strategies, which ensure the preservation of these species for future generations (McCormick et al. 2018).

The dependency of orchids on fungi plays an important role in their conservation. The presence or absence of various fungal species in a natural habitat has a strong impact on the potential of orchids' growth and their survival (McCormick et al. 2018). Endangered and threatened orchid species reside in microhabitats like shaded forest and wetlands, where the environmental conditions are beneficial not only for the orchids themselves but also for their symbiotic fungal partners (Swarts et al. 2010). Therefore, conservation must go beyond orchid protection to include the conservation of fungal diversity and of the soil microbial networks upon which they depend. Application of OMF in conservation is demonstrating an increasing trend toward success. Symbiotic propagation methods, where orchid seeds are germinated together with their fungal associates, produce seedlings that are more robust and resilient for successful reintroduction (Zhao et al. 2021). These approaches stimulate natural processes and maximize long-term viability compared to traditional tissue culture methods. Furthermore, recent restoration initiative includes the inoculation of poor-quality soils with beneficial fungi alongside their mycorrhizal partners to promote establishment rates. Preserving living fungal collections within botanical institutions represents a developing approach to ensure the availability of fungal resources for future propagation initiatives (Asmelash et al. 2016).

OMF has an important role as a principal partner in the conservation of orchids. They involve ex situ propagation and in situ restoration processes by ensuring that orchids are ecologically prepared to survive outside controlled environments (Zhao et al. 2021). An integrated conservation plan incorporating both the orchids and their mycorrhizal fungi offers the best way for orchid conservation in the wild environment (McCormick et al. 2018).

8.7 Conclusion

Orchids are subjected to significant threats due to habitat loss, overexploitation, and climate change, emphasizing the desperate need for effective conservation strategies. Survival for them relies on an obligate symbiotic relationship with orchid mycorrhizal fungi (OMF), which belong to the families Tulasnellaceae, Ceratobasidiaceae, and Serendipitaceae. The fungi established colonies in the form of pelotons in the cortical cells of orchid roots, providing the necessary carbon, nitrogen, phosphorus, and water needed for seed germination, seedling establishment, and overall plant development.

Aside from their nutritional value, OMF also enhances the resilience of orchids, modulates their defense system, and induces growth through bioactive compounds, making this symbiotic association one of the most specialized plant–fungal interactions in nature. Significantly, OMF integrates ex situ propagation with in situ restoration, preparing orchids to thrive naturally. Therefore, the preservation of orchids needed the protection of their fungal partners. The long-term sustainability of the orchids was relying on their mutualistic development, enhances soil applications, and enhances habitat restoration.

References

Agarwal M, Shrivastava N, Padh H (2008) Advances in molecular marker techniques and their applications in plant sciences. Plant Cell Rep 27:617–631. https://doi.org/10.1007/s00299-008-0507-z

Asmelash F, Bekele T, Birhane E (2016) The potential role of arbuscular mycorrhizal fungi in the restoration of degraded lands. Front Microbiol 7:1095. https://doi.org/10.3389/fmicb.2016.01095

Bairu MW, Aremu AO, Van Staden J (2011) Somaclonal variation in plants: causes and detection methods. Plant Growth Regul 63:147–173. https://doi.org/10.1007/s10725-010-9554-x

Bellgard S, Williams S (2011) Response of mycorrhizal diversity to current climatic changes. Diversity 3:8–90. https://doi.org/10.3390/d3010008

Bhuiyan M, Hossain M, Hossain K, Islam MN (2021) Isolation and identification of mycorrhizal fungus from an epiphytic orchid (*Rhynchostylis retusa* L. Bl.). Bangladesh J Bot 50:85–91. https://doi.org/10.3329/bjb.v50i1.52675

Brundrett MC, Scade A, Batty AL et al (2003) Development of *in situ* and *ex situ* seed baiting techniques to detect mycorrhizal fungi from terrestrial orchid habitats. Mycol Res 107:1210–1220. https://doi.org/10.1017/S0953756203008463

Cameron DD, Johnson I, Leake JR, Read DJ (2007) Mycorrhizal acquisition of inorganic phosphorus by the green-leaved terrestrial orchid *Goodyera repens*. Ann Bot 99:831–834. https://doi.org/10.1093/aob/mcm018

Chand K, Shah S, Sharma J et al (2020) Isolation, characterization, and plant growth-promoting activities of endophytic fungi from a wild orchid Vanda cristata. Plant Signal Behav 15: 1744294. https://doi.org/10.1080/15592324.2020.1744294

Chen D-Y, Wang X-J, Li T-Q et al (2021) *In situ* seedling baiting to isolate plant growth-promoting fungi from *Dendrobium officinale*, an over-collected medicinal orchid in China. Glob Ecol Conserv 28:e01659. https://doi.org/10.1016/j.gecco.2021.e01659

Correia M, Rodríguez-Echeverría S, Timóteo S et al (2019) Integrating plant species contribution to mycorrhizal and seed dispersal mutualistic networks. Biol Lett 15:20180770. https://doi.org/10.1098/rsbl.2018.0770

De Rose S, Kuga Y, Sillo F et al (2023) Plant and fungal gene expression coupled with stable isotope labeling provide novel information on sulfur uptake and metabolism in orchid mycorrhizal protocorms. Plant J 116:416–431. https://doi.org/10.1111/tpj.16381

Frank AB (1985) Über die auf Wurzelsymbiose beruhende Ernährung gewisser Bäume durch unterirdische Pilze. Berichte der Deutschen Botanischen Gesellschaft 3:128–145

Ghirardo A, Fochi V, Lange B et al (2020) Metabolomic adjustments in the orchid mycorrhizal fungus *Tulasnella calospora* during symbiosis with *Serapias vomeracea*. New Phytol 228: 1939–1952. https://doi.org/10.1111/nph.16812

Gómez-Vásquez R, Day R, Buschmann H et al (2004) Phenylpropanoids, phenylalanine ammonia lyase and peroxidases in elicitor-challenged cassava (*Manihot esculenta*) suspension cells and leaves. Ann Bot 94:87–97. https://doi.org/10.1093/aob/mch107

Gong M, Guan Q, Lin T, Lan J, Liu S (2018) Effects of fungal elicitors on seed germination and tissue culture of Cymbidium goeringii. AIP Conf Proc 1956:020045. https://doi.org/10.1063/1.5034297

Hemanta L, Ariina M, Mathukmi K, Devi MP (2021) Micropropagation of orchids. pp 379–383

Hossain M (2019) Morpho-molecular characterization of Ceratobasidium sp.: a mycorrhizal fungi isolated from a rare epiphytic orchid *Gastrochilus calceolaris* (J. E. Sm.) D. Don. Bangladesh J Plant Taxon 26:249–257. https://doi.org/10.3329/bjpt.v26i2.44584

Hossain M (2022) Orchid mycorrhiza: isolation, culture, characterization and application. S Afr J Bot 151:365–384. https://doi.org/10.1016/j.sajb.2022.10.003

Hossain MM, Rahi P, Gulati A, Sharma M (2013) Improved *ex vitro* survival of asymbiotically raised seedlings of *Cymbidium* using mycorrhizal fungi isolated from distant orchid taxa. Sci Hortic 159:109–116

Jiang J-H, Lee Y-I, Cubeta MA, Chen L-C (2015) Characterization and colonization of endomycorrhizal Rhizoctonia fungi in the medicinal herb Anoectochilus formosanus (Orchidaceae). Mycorrhiza 25:431–445. https://doi.org/10.1007/s00572-014-0616-1

Kaur S, Deb A (2025) The role of mycorrhizal fungi in orchids. Recent Pat Biotechnol 19:210–220. https://doi.org/10.2174/0118722083312186240822051057

Kendon JP, Yokoya K, Zettler LW et al (2020) Recovery of mycorrhizal fungi from wild collected protocorms of Madagascan endemic orchid *Aerangis ellisii* (B.S. Williams) Schltr. and their use in seed germination in vitro. Mycorrhiza 30:567–576. https://doi.org/10.1007/s00572-020-00971-x

Lee E-H, Eo J-K, Ka K-H, Eom A-H (2013) Diversity of arbuscular mycorrhizal fungi and their roles in ecosystems. Mycobiology 41:121–125. https://doi.org/10.5941/MYCO.2013.41.3.121

Leng C, Hou M, Xing Y, Chen J (2024) Perspective and challenges of mycorrhizal symbiosis in orchid medicinal plants. Chin Herb Med 16:172–179. https://doi.org/10.1016/j.chmed.2024.03.001

Li T, Yang W, Wu S et al (2021) Progress and prospects of mycorrhizal fungal diversity in orchids. Front Plant Sci 12. https://doi.org/10.3389/fpls.2021.646325

Matsubara T, Yoneda M, Ishii T (2012) Fungal isolate "KMI" is a new type of orchid mycorrhizal fungus. Am J Plant Sci 3:1121–1126. https://doi.org/10.4236/ajps.2012.38135

McCormick MK, Whigham DF, Canchani-Viruet A (2018) Mycorrhizal fungi affect orchid distribution and population dynamics. New Phytol 219:1207–1215. https://doi.org/10.1111/nph.15223

Novotná A, Mennicken S, de Paula CCP et al (2023) Variability in nutrient use by orchid mycorrhizal fungi in two medium types. J Fungi (Basel) 9:88. https://doi.org/10.3390/jof9010088

Pandey M, Sharma J, Taylor DL, Yadon V (2013) A narrowly endemic photosynthetic orchid is non-specific in its mycorrhizal associations. Mol Ecol 22(8):2341–2354. https://doi.org/10.1111/mec.12249

Pecoraro L, Girlanda M, Liu Z-J et al (2015) Molecular analysis of fungi associated with the Mediterranean orchid Ophrys bertolonii Mor. Ann Microbiol 65:2001–2007. https://doi.org/10.1007/s13213-015-1038-9

Pereira MC, Vieira NM, Tótola MR, Kasuya MCM (2011) Total fatty acid composition in the characterization and identification of orchid mycorrhizal fungi Epulorhiza spp. Rev Bras Ciênc Solo 35:1159–1166. https://doi.org/10.1590/S0100-06832011000400009

Perotto S, Balestrini R (2024) At the core of the endomycorrhizal symbioses: intracellular fungal structures in orchid and arbuscular mycorrhiza. New Phytol 242:1408–1416. https://doi.org/10.1111/nph.19338

Roth-Bejerano N, Navarro-Ródenas A, Gutierrez A (2014) Types of mycorrhizal association. pp 69–80

Shah S, Shrestha R, Maharjan S et al (2019) Isolation and characterization of plant growth-promoting endophytic fungi from the roots of *Dendrobium moniliforme*. Plants 8:5. https://doi.org/10.3390/plants8010005

Shah S, Paudel MR, Thapa BB et al (2025) Extract from endophytic Fusarium isolates stimulates seed germination of the host and protocorm development of non-host orchids. Commun Integr Biol 18:2439798. https://doi.org/10.1080/19420889.2024.2439798

Shan XC, Liew ECY, Weatherhead MA, Hodgkiss IJ (2002) Characterization and taxonomic placement of Rhizoctonia-like endophytes from orchid roots. Mycologia 94:230–239

Sharma B, Rani J, Patra et al (2025) Saprophytic orchids of Asia: diversity & conservation. J Biodivers Conserv 9:41–49. https://doi.org/10.5281/zenodo.16946661

Siebers M, Brands M, Wewer V et al (2016) Lipids in plant-microbe interactions. Biochim Biophys Acta 1861:1379–1395. https://doi.org/10.1016/j.bbalip.2016.02.021

Sun X, Guo L-D (2012) Endophytic fungal diversity: review of traditional and molecular techniques. Mycology 3:65–76. https://doi.org/10.1080/21501203.2012.656724

Swarts ND, Sinclair EA, Francis A, Dixon KW (2010) Ecological specialization in mycorrhizal symbiosis leads to rarity in an endangered orchid. Mol Ecol 19:3226–3242. https://doi.org/10.1111/j.1365-294X.2010.04736.x

Tian F, Liao X-F, Wang L-H et al (2022) Isolation and identification of beneficial orchid mycorrhizal fungi in *Paphiopedilum barbigerum* (Orchidaceae). Plant Signal Behav 17:2005882. https://doi.org/10.1080/15592324.2021.2005882

Tupac Otero J, Ackerman JD, Bayman P (2002) Diversity and host specificity of endophytic Rhizoctonia-like fungi from tropical orchids. Am J Bot 89:1852–1858. https://doi.org/10.3732/ajb.89.11.1852

Xi G, Shi J, Li J, Han Z (2020) Isolation and identification of beneficial orchid mycorrhizal fungi in Bletilla striata (Thunb.) Rchb.f. (Orchidaceae). Plant Signal Behav 15:1816644. https://doi.org/10.1080/15592324.2020.1816644

Yang Q, Xu L, Xia W, Liang L, Bai X, Li L, Xu L, Liu L (2021) Mycorrhizal compatibility and germination-promoting activity of Tulasnella species in two species of orchid (Cymbidium mannii and Epidendrum radicans). Horticulturae 7(11):472. https://doi.org/10.3390/horticulturae7110472

Yeh C-M, Chung K, Liang C-K, Tsai W-C (2019) New insights into the symbiotic relationship between orchids and fungi. Appl Sci 9:585. https://doi.org/10.3390/app9030585

Yokoya K, Zettler LW, Bell J et al (2021) The diverse assemblage of fungal endophytes from orchids in Madagascar linked to abiotic factors and seasonality. Diversity 13:96. https://doi.org/10.3390/d13020096

Zettler L, Piskin K (2011) Mycorrhizal fungi from protocorms, seedlings and mature plants of the eastern prairie fringed orchid, *Platanthera leucophaea* (Nutt.) Lindley: a comprehensive list to augment conservation. Am Midl Nat:29–39. https://doi.org/10.1674/0003-0031-166.1.29

Zhang F, Lv Y, Zhao Y, Guo S (2013) Promoting role of an endophyte on the growth and contents of kinsenosides and flavonoids of *Anoectochilus formosanus* Hayata, a rare and threatened medicinal Orchidaceae plant. J Zhejiang Univ Sci B 14:785–792. https://doi.org/10.1631/jzus.B1300056

Zhao JN, Liu HX (2008) Effects of fungal elicitors on the protocorm of Cymbidium eburneum. Eco Sci 27(2):134–137

Zhao D-K, Selosse M-A, Wu L et al (2021) Orchid reintroduction based on seed germination-promoting mycorrhizal fungi derived from protocorms or seedlings. Front Plant Sci 12. https://doi.org/10.3389/fpls.2021.701152

Zhou Z, Shi R, Zhang Y et al (2021) Orchid conservation in China from 2000 to 2020: achievements and perspectives. Plant Divers 43:343–349. https://doi.org/10.1016/j.pld.2021.06.003

Zhu GS, Yu ZN, Gui Y, Liu ZY (2008) A novel technique for isolating orchid mycorrhizal fungi. Fungal Divers 33:123–137

DNA Barcoding in Orchids as an Effective Tool for Species Identification

9

Pawan Singh Rana and Swetnisha

9.1 Introduction

The family Orchidaceae is one of the most evolutionarily advanced and species-rich families among the angiosperms, comprising more than 28,000 species distributed in over 850 genera worldwide (Hebert et al. 2003; Husen and Pant 2024). Members of this family occur in almost all habitats—from the humid tropics to alpine regions—displaying remarkable adaptive radiation and morphological plasticity (CB Plant 2009). Their sophisticated floral architecture, co-evolution with pollinators, and dependency on specific mycorrhizal fungi have long attracted the interest of taxonomists, conservation biologists, and evolutionary ecologists (Kress and Erickson 2012a, b). India alone harbors more than 1300 orchid species, a majority of which are endemic and ecologically specialized (Chase et al. 2015).

Though one of the most diverse families of plant kingdom, accurate species identification in orchid is a notoriously arduous task; the reason being the dependency of traditional morphological taxonomy on floral characteristics, often affected by phenotypic plasticity, convergent evolution, and natural hybridization, leading to misidentifications or unresolved species complexes (Dressler 1993; Bateman et al. 2018; Freudenstein 2025). Moreover, many orchids flower infrequently, and vegetative characters are often insufficient for reliable discrimination. These challenges are compounded by the illegal trade of rare and medicinally important orchids, which threatens wild populations and underscores the necessity of robust identification tools for conservation and regulatory enforcement (Kothakota et al. 2024).

P. S. Rana (✉)
Department of Biotechnology, Hemvati Nandan Bahuguna Garhwal University (A Central University), Srinagar Garhwal, India

Swetnisha
Department of Microbiology, Patna Women's College, Patna, India

A. Husen, M. Pant (eds.), *Forest-based Orchids*,
https://doi.org/10.1007/978-981-92-0096-2_9

In modern era, the discipline of biological systematics has taken a novel course of species identification based on molecular approaches of determining specialized short DNA sequence, often known as DNA barcoding (Hubert and Hanner 2015; Adhikari et al., 2017; Antil et al. 2023a, b). Originally proposed by Hebert et al. (2003) for animal species identification using the mitochondrial cytochrome c oxidase I (COI) gene (Hebert et al. 2003; Tobe and Linacre 2009), the concept was later adapted for plants through plastid and nuclear markers due to the slow evolution of plant mitochondrial genes (Dawnay et al. 2007; Kim and Jung 2018). The Consortium for the Barcode of Life (CBOL) Plant Working Group (2009) recommended two chloroplast loci—rbcL and matK—as the core plant barcode regions (Haider 2011a, b; Chac and Thinh 2023). Subsequent studies demonstrated the utility of nuclear internal transcribed spacer regions (ITS, ITS2) and intergenic spacers such as trnH–psbA in improving discriminatory power among closely related taxa (Pang et al. 2012; Tripathi et al. 2013; Vu et al. 2017; Srivastava and Manjunath 2020). In orchids, DNA barcoding has proven particularly valuable. Research across diverse genera such as *Dendrobium*, *Bulbophyllum*, *Paphiopedilum*, *Vanda*, *and Vanilla* revealed that barcode combinations (matK + ITS2 or rbcL + ITS2) significantly enhance resolution and enable the identification of cryptic or morphologically similar species (Xu et al. 2015; Praveen et al. 2017; Rajaram et al. 2019; Raskoti and Ale 2021; Su'udi et al. 2025). Furthermore, DNA barcoding assists in monitoring illegal trade, authenticating medicinal orchids like *Gastrodia elata* and *Vanilla planifolia*, and supporting conservation programs by enabling precise mapping of species distributions (Veldman et al. 2018; Tiwari et al. 2024).

Despite its effectiveness, DNA barcoding in orchids is not without challenges. Issues such as low interspecific sequence divergence, incomplete reference databases, and amplification difficulties due to secondary metabolites in orchid tissues persist (Chattopadhyay et al. 2017; Bhatt 2019; Raskoti and Ale 2021). Nevertheless, with advances in next-generation sequencing (NGS) and multilocus or whole-plastome barcoding, the resolution of species boundaries is continually improving (Ji et al. 2021; Fu et al. 2024a, b).

This chapter explores the principles, methodologies, applications, and future prospects of DNA barcoding in orchids, emphasizing its role as a reliable tool for accurate species identification and conservation. By integrating molecular and morphological data, DNA barcoding contributes to a more stable and informed taxonomy of one of the planet's most charismatic plant families.

9.2 Diversity and Distribution of Orchids

The Orchidaceae is among the largest families of angiosperms, with an estimated 25,000–30,000 species in over 800 genera globally Christenhusz and Byng (2016). Orchids occupy a wide range of habitats—from tropical rainforests and cloud forests to temperate zones and even alpine regions—and display remarkable morphological, ecological, and physiological diversity (Zhang et al. 2017). With remarkable diversity and ecological specilization orchhids are widely distributed in varying geograpical specification (Amom and Nongdam 2017). In India alone, more than 1300 species have been recorded, many of which are endemic to hot spots such as the Western Ghats and the northeastern Himalayas (Chatterjee 2006; Chitale et al. 2014). Orchids are also prominently represented in the Convention on International Trade in Endangered Species of Wild Fauna and Flora (CITES); they constitute a significant proportion of the appendix-listed taxa, reflecting pressures from trade and habitat loss (Secretariat CI 2019).

Recent global-scale mapping using AI models has further elucidated patterns of threat across orchid assemblages. Estopinan et al. (2024) used deep learning models on one million occurrence records of ~14,000 orchid species and demonstrated sharp spatial variation in conservation status, highlighting regions (e.g., Madagascar and Sumatra) with high proportions of threatened species (Zizka et al. 2021; Estopinan et al. 2024). Such analyses underscore not only orchid diversity but also their vulnerability and the need for robust identification tools across vast spatial scales.

9.3 Taxonomic Complexities in Orchids

Despite their charismatic diversity, orchids pose especially difficult challenges for taxonomists:

9.3.1 Phenotypic Plasticity and Convergence

Many orchid species show wide morphological variation depending on environmental factors (light, humidity, and elevation), which may blur species boundaries. Moreover, convergent evolution—especially in floral traits aimed at pollinator attraction—often leads to unrelated taxa evolving similar morphologies, complicating purely morphological classification (Papadopulos et al. 2013; Brundrett 2019; Zhang et al. 2022).

9.3.2 Natural Hybridization and Cryptic Species

Some orchids readily hybridize in nature, producing intermediates that are hard to place. Additionally, cryptic species—genetically distinct but morphologically very similar—are frequent in speciose genera, making morphological delimitations unreliable in many cases (Surveswaran et al. 2018; Evans et al. 2023).

9.3.3 Seasonality of Floral Characters

Most of the taxonomic keys rely heavily on floral structures (labellum shape, column, and pollinia), which are evident only during flowering. Specimens collected during the vegetative phase or dormant stage lack these diagnostic features, making morphological identification impossible (Bateman and Rudall 2023).

9.3.4 Limited Herbarium and Reference Material

Morphological features of Herbarium specimens may degrade over time (color fading and shape shrinkage), and for many orchids, especially rarer taxa, adequate reference specimens are lacking, thereby impeding comparative identification.

9.3.5 Geographic Variation and Subspecies Complexity

In many genera, geographically discrete populations may exhibit subtle morphological differences (e.g., in size and coloration) that may or may not correspond to distinct taxa, leading to taxonomic uncertainty. Added to this the morphological identification remain challenging in the closely related taxa (Al-juhani, 2019).

9.3.6 Molecular Interference from Secondary Compounds

The high concentration of secondary metabolites (phenolics and polysaccharides) in orchids often in leaf tissues inhibits PCR amplification or degrade DNA quality if extraction protocols are not optimized. Owing to these challenges, integrative taxonomy—blending molecular, morphological, ecological, and biogeographic data—has become the norm in recent orchid systematic studies (Pessoa et al. 2012; Burzacka-Hinz et al. 2025). DNA barcoding is a key molecular component of this integrative toolkit, providing molecular "anchors" for species delimitation where morphology or geography fails (Li et al. 2015).

9.4 Principles of DNA Barcoding

9.4.1 Concept and Evolution of DNA Barcoding

DNA barcoding is a molecular diagnostic approach that uses a short, standardized region of the genome to identify species through sequence comparison against reference databases. The concept was first introduced by Hebert et al. (2003) for animals using the mitochondrial cytochrome c oxidase I (COI) gene, providing a simple, universal framework for distinguishing species across large taxonomic groups (Kress and Erickson 2012a, b). However, due to the slow evolutionary rate and structural complexity of plant mitochondrial DNA, this marker proved unsuitable for plants (Roy et al. 2010; Nakayama et al. 2015).

In response, the Consortium for the Barcode of Life (CBOL) Plant Working Group (2009) proposed the use of two plastid genes—rbcL and matK—as the core barcoding loci for land plants (CBOL Plant Working Group 1 2009). CBOL plant working group recommendds these two as the core plant barcode regions (CBOL plant working group, 2024). These markers offer high universality and ease of amplification, although their discriminatory power is sometimes limited among closely related taxa. To overcome these limitations, several supplementary loci—including the nuclear ribosomal internal transcribed spacer (ITS and ITS2), and the plastid intergenic spacer trnH–psbA—have been introduced to enhance resolution (Schoch et al. 2012; Dong et al. 2015; Mishra et al. 2016).

DNA barcoding operates under the principle that interspecific genetic divergence exceeds intraspecific variation, often visualized as the "barcode gap." Species identification is performed through sequence similarity searches (e.g., BLAST, BOLD Identification Engine) or tree-based clustering approaches such as neighbor-joining (NJ) or maximum likelihood (ML) phylogenetic analyses (Kress 2017; Phillips et al. 2022; Wilson et al. 2023). Recent methodological improvements—like next-generation sequencing (NGS) and DNA metabarcoding—allow simultaneous analysis of multiple species from bulk or environmental samples, greatly expanding the applicability of barcoding in ecology and conservation (Taberlet et al. 2012; Abdelaziz et al. 2024; Letsiou et al. 2024). Understandng of the molecular level diversity helps in biodiversity conservation which is a sustainable developmental goal (FAO 2025) helps in maitaining the species richness.

9.4.2 Criteria for an Ideal Plant Barcode

An ideal barcode locus must satisfy several criteria:

1. Universality—Conserved primer binding sites enabling amplification across a wide range of taxa.
2. Sequence Quality and Alignment—High-quality sequences that can be reliably aligned across species.
3. Discriminatory Power—Sufficient interspecific divergence to separate species.
4. Short Length—Easy amplification even from degraded DNA samples (important for herbarium or traded materials).
5. Standardization and Reproducibility—Global acceptance and consistency across laboratories (Kress and Erickson 2008; Lucas et al. 2012).

Among available markers, chloroplast genes (matK, rbcL) are the most universally used due to stable structure and low recombination rates, while nuclear markers (ITS2) provide higher species-level resolution because of faster mutation rates (Praveen et al. 2017; Fu et al. 2024a, b).

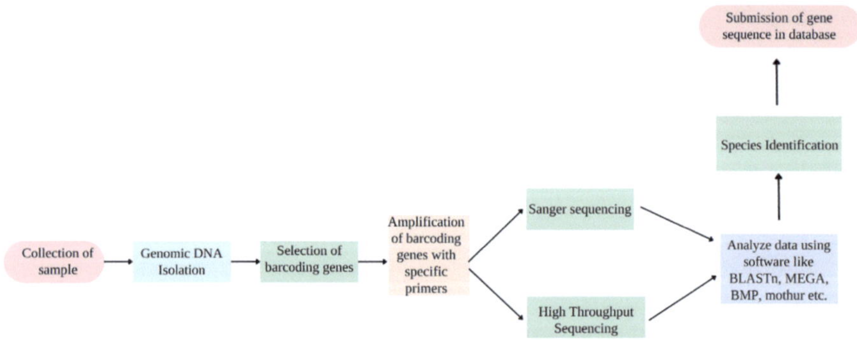

Fig. 9.1 Schematic representation of the DNA barcoding workflow in orchids—from sample collection to species identification through sequencing and database comparison

9.4.3 DNA Barcoding Workflow

The general workflow of DNA barcoding in orchids comprises four main steps (Fig. 9.1):

1. ***Sample Collection and Preservation***
   Fresh leaf or floral tissues are collected from authenticated specimens or field populations. Silica gel-dried or ethanol-preserved samples are preferred to maintain DNA integrity. Voucher specimens are deposited in recognized herbaria for future reference (Prendini et al. 2013).
2. ***DNA Extraction***
   Genomic DNA is extracted using modified cetyltrimethylammonium bromide (CTAB) protocols or commercial plant DNA kits optimized for orchids rich in polysaccharides and phenolics (Kamba and Deb 2018).
3. ***PCR Amplification and Sequencing***
   Target regions such as matK, rbcL, ITS2, and trnH–psbA are amplified using universal primers. Amplified products are visualized by gel electrophoresis and sequenced using Sanger sequencing or, increasingly, Illumina/ONT-based NGS for multilocus analysis (Gostel et al. 2020).
4. ***Sequence Alignment and Identification***
   Raw sequences are trimmed and aligned using tools such as ClustalW, MUSCLE, or MEGA X. Species identity is determined through comparison with databases such as NCBI GenBank or the Barcode of Life Data Systems (BOLD) (Ismail 2022; Obembe et al. 2025). Phylogenetic trees (NJ, ML, Bayesian) are constructed to confirm monophyletic clustering of conspecific sequences (Rannala and Yang 2017; Balaban et al. 2019).

9.4.4 Current Advances and Standards

The BOLD Systems v4 platform (2023 update) and GenBank Plant Reference Dataset (RefSeq 2024) now integrate more than 450,000 plant barcode records, including over 3000 orchid species (Ratnasingham et al. 2024). Recent developments such as whole-plastome barcoding, target enrichment sequencing, and AI-assisted taxonomic prediction have improved accuracy and processing speed (Song et al. 2020; Ahmed et al., 2022; Kutuzova et al. 2024). For example, Wang et al. (2023) demonstrated that machine learning algorithms trained on ITS2 + matK sequences could predict orchid species identity with >98% accuracy (de Medeiros et al. 2025; Shelke et al. 2025).

Thus, DNA barcoding represents a robust, scalable, and evolving system bridging molecular genetics and classical taxonomy—providing an indispensable foundation for orchid research and conservation in the genomic era.

9.5 Methodology

9.5.1 Sample Collection and Preservation

Accurate DNA barcoding begins with authentic and traceable sampling. Orchid specimens should be collected from verified field populations, botanical gardens, or herbaria. Each specimen must be accompanied by detailed metadata—collection site, GPS coordinates, habitat type, collector name, and voucher number—to ensure reproducibility and traceability (Howard et al. 2020; Kress and Gostel 2022). Tissues such as young leaves or floral parts are ideal, as they contain high-quality DNA with fewer secondary metabolites. Samples are typically preserved in silica gel or 95% ethanol for field collection, whereas laboratory samples can be stored at −20 °C for short-term or −80 °C for long-term preservation (Čertner et al. 2022). Voucher specimens should be prepared and deposited in a recognized herbarium (e.g., Kew, CAL, or CFTRI Herbarium) following standard taxonomic procedures. Linking molecular data to physical vouchers ensures long-term validation and traceability of barcodes (Sullivan and Nazaire 2022; Shweta et al. 2024).

9.5.2 DNA Extraction

DNA extraction from orchid tissues poses unique challenges due to high concentrations of polysaccharides, phenolic compounds, and secondary metabolites, which can co-precipitate with DNA and inhibit PCR. The CTAB method remains the most widely used approach, with modifications involving polyvinylpyrrolidone (PVP) or β-mercaptoethanol to remove phenolics (Heikrujam et al. 2020; Lal et al. 2023). Alternative commercial kits (e.g., Qiagen DNeasy Plant Mini Kit and NucleoSpin Plant II Kit) are also effective, particularly for processed or dried samples. Extracted DNA quality is verified using agarose gel electrophoresis (1%)

Table 9.1 Commonly used barcode loci and primer sequences in orchids

Marker	Primer name	Sequence (5′–3′)
rbcL	rbcLa-F	ATGTCACCACAAACGAACAAG
	rbcLa-R	GTAAAATCAAGTCCACCRCG
matK	matK-390F	CGATCTATTCATTCAATATTC
	matK-1326R	TCTAGCACACGAAAGTCGAAGT
trnH-psbA	trnH-F	CGCGCATGGTGGATTCACAATCC
	psbA-R	GTTATGCATGAACGTAATGCTC
ITS2	ITS-S2F	ATGCGATACTTGGTGTGAAT
	ITS-S3R	GACGCTTCTCCAGACTACAAT

and quantified by spectrophotometric ratios (A260/A280 ≈ 1.8) using a Nanodrop or Qubit fluorometer (Lee et al. 2012).

9.5.3 PCR Amplification

DNA barcoding relies on polymerase chain reaction (PCR) amplification of target loci using universal primer pairs (Table 9.1).

Amplification success is checked on 1.5% agarose gels stained with ethidium bromide or SYBR Safe and visualized under UV light. PCR products are purified using ExoSAP-IT or commercial column kits prior to sequencing (Gabriyan and Avashia 2013).

9.5.4 Sequencing and Data Processing

Purified PCR products are subjected to bidirectional Sanger sequencing, ensuring sequence accuracy through consensus alignment. In large-scale projects, high-throughput Illumina or Oxford Nanopore sequencing may be employed for multilocus datasets or plastome-wide barcoding (Shendure et al. 2011; Namias et al. 2023).

Raw chromatograms are visualized and trimmed using Chromas, Geneious Prime, or MEGA X. Consensus sequences are assembled from forward and reverse reads, aligned with reference sequences using ClustalW or MUSCLE, and manually corrected where necessary (Smith 2015; Al-Shuhaib and Hashim 2023).

9.5.5 Sequence Analysis and Species Identification

Sequence identity is confirmed using BLASTn (NCBI) and BOLD Identification System searches. Species-level resolution is assessed via:

Table 9.2 Tools for genomic analysis

Application	Tool/database	Description
Sequence editing and alignment	MEGA X, MUSCLE, Geneious	Quality check and multiple sequence alignment
Distance analysis	MEGA X, PAUP*, Kimura-2-P	Calculate pairwise divergences
Tree building	IQ-TREE, RAxML, MrBayes	Phylogenetic inference
Species identification	BOLD IDS, BLASTn	Match barcode sequences to reference datasets
Data deposition	GenBank, BOLD Systems v4	Public repository with metadata integration

- Genetic distance analysis (Kimura-2-Parameter model)
- Phylogenetic reconstruction (Neighbor-Joining, Maximum Likelihood, or Bayesian methods)
- Barcode gap analysis to differentiate intra- vs. interspecific variation (Mahadani et al. 2016; Sananka 2024)

For visualization, phylogenetic trees are constructed in MEGA X, IQ-TREE, or RAxML, with bootstrap support $\geq$70% considered statistically robust. Sequences and metadata are deposited in GenBank and BOLD repositories following FAIR data principles (Li et al. 2016) (Table 9.2).

9.5.6 Quality Assurance

To ensure reliability and reproducibility:

- Include negative controls in each PCR batch to detect contamination.
- Perform duplicate amplifications for critical specimens.
- Deposit voucher and sequence metadata together in global databases (GBIF, BOLD).
- Follow MIxS and CBOL data submission standards.

9.6 Case Studies and Applications in Orchids

9.6.1 Authentication of Medicinal and Commercial Orchids

Many orchid species are used in traditional medicine and commercial trade, yet they are frequently misidentified or adulterated. Molecular markers have improved taxonomica resilution (Ge et al. 2021). DNA barcoding has proven invaluable for verifying authenticity and ensuring traceability of raw materials. A recent study by

Wang et al. (2017) employed ITS2 and matK barcoding to authenticate Dendrobium species sold in herbal markets in China. The analysis identified several cases of substitution with non-medicinal taxa, highlighting the economic and safety implications of accurate molecular identification (Xu et al. 2015). Similarly, Sayem et al. (2024) successfully used rbcL + ITS2 barcodes to authenticate *Vanda tessellata* and *Rhynchostylis retusa*, two valuable Indian medicinal orchids, from adulterants and morphologically similar species. Another noteworthy application involves the authentication of *Vanilla planifolia*, the primary source of natural vanilla flavor. Barcoding using trnH–psbA and ITS2 markers has effectively differentiated genuine *V. planifolia* from related species and hybrids (*V. tahitensis*, *V. pompona*), ensuring quality control in the flavor industry (Besse et al. 2020; Morton 2022). A more recent study by Lal et al. (2024) has established the use of DNA barcoding for species-level identification in two medicinally significant, epiphytic orchids *Aerides multiflora* and *Rhyncostylis retusa*.

9.6.2 Conservation and CITES Regulation

Due to their ornamental and medicinal value, orchids constitute nearly 70% of all plant species listed under CITES Appendix II (Hinsley et al. 2018). Illegal trade in wild-collected orchids has placed several species at risk of extinction. DNA barcoding enables enforcement agencies to identify confiscated or traded specimens, even when floral structures are absent. For instance, Phelps et al. (2016) demonstrated the use of matK + ITS2 barcodes to confirm the identity of illegally traded Paphiopedilum specimens in Southeast Asia, assisting authorities in enforcement actions and repatriation of confiscated material (Phillips et al. 2020; Nurse 2024). Similarly, Mahadani et al. (2022) used barcoding to identify wild Dendrobium species from tissue cultures in Indian nurseries, facilitating traceability and legal certification for commercial propagation (Tiwari et al. 2023). Recent developments in DNA mini-barcoding and metabarcoding approaches have further enhanced the applicability of molecular identification in forensic contexts. Mini-barcodes—short fragments (~150–300 bp)—are particularly useful for processed or degraded samples commonly found in illegal trade (Staats et al. 2016; Gao et al. 2019; Abdi et al. 2024).

9.6.3 Orchid Phylogenetics and Species Discovery

DNA barcoding has not only aided identification but has also illuminated phylogenetic relationships and cryptic diversity. Using multilocus barcodes (matK, rbcL, trnH–psbA, ITS2), Chase et al. (2015) reconstructed the phylogeny of 300 orchid taxa, resolving several ambiguous clades and supporting the reclassification of *Bulbophyllum* and *Coelogyne* groups (Raskoti and Ale 2021). In another landmark study, Averyanov et al. (2021) discovered new cryptic species of Dendrobium in Vietnam by combining DNA barcoding with morphometric analyses. Molecular

Table 9.3 Applications of DNA barcoding in Orchidaceae (2020–2025)

Application	Marker(s) used	Genus/species studied	Key outcome
Medicinal authentication	*ITS2, matK*	*Dendrobium officinale*	Detection of adulteration in herbal markets
Trade and CITES enforcement	*matK + ITS2*	*Paphiopedilum* spp.	Identification of illegally traded orchids
Species discovery	*ITS2 + trnH–psbA*	*Dendrobium* sp. nov.	Discovery of cryptic taxa
Phylogenetic resolution	*matK + rbcL + ITS2*	*Bulbophyllum, Vanda*	Clarified generic relationships
Conservation planning (India)	*rbcL + matK*	Himalayan orchids	Resolved misidentified taxa; established database

data revealed >3% divergence in ITS2 sequences compared to their nearest relatives, confirming their distinct taxonomic status (Nguyen et al. 2020). Further, DNA metabarcoding—an NGS-based extension—has been used to characterize orchid mycorrhizal associations from soil and root samples. Jacquemyn et al. (2017) demonstrated that specific fungal communities underpin the germination success of rare Epipactis orchids, revealing crucial insights into species ecology and conservation genetics (Davis et al. 2015).

9.6.4 Regional Applications in India

India, being a megadiversity hot spot for orchids, has seen growing adoption of barcoding tools for conservation management. Many Himalayan orchid species have been barcoded using rbcL, matK, and ITS2, producing the first region-wide molecular reference database (Srivastava and Manjunath 2020; Jyothsna et al. 2023; Mahadani et al. 2022). These results revealed multiple mislabeled herbarium specimens and resolved several taxonomic ambiguities in Calanthe and Habenaria (Zhai et al. 2014; Pandey and Jin 2021; Mahadani et al. 2022). Additionally, combining DNA barcoding with ecological niche modeling to map the potential distribution of threatened orchids in Western ghats of India, aided in identifying priority conservation zones (Jalal and Singh 2017). This integrative approach illustrates the practical role of barcoding in biodiversity planning under climate change scenarios (Table 9.3).

9.6.5 Implications and Future Relevance

These applications demonstrate that DNA barcoding is not limited to taxonomy—it is a cross-disciplinary tool underpinning conservation biology, phytochemistry, forensic botany, and sustainable trade. With the ongoing development of AI-based barcode classifiers, portable sequencing platforms (e.g., MinION), and blockchain-integrated biodiversity databases, orchid identification is entering an era of real-time molecular verification (Kothakota et al. 2024). By integrating traditional taxonomy with molecular evidence, barcoding ensures that orchid biodiversity is accurately documented, conserved, and legally protected, fulfilling both scientific and societal imperatives.

9.7 Advantages and Limitations of DNA Barcoding in Orchids

9.7.1 Advantages

DNA barcoding represents one of the most transformative molecular tools in modern plant systematics and biodiversity monitoring. For the Orchidaceae, where morphological and ecological variability are exceptionally high, the method offers several unique advantages:

9.7.1.1 Accurate and Rapid Species Identification

DNA barcoding allows reliable discrimination even among morphologically similar or cryptic species. It is particularly effective for sterile or juvenile orchids, where traditional morphology-based identification is not possible. Studies such as Lal et al. (2025) and Ho et al. (2021) demonstrated ultra-high accuracy in species-level discrimination using ITS2 and matK combinations (Srivastava and Manjunath 2020; Raskoti and Ale 2021).

9.7.1.2 Applicability Across Developmental and Processed Material

Unlike morphological keys that require flowering material, DNA barcoding can be performed using any tissue type—leaves, roots, pollen, or even processed herbal samples. This enables authentication of traded or degraded material, crucial for both CITES enforcement and pharmaceutical quality control (Hollingsworth 2009; Techen et al. 2014; Zhu et al. 2022).

9.7.1.3 Contribution to Conservation and Trade Regulation

Barcoding supports biodiversity conservation through species verification, monitoring of illegal trade, and detection of invasive or misdeclared taxa. Its integration into CITES enforcement workflows has made orchid identification feasible for customs authorities, even from minute or mixed tissue samples (Veldman et al. 2014; de Boer et al. 2017; Hinsley et al. 2018).

9.7.1.4 Enhancement of Phylogenetic and Evolutionary Studies

Molecular data from barcodes provide baseline information for constructing phylogenetic relationships. Combined with plastid genomes or nuclear data, they contribute to resolving complex evolutionary patterns and verifying generic delimitations within orchids (Kim et al. 2020; Freudenstein 2025).

9.7.1.5 Facilitation of Data Standardization and Sharing

The Barcode of Life Data Systems (BOLD v4, 2023) and GenBank RefSeq (2024) allow seamless global data sharing, with standardized metadata including voucher information, geolocation, and sequence trace files (Meiklejohn et al. 2019; Nakazato and Jinbo 2022). GBIF provides large-scale biodiversity data (GBIF 2024). This ensures transparency, reproducibility, and long-term accessibility of orchid genetic information.

9.7.1.6 Integration with Emerging Technologies

Recent advancements such as portable sequencing platforms (Oxford Nanopore MinION), AI-assisted species prediction models, and metabarcoding of environmental DNA (eDNA) samples have revolutionized field-based identification. These approaches enable real-time biodiversity assessment and community-level monitoring of orchids (Estopinan et al. 2024; Chang et al. 2025; Ullah et al. 2025).

9.8 Limitations and Challenges

While powerful, DNA barcoding also faces several technical, biological, and logistical limitations when applied to orchids:

9.8.1 Low Inter-Specific Divergence in Chloroplast Genes

Chloroplast barcodes (rbcL, matK) evolve slowly in orchids, resulting in insufficient sequence divergence between closely related species (Li et al. 2021). As a result, multiple loci or nuclear markers such as ITS2 are often required for species-level resolution (Dupuis et al. 2012). Despite its all utilities in some cases DNA barccoding faces challenges in resolving colosely related yet different species (Gueuning et al. 2019).

9.8.2 Hybridization and Polyploidy

Orchids frequently undergo natural hybridization, introgression, and polyploidization—processes that complicate barcode interpretation by producing incongruent plastid and nuclear signals. Hybrid taxa often cluster with either parent species in phylogenetic analyses, leading to ambiguous identification (Bersweden 2021; Evans et al. 2023).

9.8.3 Limited Reference Databases

Although over 3000 orchid species are represented in BOLD and GenBank, many remain unsequenced or misidentified. This reference gap reduces the reliability of barcode-based identification for under-studied genera, particularly in tropical regions (Ratnasingham and Hebert 2007; Srivastava and Manjunath 2020).

9.8.4 Technical Constraints in DNA Extraction and Amplification

High levels of polysaccharides and secondary metabolites (phenolics and alkaloids) in orchid tissues can inhibit PCR amplification or degrade DNA. Optimized extraction protocols are necessary but increase cost and complexity (Hong et al. 2024).

9.8.5 Intraspecific Variation and Database Errors

In some orchids, especially those with wide geographic ranges, intraspecific variation may overlap with interspecific divergence, confounding the "barcode gap." Moreover, erroneous sequences in public databases due to mislabeling or contamination can propagate identification errors (Ahmed 2022).

9.8.6 Cost, Expertise, and Infrastructure

Although costs have declined, molecular barcoding still requires laboratory infrastructure, trained personnel, and bioinformatics expertise, which may not be readily available in many biodiversity-rich but resource-limited regions (Zhivkoplias et al. 2025).

9.9 Addressing Limitations Through Integrative Approaches

To overcome these challenges, researchers are adopting integrative taxonomy, combining molecular, morphological, ecological, and geographical data. Multilocus or genome-scale barcoding (e.g., whole-plastome sequencing, target-capture) has improved resolution dramatically (Barrett et al. 2022; Karbstein et al. 2024). The Plant Working Group of CBOL (2024 update) recommends combining rbcL, matK, ITS2, and trnH–psbA as a multi-marker barcode panel for orchids, offering the best balance between universality and specificity (Chen et al. 2024). Furthermore, AI-driven classification models (e.g., neural networks trained on barcode datasets) are now capable of distinguishing subtle sequence patterns beyond human perception, providing near-complete automation of species identification (Fan et al. 2025) (Table 9.4).

Table 9.4 Comparative analysis of the applicability of DNA barcoding in orchid species identification

Strengths	Limitations	Future solutions
Accurate, standardized species identification	Low sequence divergence in chloroplast loci	Use multi-locus or genome-wide barcodes
Applicable to sterile, processed, or traded samples	Hybridization causes ambiguous results	Integrative taxonomy with morphological and ecological data
Supports conservation and trade enforcement	Incomplete reference databases	Expand global sequencing efforts
Compatible with AI and portable devices	PCR inhibition in secondary metabolite-rich tissues	Use DNA-cleaning reagents and robust polymerases

9.10 Future Prospects

9.10.1 Emerging Trends in DNA Barcoding of Orchids

The rapid evolution of molecular and computational tools has greatly expanded the scope of DNA barcoding beyond single-locus approaches. Recent years (2022–2025) have witnessed a paradigm shift toward integrative, high-throughput, and real-time identification systems, particularly relevant for a complex plant family like Orchidaceae.

9.10.2 Next-Generation and Whole-Plastome Barcoding

The development of next-generation sequencing (NGS) technologies has enabled complete plastome sequencing, providing hundreds of loci for species discrimination and phylogenetic resolution. Demonstrated that plastome-scale barcoding in Dendrobium yielded >99% discrimination accuracy among 78 species, outperforming traditional markers (Xu et al. 2015). Similarly, Lei et al. (2024) used target-capture sequencing of nuclear exons to resolve the long-standing Vanda–Aerides taxonomic complex, illustrating the power of genome-informed taxonomy (Tsai et al. 2017).

9.10.3 Environmental DNA (eDNA) and Metabarcoding

DNA metabarcoding, based on bulk environmental samples (soil, root microbiome, and air), is revolutionizing orchid ecology and conservation. Using high-throughput sequencing of ITS2 and trnH–psbA, Jacquemyn et al. (2015) characterized orchid–mycorrhizal fungal networks across 60 European sites, revealing fine-scale species–fungus specificity critical for habitat restoration (Wang et al. 2024). Such approaches also facilitate non-invasive monitoring of orchid diversity in inaccessible or protected ecosystems (Rasmussen and Rasmussen 2018).

9.10.4 Portable Sequencing and Real-Time Field Identification

The advent of portable sequencers like the Oxford Nanopore MinION and Flongle platforms has brought DNA barcoding to the field. These devices enable on-site sequencing and identification within hours, eliminating dependence on centralized laboratories. Srivathsan et al. (2018) successfully applied MinION barcoding to identify *Bulbophyllum* species in Borneo rainforests in real time, setting a precedent for molecular ecology expeditions (Srivathsan et al. 2018; Su'udi et al. 2025).

9.10.5 AI- and Machine Learning-Assisted Taxonomy

Machine learning models trained on large DNA-barcode datasets (e.g., matK, ITS2) are now being explored for orchid species classification and have shown high accuracy in pilot studies—though genus-wide datasets of 200+ species remain rare and published accuracy figures vary. Such systems hold promise to automatically flag anomalous sequences or detect hybridization signatures, enhancing reproducibility and robustness (Yang et al. 2021; Kothakota et al. 2024; Nanni et al. 2024).

9.10.6 Digital Databases and Policy Integration

Global repositories like BOLD Systems v4 (2024), GenBank RefSeq Plant Dataset, and GBIF Molecular Portal are integrating genomic, geographic, and phenotypic data into unified bioinformatics pipelines (Ratnasingham et al. 2024). Linking these with CITES trade databases and biodiversity policy frameworks can enable real-time monitoring of trade and conservation compliance (Nakazato and Jinbo 2022). Projects like Global Orchid Genome Initiative (GOGI 2025) are building open-access barcode libraries covering >10,000 orchid taxa, forming the backbone for legal and scientific verification worldwide (Andersson et al. 2021).

9.10.7 Integrative Taxonomy: The Future Paradigm

The consensus emerging among taxonomists is that no single dataset can define species boundaries. Hence, integrative taxonomy—which combines molecular barcodes with morphological, ecological, and chemotaxonomic data—is now the gold standard for orchid systematics (Dayrat 2005; Li et al. 2012; Chase et al. 2015. In this framework, DNA barcoding serves as the molecular anchor, enabling cross-validation of morphological and ecological observations. Such synergy between disciplines ensures a more holistic understanding of orchid biodiversity. Moreover, citizen science initiatives and mobile apps using barcode-linked QR codes can democratize biodiversity monitoring. With portable sequencing, field ecologists, forest departments, and even trained volunteers can contribute validated genetic data to global repositories, fostering participatory conservation (Germain et al. 2021; Odah 2024).

9.11 Future Challenges and Opportunities

While the future of orchid DNA barcoding is promising, several key challenges must be addressed:

- Expansion of reference libraries for underrepresented genera and regions (Fay 2018; Raskoti and Ale 2021).
- Development of cost-effective, field-ready extraction kits, and sequencing workflows (Watsa et al. 2020).
- Integration of barcode data with ecological modeling to predict species distribution under climate change (Phillips et al. 2017).
- Ethical and data governance frameworks ensuring benefit-sharing under Nagoya Protocol compliance (Buck and Hamilton 2011; McManis and Kumar 2017).

By overcoming these barriers, DNA barcoding can evolve into a universal biodiversity language—linking taxonomy, ecology, trade regulation, and conservation biology in a unified digital ecosystem.

9.12 Conclusion

DNA barcoding has unequivocally transformed the way orchid species are identified, classified, and conserved. Its ability to deliver rapid, reliable, and reproducible molecular identification across life stages, and sample types make it indispensable in both research and regulatory domains. For orchids—one of nature's most intricate plant families—barcoding bridges the long-standing gap between morphological ambiguity and molecular precision. When integrated with next-generation sequencing, artificial intelligence, and open-access databases, it paves the way for real-time biodiversity authentication and data-driven conservation. Thus, the future of orchid taxonomy is not merely in herbarium drawers or greenhouses but also within digital genomes and global data networks. As we advance toward an era of genomic conservation, DNA barcoding will remain the cornerstone of species identity, ensuring that the elegance of orchids continues to be scientifically understood and ecologically preserved.

References

Abdelaziz SA, Khaled KA, Younis RA, Al-Kordy MA, El-Domyati FM, Moghazee MM (2024) Comparison of four DNA barcoding loci to distinguish between some Apiaceae family species. Beni-Suef Univ J Basic Appl Sci 13(1):12

Abdi G, Singh S, Selvakumar S, Dhar SK, Mudgal G, Swaminathan P, Jain M, Tarighat MA (2024) DNA barcoding and its applications. In: Advances in genomics: methods and applications. Springer Nature Singapore, Singapore, pp 91–117

Adhikari S, Saha S, Biswas A, Rana TS, Bandyopadhyay TK, Ghosh P (2017) Application of molecular markers in plant genome analysis: a review. Nucleus 60(3):283–297

Ahmed SS (2022) DNA barcoding in plants and animals: a critical review

Ahmed S, Ibrahim M, Nantasenamat C, Nisar MF, Malik AA, Waheed R, Ahmed MZ, Ojha SC, Alam MK (2022) Pragmatic applications and universality of DNA barcoding for substantial organisms at species level: a review to explore a way forward. Biomed Res Int 2022(1):1846485

Al-Juhani WS (2019) Evaluation of the capacity of the DNA barcode ITS2 for identifying and discriminating dryland plants. Genet Mol Res 18(1):32–46

Al-Shuhaib MB, Hashim HO (2023) Mastering DNA chromatogram analysis in Sanger sequencing for reliable clinical analysis. J Genet Eng Biotechnol 21(1):115

Amom T, Nongdam P (2017) The use of molecular marker methods in plants: a review. Int J Curr Res Rev 9(17):1–7

Andersson AA, Tilley HB, Lau W, Dudgeon D, Bonebrake TC, Dingle C (2021) CITES and beyond: illuminating 20 years of global, legal wildlife trade. Glob Ecol Conserv 26:e01455

Antil S, Abraham JS, Sripoorna S, Maurya S, Dagar J, Makhija S, Bhagat P, Gupta R, Sood U, Lal R, Toteja R (2023a) DNA barcoding, an effective tool for species identification: a review. Mol Biol Rep 50(1):761–775

Antil S, Malik S, Chhillar AK (2023b) DNA barcoding and phylogenetic assessment of medicinal orchids from northern India using ITS and matK markers. Plant Gene 34:100389

Averyanov LV, Maisak TV, Nguyen TL, Nguyen VK, Truong BV (2021) Dendrobium parahendersonii, a new orchid species (Orchidaceae) from Southern Vietnam. Taiwania 66(3): 360–363

Balaban M, Moshiri N, Mai U, Jia X, Mirarab S (2019) TreeCluster: clustering biological sequences using phylogenetic trees. PLoS One 14(8):e0221068

Barrett CF, Santee MV, Fama NM, Freudenstein JV, Simon SJ, Sinn BT (2022) Lineage and role in integrative taxonomy of a heterotrophic orchid complex. Mol Ecol 31(18):4762–4781

Bateman RM, Rudall PJ (2023) Morphological continua make poor species: genus-wide morphometric survey of the European bee orchids (Ophrys L.). Biology 12(1):136

Bateman RM, Sramkó G, Paun O (2018) Integrating restriction site-associated DNA sequencing (RAD-seq) with morphological cladistic analysis clarifies evolutionary relationships among major species groups of bee orchids. Ann Bot 121(1):85–105

Bersweden L. Species integrity and hybridization in anthropomorphic Orchis species (Orchidaceae). 2021

Besse P, Da Silva D, Grisoni M (2020) Plant DNA barcoding principles and limits: a case study in the genus Vanilla. In: Molecular plant taxonomy: methods and protocols. Springer US, New York, pp 131–148

Bhatt ID (2019) DNA barcoding in Himalayan orchids: progress, prospects, and limitations. Curr Sci 116(6):948–954

Brundrett MC (2019) The evolution and ecology of mycorrhizal associations in orchids. Ann Bot 123(6):1087–1102

Buck M, Hamilton C (2011) The Nagoya Protocol on access to genetic resources and the fair and equitable sharing of benefits arising from their utilization to the Convention on Biological Diversity. Rev Eur Comp Int Environ Law 20(1):47–61

Burzacka-Hinz A, Dudek M, Olędrzyńska N, Naczk AM, Szlachetko DL (2025) Evolution of morphological traits of Dendrobium sensu lato (Orchidaceae)—an attempt to resolve

phylogenetic relationships in nominal and morphologically convergent sections. BMC Plant Biol 25(1):239
CBOL (2024) Plant Working Group: updated standards for plant DNA barcoding. Consortium for the Barcode of Life (CBOL), Washington, DC
CBOL Plant Working Group (2009) A DNA barcode for land plants. Proc Natl Acad Sci USA 106(31):12794–12797
CBOL Plant Working Group (2024) Multi-locus plant barcoding strategy for orchids. Consortium for the Barcode of Life (CBOL), Washington, DC
CBOL Plant Working Group 1, Hollingsworth PM, Forrest LL, Spouge JL, Hajibabaei M, Ratnasingham S, van der Bank M, Chase MW, Cowan RS, Erickson DL, Fazekas AJ (2009) A DNA barcode for land plants. Proc Natl Acad Sci 106(31):12794–12797
Čertner M, Lučanová M, Sliwinska E, Kolář F, Loureiro J (2022) Plant material selection, collection, preservation, and storage for nuclear DNA content estimation. Cytometry A 101(9):737–748
Chac VT, Thinh TT (2023) Efficiency of DNA barcoding markers rbcL, matK, and ITS2 for identification of selected medicinal plant species. Biodiversitas 24(5):2579–2588
Chang H, Ye T, Xie Z, Liu X (2025) Application of environmental DNA in aquatic ecosystem monitoring: opportunities, challenges and prospects. Water 17(5):20734441
Chase MW, Cameron KM, Freudenstein JV, Pridgeon AM, Salazar G, Van den Berg C, Schuiteman A (2015) An updated classification of Orchidaceae. Bot J Linn Soc 177(2):151–174
Chatterjee D (2006) Endemic flora of India and its significance. Bull Bot Surv India 47(1–4):1–20
Chattopadhyay P, Sa G, Chakraborty A, Das S (2017) Challenges in DNA barcoding of orchids: influence of secondary metabolites on PCR amplification. J Plant Biochem Biotechnol 26(4): 392–399
Chen Z, Gao L, Wang H, Feng S (2024) Molecular identification and phylogenetic analysis of Cymbidium species (Orchidaceae) based on the potential DNA barcodes matK, rbcL, psbA-trnH, and internal transcribed spacer. Agronomy 14(5):933
Chitale VS, Behera MD, Roy PS (2014) Orchid distribution in India: conservation implications for endemic and rare species. J For Res 25(3):557–564
Christenhusz MJM, Byng JW (2016) The number of known plants species in the world and its annual increase. Phytotaxa 261(3):201–217
CITES Secretariat (2019) The CITES appendices I, II and III. Convention on International Trade in Endangered Species of Wild Fauna and Flora, Geneva
Davis BJ, Phillips RD, Wright M, Linde CC, Dixon KW (2015) Continent-wide distribution in mycorrhizal fungi: implications for the biogeography of specialized orchids. Ann Bot 116(3): 413–421
Dawnay N, Ogden R, McEwing R, Carvalho GR, Thorpe RS (2007) Validation of the COI gene for DNA barcoding across multiple animal phyla. PLoS One 2(10):e105
Dayrat B (2005) Towards integrative taxonomy. Biol J Linn Soc 85(3):407–417
de Boer HJ, Ghorbani A, Manzanilla V, Raclariu AC, Kreziou A, Ounjai S, Osathanunkul M, Gravendeel B (2017) DNA metabarcoding of orchid-derived products reveals widespread illegal orchid trade. Proc R Soc B Biol Sci 284(1863):20171182
de Medeiros BA, Cai L, Flynn PJ, Yan Y, Duan X, Marinho LC, Anderson C, Davis CC (2025) A composite universal DNA signature for the Tree of Life. Nat Ecol Evol 25:1–5
Dong W, Xu C, Li C, Sun J, Zuo Y, Shi S, Cheng T, Guo J, Zhou S (2015) ycf1, the most promising plastid DNA barcode of land plants. Sci Rep 5(1):8348
Dressler RL (1993) Phylogeny and classification of the orchid family. Cambridge University Press
Dupuis JR, Roe AD, Sperling FA (2012) Multi-locus species delimitation in closely related animals and fungi: one marker is not enough. Mol Ecol 21(18):4422–4436
Estopinan J, Servajean M, Bonnet P, Joly A, Munoz F (2024) AI-based mapping of the conservation status of orchid assemblages at global scale. arXiv preprint arXiv:2401.04691
Evans SA, Whigham DF, Hartvig I, McCormick MK (2023) Hybridization in the fringed orchids: an analysis of species boundaries in the face of gene flow. Diversity 15(3):384

Fan Z, Wang T, Huang K, Ying B, Zhou X (2025) Unleashing the power of computational insights in revealing the complexity of biological systems in the new era of spatial multi-omics. arXiv preprint arXiv:2509.13376

FAO (2025) Digital biodiversity governance and Nagoya Protocol compliance framework. FAO Biodivers Pol Pap 14:33–46

Fay MF (2018) Orchid conservation: how can we meet the challenges in the twenty-first century? Bot Stud 59(1):16

Freudenstein JV (2025) Orchid phylogenetics and evolution: history, current status and prospects. Ann Bot 135(5):805–822

Fu N, Xu Y, Jin L, Xiao TW, Song F, Yan HF, Chen YS, Ge XJ (2024a) Testing plastomes and nuclear ribosomal DNA sequences as the next-generation DNA barcodes for species identification and phylogenetic analysis in Acer. BMC Plant Biol 24(1):445

Fu Y, Luo Y, Zhao Y, Li Q, Jin X, Zhang L et al (2024b) Genome skimming and plastome-scale barcoding improve phylogenetic resolution in Dendrobium species. BMC Plant Biol 24(1):178

Gabriyan L, Avashia N (2013) Research techniques made simple: polymerase chain reaction (PCR). J Invest Dermatol 133(3):e6

Gao Z, Liu Y, Wang X, Wei X, Han J (2019) DNA mini-barcoding: a derived barcoding method for herbal molecular identification. Front Plant Sci 10:987

Ge Y, Xia C, Wang J, Zhang X, Ma X, Zhou Q (2021) The efficacy of DNA barcoding in the classification, genetic differentiation, and biodiversity assessment of benthic macroinvertebrates. Ecol Evol 11(10):5669–5681

Germain RM, Hart SP, Turcotte MM, Otto SP, Sakarchi J, Rolland J, Usui T, Angert AL, Schluter D, Bassar RD, Waters MT (2021) On the origin of coexisting species. Trends Ecol Evol 36(4):284–293

Global Biodiversity Information Facility (GBIF) (2024) Molecular data portal for Orchidaceae GBIF Dataset Report v1.2

Global Orchid Genome Initiative (GOGI) (2025) Global open-access orchid barcode database GOGI Technical Bulletin, v2.1

Gostel MR, Zúñiga JD, Kress WJ, Funk VA, Puente-Lelievre C (2020) Microfluidic Enrichment Barcoding (MEBarcoding): a new method for high throughput plant DNA barcoding. Sci Rep 10(1):8701

Gueuning M, Ganser D, Blaser S, Albrecht M, Knop E, Praz C, Frey JE (2019) Evaluating next-generation sequencing (NGS) methods for routine monitoring of wild bees: metabarcoding, mitogenomics or NGS barcoding. Mol Ecol Resour 19(4):847–862

Haider N (2011a) Identification of plant species using traditional and molecular-based methods. In: Davis RE (ed) Wild plants: identification, uses and conservation. Nova Science Publishers, Inc., New York, pp 1–62

Haider N (2011b) Plant DNA barcoding: from gene to genome. Biochem Syst Ecol 39(4–6): 188–191

Hebert PDN, Cywinska A, Ball SL, deWaard JR (2003) Biological identifications through DNA barcodes. Proc Natl Acad Sci USA 100(9):313–317

Heikrujam J, Kishor R, Mazumder PB (2020) The chemistry behind plant DNA isolation protocols. In: Biochemical analysis tools—methods for bio-molecules studies. IntechOpen

Hinsley A, De Boer HJ, Fay MF, Gale SW, Gardiner LM, Gunasekara RS, Kumar P, Masters S, Metusala D, Roberts DL, Veldman S (2018) A review of the trade in orchids and its implications for conservation. Bot J Linn Soc 186(4):435–455

Ho VT, Tran TK, Vu TT, Widiarsih S (2021) Comparison of matK and rbcL DNA barcodes for genetic classification of jewel orchid accessions in Vietnam. J Genet Eng Biotechnol 19(1):93

Hollingsworth PM (2009) A DNA barcode for land plants. Proc Natl Acad Sci USA 106:12569–12567

Hong BY, Driscoll M, Gratalo D, Jarvie T, Weinstock GM (2024) Improved DNA extraction and amplification strategy for 16S rRNA gene amplicon-based microbiome studies. Int J Mol Sci 25(5):2966

Howard C, Lockie-Williams C, Slater A (2020) Applied barcoding: the practicalities of DNA testing for herbals. Plants 9(9):1150
Hubert N, Hanner R (2015) DNA barcoding, species delineation and taxonomy: a historical perspective. DNA barcodes 3(1)
Husen H, Pant M (2024) Exploring medicinal orchids. Taylor & Francis Group, LLC, Boca Raton, FL. https://doi.org/10.1201/b23375
Ismail HD (2022) Bioinformatics: a practical guide to NCBI databases and sequence alignments. Chapman and Hall/CRC
Jacquemyn H, Brys R, Waud M, Busschaert P, Lievens B (2015) Mycorrhizal networks and coexistence in species-rich orchid communities. New Phytol 206(3):1127–1134
Jacquemyn H, Duffy KJ, Selosse MA (2017) Biogeography of orchid mycorrhizas. In: Biogeography of mycorrhizal symbiosis. Springer International Publishing, Cham, pp 159–177
Jalal JS, Singh P (2017) Ecological niche modelling for conservation of Habenaria Suaveolens Dalzell, an endangered orchid species endemic to Western Ghats: a case study. J Orchid Soc India 31:77–83
Ji Y, Liu DK, Yang JB, Jin WT, Xiang XG, Li DZ et al (2021) Plastid phylogenomics and super-barcode analyses resolve complex relationships within Orchidaceae. Front Plant Sci 12:692365
Jyothsna B, Sharma K, Srivastava D (2023) DNA barcoding of orchids in India: a systematic survey. J Environ Biosci 37(2)
Kamba JO, Deb CR (2018) A new simple and efficient DNA extraction protocol for orchid without liquid nitrogen and phenol. Plant Cell Biotechnol Mol Biol 19(3–4):143–147
Karbstein K, Kösters L, Hodač L, Hofmann M, Hörandl E, Tomasello S, Wagner ND, Emerson BC, Albach DC, Scheu S, Bradler S (2024) Species delimitation 4.0: integrative taxonomy meets artificial intelligence. Trends Ecol Evol 39(8):771–784
Kim HM, Jung J (2018) Evaluation of universal plant DNA barcode loci for identification of Korean plant species. Plants (Basel) 7(3):64
Kim YK, Jo S, Cheon SH, Joo MJ, Hong JR, Kwak M, Kim KJ (2020) Plastome evolution and phylogeny of Orchidaceae, with 24 new sequences. Front Plant Sci 11:22
Kothakota N, Behera B, Teja S, Routray S, Babu R, Behura N (2024) Advancements in Orchidaceae species identification: a comprehensive review of traditional and molecular methods. Int J Exp Res Rev 43:229–252
Kress WJ (2017) Plant DNA barcodes: applications today and in the future. J Syst Evol 55(4): 291–307
Kress WJ, Erickson DL (2008) DNA barcodes: genes, genomics, and bioinformatics. Proc Natl Acad Sci 105(8):2761–2762
Kress WJ, Erickson DL (2012a) DNA barcodes: methods and protocols. Methods Mol Biol 858:3–8
Kress WJ, Erickson DL (2012b) DNA barcodes: methods and protocols. In: DNA barcodes: methods and protocols. Humana Press, Totowa, NJ, pp 3–8
Kress WJ, Gostel MR (2022) Plant DNA barcodes, community ecology, and species interactions. Diversity 14(6):453
Kutuzova S, Nielsen M, Piera P, Nissen JN, Rasmussen S (2024) Taxometer: improving taxonomic classification of metagenomics contigs. Nat Commun 15(1):8357
Lal A, Pant M, Pant G, Palni LMS, Datta A, Kumar G (2023) ISSR marker assisted genetic diversity assessment in natural populations of two endemic orchids Aerides multiflora and Rhynchostylis retusa from Uttarakhand, India. S Afr J Bot 157:151–158. https://doi.org/10.1016/j.sajb.2023.03.031
Lal A, Pant M, Kumar A, Palni LMS, Singh A, Siddiqui MR, Husen A (2024) DNA barcoding of epiphytic orchids for species level identification. J Hortic Sci Biotechnol:1–11. https://doi.org/10.1080/14620316.2024.2371591
Lal A, Pant M, Kumar A, Palni LM, Singh A, Siddiqui MR, Husen A (2025) DNA barcoding of epiphytic orchids for species level identification. J Hortic Sci Biotechnol 100(1):127–137
Lee PY, Costumbrado J, Hsu CY, Kim YH (2012) Agarose gel electrophoresis for the separation of DNA fragments. J Vis Exp 62:3923

Lei W, Zhou P, Pei Z, Liu Y, Luo Y, Xiang X (2024) Plastome evolution and comparative analyses of a recently radiated genus Vanda (Aeridinae, Orchidaceae). Int J Mol Sci 25(17):9538

Letsiou S, Madesis P, Vasdekis E, Montemurro C, Grigoriou ME, Skavdis G, Moussis V, Koutelidakis AE, Tzakos AG (2024) DNA barcoding as a plant identification method. Appl Sci 14(4):1415

Li P, Pemberton R, Zheng G, Luo Y (2012) Fly pollination in Cypripedium: a case study of sympatric C. sichuanense and C. micranthum. Bot J Linn Soc 170(1):50–58

Li X, Yang Y, Henry RJ, Rossetto M, Wang Y, Chen S (2015) Plant DNA barcoding: from gene to genome. Biol Rev 90(1):157–166

Li MH, Zhang GQ, Lan SR, Liu ZJ, China Phylogeny Consortium (2016) A molecular phylogeny of Chinese orchids. J Syst Evol 54(4):349–362

Li H, Xiao W, Tong T, Li Y, Zhang M, Lin X, Zou X, Wu Q, Guo X (2021) The specific DNA barcodes based on chloroplast genes for species identification of Orchidaceae plants. Sci Rep 11(1):1424

Lucas C, Thangaradjou T, Papenbrock J (2012) Development of a DNA barcoding system for seagrasses: successful but not simple. PLoS One 7(1):e29987

Mahadani P, Trivedi S, Rehman H, Saggu S (2016) Bioinformatics tools in marine DNA barcoding. In: DNA barcoding in marine perspectives: assessment and conservation of biodiversity. Springer International Publishing, Cham, pp 71–85

Mahadani P, Chakrabarti S, Pal R, Jain SK, Sevanthi AM (2022) Application of DNA barcoding tool for authentication of traditionally important medicinal Dendrobium (Orchidaceae) species from Sikkim Himalaya. S Afr J Bot 146:905–911

McManis C, Kumar R (2017) Access and benefit sharing of genetic resources: issues and solutions. Asian Biotechnol Dev Rev 19(3):35–54

Meiklejohn KA, Damaso N, Robertson JM (2019) Assessment of BOLD and GenBank—their accuracy and reliability for the identification of biological materials. PLoS One 14(6):e0217084

Mishra P, Kumar A, Nagireddy A, Mani DN, Shukla AK, Tiwari R, Sundaresan V (2016) DNA barcoding: an efficient tool to overcome authentication challenges in the herbal market. Plant Biotechnol J 14(1):8–21

Morton WG (2022) Conservation of North America's Native Vanilla (Orchidaceae) species: population structure, isolation & identification of fungal endophytes, and mycobiome metabarcoding. The University of Wisconsin-Madison

Nakayama J, Watanabe K, Jiang J, Matsuda K, Chao SH, Haryono P, La-Ongkham O, Sarwoko MA, Sujaya IN, Zhao L, Chen KT (2015) Diversity in gut bacterial community of school-age children in Asia. Sci Rep 5(1):8397

Nakazato T, Jinbo U (2022) Cross-sectional use of barcode of life data system and GenBank as DNA barcoding databases for the advancement of museomics. Front Ecol Evol 10:966605

Namias A, Sahlin K, Makoundou P, Bonnici I, Sicard M, Belkhir K, Weill M (2023) Nanopore sequencing of PCR products enables multicopy gene family reconstruction. Comput Struct Biotechnol J 21:3656–3664

Nanni L, Cuza D, Brahnam S (2024) AI-powered biodiversity assessment: species classification via DNA barcoding and deep learning. Technologies 12(12):240

Nguyen NH, Vu HT, Le ND, Nguyen TD, Duong HX, Tran HD (2020) Molecular identification and evaluation of the genetic diversity of Dendrobium species collected in Southern Vietnam. Biology 9(4):76

Nurse A (2024) Illegal trafficking in plants and timber. In: Research handbook on environmental crimes and criminal enforcement. Edward Elgar Publishing, pp 227–245

Obembe OO, Obadire TT, Bello OA, Aworunse OS (2025) Phylogenetic analyses of selected underutilized members of the Fabaceae family based on DNA barcodes. In: Harnessing biotechnology tools for product development: selected papers from the international biotechnology conference exhibition and workshop (IBCEW 2024). Springer Nature Switzerland, Cham, pp 111–123

Odah M (2024) Advancements in DNA barcoding: revolutionizing taxonomy and biodiversity studies
Pandey TR, Jin XH (2021) Taxonomic revision of Habenaria josephi group (sect. Diphyllae sl) in the Pan-Himalaya. PhytoKeys 175:109
Pang X, Liu C, Shi L, Liu R, Liang D, Li H et al (2012) Utility of the trnH-psbA intergenic spacer region and its combinations as plant DNA barcodes: a meta-analysis. PLoS One 7(11):e48833
Papadopulos AST, Hollingsworth PM, Watson LE, Paun O (2013) Convergent evolution of floral signals underlies the success of Neotropical orchids. New Phytol 200(1):198–210
Pessoa EM, Alves M, Alves-Araújo A, Palma-Silva C, Pinheiro F (2012) Integrating different tools to disentangle species complexes: a case study in Epidendrum (Orchidaceae). Taxon 61(4): 721–734
Phelps J, Biggs D, Webb EL (2016) Tools and terms for understanding illegal wildlife trade. Front Ecol Environ 14(9):479–489
Phillips SJ, Anderson RP, Dudík M, Schapire RE, Blair ME (2017) Opening the black box: an open-source release of Maxent. Ecography 40(7):887–893
Phillips RD, Reiter N, Peakall R (2020) Orchid conservation: from theory to practice. Ann Bot 126(3):345–362
Phillips JD, Gillis DJ, Hanner RH (2022) Lack of statistical rigor in DNA barcoding likely invalidates the presence of a true species' barcode gap. Front Ecol Evol 10:859099
Praveen M, Parvathy VA, Sheeja TE, Sasikumar B (2017) DNA barcoding of four important Vanilla species in India: a step towards identification and conservation. Plant Gene 9:65–70
Prendini L, Hanner R, DeSalle R (2013) 11 Obtaining, storing and archiving. Techn Mol Systemat Evol 227(233):176
Rajaram R, Thamilarasan SK, Rajasekaran S, Manickavasagam M (2019) DNA barcoding and species delimitation of Bulbophyllum species from South India. PhytoKeys 124:37–52
Rannala B, Yang Z (2017) Efficient Bayesian species tree inference under the multispecies coalescent. Syst Biol 66(5):823–842
Raskoti BB, Ale R (2021) DNA barcoding of medicinal orchids in Asia. Sci Rep 11(1):23651
Rasmussen HN, Rasmussen FN (2018) The epiphytic habitat on a living host: reflections on the orchid–tree relationship. Bot J Linn Soc 186(4):456–472
Ratnasingham S, Hebert PD (2007) BOLD: the barcode of life data system (http://www.barcodinglife.org). Mol Ecol Notes 7(3):355–364
Ratnasingham S, Wei C, Chan D, Agda J, Agda J, Ballesteros-Mejia L, Boutou HA, El Bastami ZM, Ma E, Manjunath R, Rea D (2024) BOLD v4: a centralized bioinformatics platform for DNA-based biodiversity data. In: DNA barcoding: methods and protocols. Springer US, New York, pp 403–441
Roy S, Tyagi A, Shukla V, Kumar A, Singh UM, Chaudhary LB, Datt B, Bag SK, Singh PK, Nair NK, Husain T (2010) Universal plant DNA barcode loci may not work in complex groups: a case study with Indian Berberis species. PLoS One 5(10):e13674
Sananka NS (2024) Molecular identification and phylogenetic analysis of African nightshade using its 2 DNA barcoding. Doctoral dissertation, University of Nairobi
Sayem MA, Miah MS, Hossain MM (2024) Isolation and identification of endophytic fungi from the roots of two epiphytic orchids: Rhynchostylis retusa (L.) Bl. and Vanda tessellata (L.) Bl
Schoch CL, Seifert KA, Huhndorf S, Robert V, Spouge JL, Levesque CA, Chen W, Fungal Barcoding Consortium, Fungal Barcoding Consortium Author List, Bolchacova E, Voigt K (2012) Nuclear ribosomal internal transcribed spacer (ITS) region as a universal DNA barcode marker for Fungi. Proc Natl Acad Sci USA 109(16):6241–6246
Shelke RG, Lahane Y, Ingle P, Pusalkar P, Chauhan S, Pardhi P, Nirmal RC, Prajapati SK (2025) Advances in plant molecular taxonomy: techniques, challenges, and applications. Plant Cell Biotechnol Mol Biol 26(3–4):89–110
Shendure JA, Porreca GJ, Church GM, Gardner AF, Hendrickson CL, Kieleczawa J, Slatko BE (2011) Overview of DNA sequencing strategies. Curr Protoc Mol Biol 96(1):7–1

Shweta S, Dwivedi A, Subramaniam B, Kaushik S, Sahu N (2024) Herbaria: a valuable resource of the time treasured historic plant specimens with boundless research potential for environmental sustainability. Environ Dev Sustain:1–41

Smith DR (2015) Buying in to bioinformatics: an introduction to commercial sequence analysis software. Brief Bioinform 16(4):700–709

Song F, Li T, Burgess KS, Feng Y, Ge XJ (2020) Complete plastome sequencing resolves taxonomic relationships among species of Calligonum L. (Polygonaceae) in China. BMC Plant Biol 20(1):261

Srivastava D, Manjunath K (2020) DNA barcoding of endemic and endangered orchids of India: a molecular method of species identification. Pharmacogn Mag 16(70)

Srivathsan A, Baloğlu B, Wang W, Tan WX, Bertrand D, Ng AH, Boey EJ, Koh JJ, Nagarajan N, Meier R (2018) A Min ION™-based pipeline for fast and cost-effective DNA barcoding. Mol Ecol Resour 18(5):1035–1049

Staats M, Arulandhu AJ, Gravendeel B, Holst-Jensen A, Scholtens I, Peelen T, Prins TW, Kok E (2016) Advances in DNA metabarcoding for food and wildlife forensic species identification. Anal Bioanal Chem 408(17):4615–4630

Su'udi M, Ulum FB, Setyati D (2025) Direct PCR for DNA barcoding of Bulbophyllum lobbii Lindl. based on rbcL sequence. In: BIO web of conferences, vol 155. EDP Sciences, p 10006

Sullivan JR, Nazaire M (2022) Specimen collection and preparation for a changing flora. Rhodora 123(993):50–66

Surveswaran S, Gowda V, Sun M (2018) Using an integrated approach to identify cryptic species, divergence patterns and hybrid species in Asian ladies' tresses orchids (Spiranthes, Orchidaceae). Mol Phylogenet Evol 124:106–121

Taberlet P, Coissac E, Pompanon F, Brochmann C, Willerslev E (2012) Towards next-generation biodiversity assessment using DNA metabarcoding. Mol Ecol 21(8):2045–2050

Techen N, Parveen I, Pan Z, Khan IA (2014) DNA barcoding of medicinal plant material for identification. Curr Opin Biotechnol 25:103–110

Tiwari P, Bose SK, Gautam A, Chen JT (2023) Emerging trends and insights into the cultivation strategies, ethnomedicinal uses, and socio-economic attributes of orchids. J Hortic Sci Biotechnol 98(3):273–298

Tiwari R, Singh P, Verma SK et al (2024) Integrative taxonomy and molecular systematics of Indian orchids: resolving species complexes and supporting conservation through DNA barcoding. Plant Syst Evol 310(2):157–172

Tobe SS, Linacre A (2009) DNA typing of non-human (animal) DNA in forensic science. Forensic Sci Med Pathol 5(3):206–213

Tripathi AM, Tyagi A, Kumar A, Singh A, Singh S, Roy S et al (2013) The internal transcribed spacer (ITS) region and trnH–psbA are suitable DNA barcodes to identify threatened medicinal plants of India. Genome 56(9):491–496

Tsai WC, Dievart A, Hsu CC, Hsiao YY, Chiou SY, Huang H, Chen HH (2017) Post genomics era for orchid research. Bot Stud 58(1):61

Ullah F, Saqib S, Xiong YC (2025) Integrating artificial intelligence in biodiversity conservation: bridging classical and modern approaches. Biodivers Conserv 34(1):45–65

Veldman S, Otieno J, Gravendeel B, Andel TV, Boer HD (2014) Conservation of endangered wild harvested medicinal plants: use of DNA barcoding. Novel Plant Bioresour Applic Food Med Cosmet:81–88

Veldman S, Gravendeel B, Otieno JN, van Andel T (2018) DNA barcoding for identification of traded medicinal orchids from Africa and Asia. Bot J Linn Soc 186(4):725–736

Vu HT, Tran HT, Tran N, Nguyen TT, Le HT, Bui TTM et al (2017) Assessment of candidate DNA barcodes for identification of some tropical tree species of Dipterocarpaceae. Plant Gene 11:90–97

Wang X, Chen X, Yang P, Wang L, Han J (2017) Barcoding the Dendrobium (Orchidaceae) species and analysis of the intragenomic variation based on the internal transcribed spacer 2. Biomed Res Int 2017(1):2734960

Wang J, Fu CN, Mo ZQ, Möller M, Yang JB, Zhang ZR, Gao LM (2022) Testing the complete plastome for species discrimination, cryptic species discovery and phylogenetic resolution in Cephalotaxus (Cephalotaxaceae). Front Plant Sci 13:768810

Wang Z, Cui J, Zhu Y (2023) Review of plant leaf recognition. Artif Intell Rev 56(5):4217–4253

Wang D, Trimbos KB, Gomes SI, Jacquemyn H, Merckx VS (2024) Metabarcoding read abundances of orchid mycorrhizal fungi are correlated to copy numbers estimated using ddPCR. New Phytol 242(4):1825–1834

Watsa M, Erkenswick GA, Pomerantz A, Prost S (2020) Portable sequencing as a teaching tool in conservation and biodiversity research. PLoS Biol 18(4):e3000667

Wilson AW, Eberhardt U, Nguyen N, Noffsinger CR, Swenie RA, Loucks JL, Perry BA, Herrera M, Osmundson TW, DeLong-Duhon S, Beker HJ (2023) Does one size fit all? Variations in the DNA barcode gaps of macrofungal genera. J Fungi 9(8):788

Xu S, Li D, Li J, Xiang X, Jin W, Huang W, Jin X, Huang L (2015) Evaluation of the DNA barcodes in Dendrobium (Orchidaceae) from mainland Asia. PLoS One 10(1):e0115168

Yang CH, Wu KC, Chuang LY, Chang HW (2021) DeepBarcoding: deep learning for species classification using DNA barcoding. IEEE/ACM Trans Comput Biol Bioinformatics 19(4): 2158–2165

Zhai JW, Zhang GQ, Li L, Wang M, Chen LJ, Chung SW, Rodríguez FJ, Francisco-Ortega J, Lan SR, Xing FW, Liu ZJ (2014) A new phylogenetic analysis sheds new light on the relationships in the Calanthe alliance (Orchidaceae) in China. Mol Phylogenet Evol 77:216–222

Zhang GQ, Liu KW, Li Z, Lohaus R, Hsiao YY, Niu SC et al (2017) The Apostasia genome and the evolution of orchids. Nature 549(7672):379–383

Zhang GQ, Liu KW, Li Z, Lohaus R, Hsiao YY, Niu SC et al (2022) Advances and prospects of orchid research and industrialization. Hortic Res 9:uhac220

Zhivkoplias E, da Silva JM, Blasiak R (2025) How transdisciplinarity can help biotech-driven biodiversity research. Trends Biotechnol

Zhu S, Liu Q, Qiu S, Dai J, Gao X (2022) DNA barcoding: an efficient technology to authenticate plant species of traditional Chinese medicine and recent advances. Chin Med 17(1):112

Zizka A, Silvestro D, Vitt P, Knight TM (2021) Automated conservation assessment of the orchid family with deep learning. Conserv Biol 35(3):897–908

Recent Trends in the Global Orchid Industry and Export-Oriented Orchid Production System for Developing Nations

10

Nikhila Vaagdevi Anumala, Labdhi Dedhia, Chandan Gowda, Suman Natta, Kalaivanan N. S, S. S. Biswas, S. P. Das, Sushmita Halagali, and Usha Bharathi T

10.1 Introduction

The Orchidaceae family, often remarked as the most diverse in the plant kingdom, displays an extraordinary array of species with a wide range of adaptations and ecological roles. Encompassing approximately 29,524 species across 763 genera (Chase et al. 2015; Vendrame and Khoddamzadeh 2017; Govaerts et al. 2021), along with over 100,000 cultivated hybrids, orchids are a testament to nature's evolutionary ingenuity. Orchids have successfully colonized a vast range of habitats found in over 200 countries, inhabiting every continent except Antarctica (Chase et al. 2003; Pridgeon et al. 2005), growing as epiphytes, lithophytes, or terrestrials. Their flowers, ranging in shape, size, fragrance, and colour, reflect this diversity, from minute blossoms just 0.083 in. wide to the colossal 34-in. blooms of *Paphiopedilum sanderianum* 'Rapunzel', the world's largest orchid flower.

Orchids' adaptability and aesthetic appeal have made them a focal point in both horticulture and conservation biology. However, their complex life histories are marked by dependencies on specific mycorrhizal fungi, pollinators, and substrates, which also render them vulnerable to environmental changes and human activities (Zhang et al. 2022). The over-collection of orchids (Koopowitz 2001; Darro and Khan 2023), driven by the demand for edible (Ghorbani et al. 2014), ornamental plants (Vermeulen et al. 2014) and traditional medicine (Liu et al. 2014), has led to significant declines in wild populations. In recognition of these threats, all wild orchid species are listed and protected under the Convention on International Trade in Endangered Species of Wild Fauna and Flora (CITES 2013) and International

N. V. Anumala (✉) · C. Gowda · S. Natta · K. N. S · S. S. Biswas · S. P. Das
ICAR-National Research Centre for Orchids, Pakyong, Sikkim, India

L. Dedhia · S. Halagali · Usha Bharathi T (✉)
ICAR-Indian Institute of Horticultural Research, Bengaluru, India

A. Husen, M. Pant (eds.), *Forest-based Orchids*,
https://doi.org/10.1007/978-981-92-0096-2_10

Union for Conservation of Nature (IUCN 2024). Recent advancements in DNA sequencing have further deepened our understanding of orchid evolution and biogeography, revealing that orchids likely originated in the northern hemisphere around 83 million years ago, rather than in Australia as previously hypothesized (Stam 2024). These findings underscore the dynamic evolutionary history of orchids, with most extant species having emerged within the last 5 million years.

The ornamental and economic significance of orchids has shifted markedly over recent decades. Orchids have become mainstream commodities, entering commercial markets as cut flowers, potted plants, and value-added products (Hinsley et al. 2018; Negi et al. 2024). Technological interventions, including micropropagation, in vitro seed germination, hybrid development, and protected cultivation, have enabled year-round production, diversification of floral traits, and reduction in costs, democratizing access to these plants worldwide (Vendrame and Khoddamzadeh 2017; Tiwari et al. 2024). These developments have transformed orchids from niche luxury items into globally traded floricultural crops, creating opportunities for both established and emerging producers.

10.2 Challenges in Commercial Production: Habitat Loss and Extinction Risk

Orchids are known for their specific temperature, humidity, and light requirements, making them vulnerable to fluctuations in climate and weather patterns. Climate change, in particular, poses a significant threat to orchid cultivation as it can alter growing conditions and impact the overall health and growth worldwide. Extreme weather events, such as droughts, floods, and storms, can also damage orchid habitats and disrupt their natural growth cycle. Additionally, deforestation and habitat destruction further endanger wild orchid populations, leading to a decline in their availability (Lal et al. 2023). To address these challenges, efforts are being made to develop more resilient orchid varieties through breeding programs and to promote sustainable cultivation practices that minimize environmental impact. Recent scientific studies indicated alarming statistics: under global climate change, the habitats of 50.63–57.81% of the studied species will be reduced, with approximately 25.31–26.88% of the species are at risk of extinction. Research from southwestern China estimates that populations of at least 15% of the orchid species will be threatened with extinction over the next two centuries, given projected climate changes (Liu et al. 2020). Current IUCN Red List assessments show that 56.5% of assessed orchid species are threatened (Swarts and Dixon 2017), though this represents only 948 species of approximately 28,000 total orchid species.

Many orchids are affected by multiple threats, including habitat loss from development, agriculture, roads, forestry, grazing, fire and illegal collection, along with other less common threats that need further investigation, including the impact of small population sizes, loss of pollinators, and climate change (PMC Research 2024). The conservation challenge is substantial, with "rapidly increasing numbers of endangered orchids and low success rates in orchid conservation efforts (Conservation Biology Studies 2023).

10.3 Technology and Production of Orchids

Orchids represent one of the most environmentally sensitive plant families, creating significant cultivation challenges for commercial production. The global orchid industry has undergone a significant transformation, shifting from traditional cultivation methods to sophisticated, technology-driven production systems. Modern orchid cultivation is defined by the use of protected cultivation systems, primarily polyhouses and greenhouses. These structures provide a controlled microclimate, insulating plants from environmental stressors such as extreme temperatures, high winds, and heavy rainfall. The true innovation, however, lies in precision climate control, where interconnected systems manage the environment with unprecedented accuracy.

10.3.1 Temperature and Humidity Regulation

Automated heating, cooling, and ventilation systems maintain a species-specific temperature range, often with a programmed day-to-night temperature differential that is crucial for inducing flowering in many commercial varieties like *Phalaenopsis* (American Orchid Society 2024). Warm-growing orchids such as *Phalaenopsis* prefer day temperatures between 18 and 29 °C with night temperatures between 13 and 16 °C, while some species like Cymbidiums require temperature drops as much as 10 °C between day and night cycles (AboutOrchids.com 2024). Humidity requirements are species-specific and demanding: virtually all orchids do best when humidity ranges from 40% to 70% (American Orchid Society 2024), though some orchids, like *Paphiopedilum*, need as little as 40% humidity and other types need as much as 80% (New York Botanical Garden 2024). Humidity is controlled through misting or fogging systems and ventilation, preventing both plant desiccation and the proliferation of fungal pathogens.

10.3.2 Light and CO_2 Management

Light requirements are particularly critical, as too little light can result in weak, leggy stems, delayed flowering, or a total failure to bloom, while too much light can cause scorching of leaves, drying out, and stress-related disease. Achieving the perfect balance is crucial, but it takes a clear knowledge of every orchid species' specific requirement for light and response to various conditions, including natural and artificial light (Cognitive Market Research 2024). Sensors measure light intensity, and automated shade cloths adjust to prevent scorching while ensuring optimal light levels for photosynthesis. Supplemental LED lighting systems are increasingly used to provide specific light spectra, a practice known as photoperiodic lighting, to accelerate growth and flowering cycles. Furthermore, controlled carbon dioxide (CO_2) enrichment enhances photosynthetic efficiency, leading to faster growth rates and more robust plants. Good air circulation is crucial for preventing disease and ensuring proper growth of plants. Fans are used to ensure constant air movement. In some advanced systems, CO_2 is injected to maintain optimal levels, which boosts the rate of photosynthesis and accelerates growth.

10.3.3 Automated Irrigation and Fertigation

Drip irrigation and nutrient delivery systems are precisely controlled by sensors and timers, supplying plants with the exact amount of water and fertilizer required at the right time for maximum productivity, thereby minimizing waste and nutrient runoff.

10.3.4 Substrate Alternatives

Orchids, being epiphytes, do not grow in soil. Their roots require exceptional aeration and drainage. Modern cultivation relies on a variety of soilless substrates, each chosen for its specific properties. The most common organic materials include pine bark, which offers excellent aeration and a stable physical support, and coconut husk chips (coir), a highly sustainable material known for its superior water retention capacity. Sphagnum moss is also widely used, especially for young plants, due to its ability to hold a large volume of water. To enhance drainage and extend the lifespan of the medium, growers often incorporate inorganic materials. Among the several inorganic substrates, perlite and pumice, volcanic materials are incorporated as they improve aeration, while expanded clay aggregate (LECA) provides a stable, porous base. Charcoal is sometimes added for its ability to filter impurities and prevent bacterial growth.

10.3.5 Integrated Pest Management (IPM)

Modern orchid cultivation employs integrated pest management (IPM) to control pests and diseases in an environmentally responsible manner. IPM is a holistic strategy that prioritizes non-chemical methods and uses pesticides only as a last resort. Cultural controls involve proactive measures like maintaining meticulous greenhouse sanitation, regulating humidity to prevent fungal pathogens, and sourcing only disease-free plant material from reputable labs. Biological controls include the introduction of beneficial organisms that prey on common orchid pests. For example, predatory mites are released to control spider mites, while certain parasitic wasps are used to manage mealybug and scale infestations. Physical and mechanical controls include using sticky traps to monitor and catch flying pests like fungus gnats and whiteflies, as well as installing fine mesh screens on vents to prevent pests from entering the greenhouse.

10.3.6 Disease Management

Viral (Cymbidium Mosaic Virus, Odontoglossum Ringspot Virus), bacterial, and fungal diseases reduce orchids yield by 15–25%, necessitating robust biosecurity measures (Lingamaneni 2024). Routine ELISA ImmunoStrip screening, costing ₹500–1000 per kit, detects viruses with 95% accuracy, while bio-pesticides like *Trichoderma viride* (₹100/kg) control fungal infections, maintaining infection rates below 5% (Lepcha et al. 2020). Subsidizing bio-pesticides can replace chemical sprays, reducing residue risks by 90%, and immediate roguing of infected plants should be standard practice to sustain infection rates below 2% (Tamang et al. 2023).

10.3.7 Mass Production of Quality Orchid Plants

Orchid propagation through conventional methods is slow, labour-intensive, and often limited by seed dormancy and low germination rates. Tissue culture provides a scalable solution for producing large numbers of genetically uniform, disease-free plants. The foundation of large-scale, high-quality orchid production is in vitro propagation, more commonly known as tissue culture or micropropagation. This technique allows for the rapid multiplication of genetically uniform and disease-free plantlets. The process begins by excising a small, sterile piece of plant tissue, or explant, which is then placed in a sterile culture medium. The medium contains essential nutrients and plant growth regulators that induce the explant to multiply and form structures known as protocorm-like bodies (PLBs). These PLBs are then asexually multiplied, producing a large number of clones.

Technological advancements have witnessed replacement of glass flasks by advanced bioreactor systems. Temporary immersion systems (TIS) are particularly effective. These systems automatically flood the plantlets with liquid nutrient medium for brief periods and then drain the medium, allowing the plantlets to be immersed and aerated intermittently. This method significantly improves nutrient uptake and reduces the risk of tissue hyperhydration, resulting in more vigorous and healthier plantlets at a much higher density. The system is commercially in high demand for the propagation of orchids—*Cattleya*, *Dendrobium*, and *Vanda*.

10.4 Global Trends

In 2024, the global floriculture sector, encompassing live trees, bulbs, roots, cut flowers, and ornamental foliage, recorded exports worth USD 26.97 billion against imports of USD 24.65 billion, resulting in a positive trade balance of USD 2.32 billion. Export growth remained robust at 5% between 2023 and 2024, with a steady

3% CAGR during 2020–2024, indicating resilience in international demand despite market fluctuations (International Trade Centre 2025). The Netherlands has maintained its position as the global leader in floriculture market with the highest trade surplus (USD 10.33 billion), followed by Colombia and Ecuador. This dominance reflects their efficient export-oriented production systems and strong integration into international value chains, serving as a model for developing nations aiming to expand floriculture exports.

Within this global framework, orchids represent a high-value segment of the floriculture market. Valued at USD 302.5 million in 2024, it is projected to grow at a CAGR of 4.0% during 2024–2031(International Trade Centre 2025). The market spans the cultivation, production, and sale of orchids as both cut flowers and potted plants, driven by their aesthetic diversity, long vase life, and association with luxury. Increasing demand from the hospitality sector, a rising interest in gardening and interior decor, and the growth of online plant retail channels are fuelling expansion. Furthermore, innovations in breeding and micropropagation have expanded species diversity and enhanced year-round availability, strengthening market appeal. A noteworthy trend is the increasing use of orchids in the wellness industry, particularly in orchid-based therapies and aromatherapy, where orchid extracts and oils are marketed for their calming and rejuvenating properties. This niche yet expanding segment aligns with the broader consumer shift towards holistic wellness and natural remedies, offering new avenues for market growth.

Among commercial orchid groups, *Phalaenopsis* dominates the global potted plant trade, supported by controlled flowering induction and large-scale propagation in Belgium, the Netherlands, Taiwan, and Thailand. *Vanda* is witnessing rapid market growth owing to their vibrant, exotic blooms and suitability for tropical climates, while *Cymbidium*, *Dendrobium*, *Oncidium*, and *Mokara* remain the leading cut flower species across markets. Regional specialization further defines trade patterns influenced by agro-climatic condition, availability, and consumers demand, such as the Netherlands and New Zealand are renowned for *Cymbidium*; the Netherlands and Taiwan for *Phalaenopsis*; Malaysia, Singapore, and Thailand for *Dendrobium* and *Mokara*; and Taiwan and Thailand for *Oncidium* and *Vanda*.

Historically, orchids were symbols of nobility and exclusivity, valued as luxury collectibles. However, the commercialization of orchids has transformed significantly over the past six decades with the development of aseptic culture and greenhouse technologies, which enabled mass production and reduced costs of production. By the early 2000s, orchids entered mainstream retail chains in Europe and North America, particularly through the Dutch and U.S. markets. This transition marked their evolution from luxury commodities to affordable consumer products, driving exponential increases in both production and global consumption.

Today, a wide array of orchids, including *Cattleya*, *Cymbidium*, *Dendrobium*, *Oncidium*, *Paphiopedilum*, *Phalaenopsis*, and flasked orchid plantlets, feature prominently in international trade, reflecting not only their horticultural and commercial value but also their expanding roles in interior decoration, wellness, and lifestyle markets. For tropical countries such as India and others in Asia, these global dynamics present a unique opportunity. With rich native orchid biodiversity, favourable climates, and relatively lower production costs, these regions are well-positioned to strengthen their participation in the global orchid trade. Strategic investments in micropropagation, controlled environment cultivation, value-added orchid products, and export-oriented supply chains can enable developing nations to capture emerging markets, particularly in the wellness and lifestyle segments. By aligning conservation with commercialization, such countries can simultaneously safeguard orchid diversity and tap into the expanding global demand.

10.5 Global Imports for Orchid Cut Flowers

The global import value of cut flowers has grown from USD 8,204,007 thousand in 2020 to USD 10,504,611 thousand in 2024, reflecting the steady expansion of international floriculture trade (International Trade Centre 2025). The United States, Germany, the Netherlands, and the United Kingdom were among the leading importers of cut flowers. Between 2023 and 2024, the United States recorded a 5% increase in import value, while the Netherlands, which had experienced a 16% decline during 2021–2022, demonstrated signs of recovery with a 7% increase in 2023–2024.

Within this broader market, cut orchids accounted for USD 190,269 thousand in 2024, representing a modest 1.81% share of total cut flower imports. Japan remained the largest importer of cut orchids, capturing 28.3% of global imports, followed by the United States (16.4%) and Vietnam (7.4%). Other important importing countries included Italy, Germany, China, France, Russia, and Australia, highlighting the wide geographical spread of orchid demand. The distribution of major importing countries and their relative shares in global orchid imports is presented in Fig. 10.1, while Fig. 10.2 shows the top ten importing countries alongside their respective annual

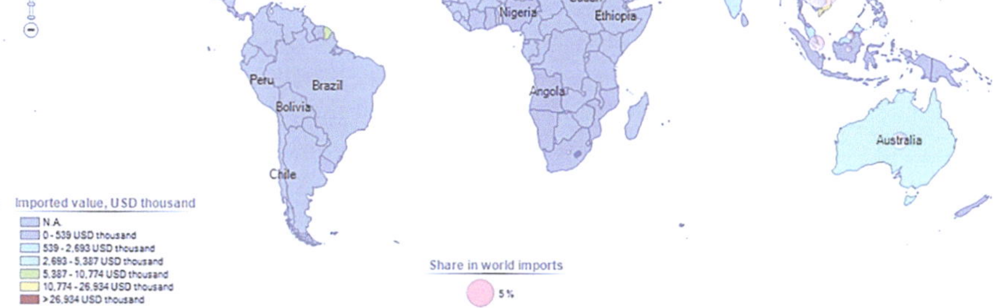

Fig. 10.1 Major importers of cut orchids and their share in global imports, 2024

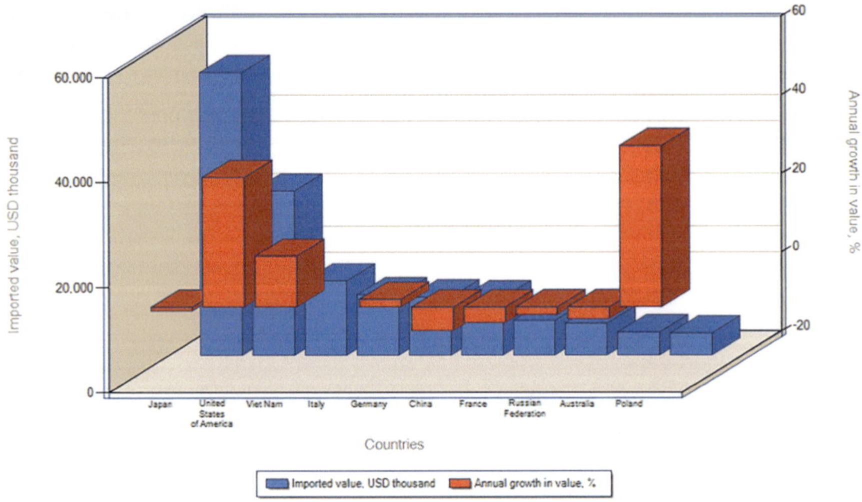

Fig. 10.2 Top ten importing countries of cut orchids and their annual growth rates during 2024

growth rates. Notably, Japan recorded a 1% decline in import value, whereas the United States and Vietnam posted strong growth of 33% and 13%, respectively.

The dynamics of trade flows are further highlighted in Fig. 10.3, which depicts the concentration of importing countries and their average distance from supplying partners. On an average, orchids travelled 4182 km from supplier to importer, though the distances varied considerably—from as little as 343 km for Germany to as much as 11,238 km for the United States. This underscores both the localized sourcing advantages within certain regions and the highly globalized nature of orchid trade.

10.6 Global Exports for Orchid Cut Flowers

The global export value of cut flowers expanded from USD 8,668,856 thousand in 2020 to USD 11,264,647 thousand in 2024, registering an average annual growth rate of 5% over the period (International Trade Centre 2025). With an average supply distance of 3354 km, this trade reflects both its globalized nature and the increasing international demand for floricultural products. The Netherlands, Colombia, Ecuador, Kenya, and Ethiopia maintained their positions as the world's leading exporters of total cut flowers (Fig. 10.4).

Within this sector, cut orchids represented a smaller but significant segment, with exports valued at USD 186,830 thousand in 2024, accounting for 2.16% of total cut flower exports. The trade was concentrated among a few major players, with the Netherlands (37.1%), Thailand (33.1%), and Taipei, Chinese (18%) together contributing nearly 90% of global orchid exports (Fig. 10.4). Other notable exporters

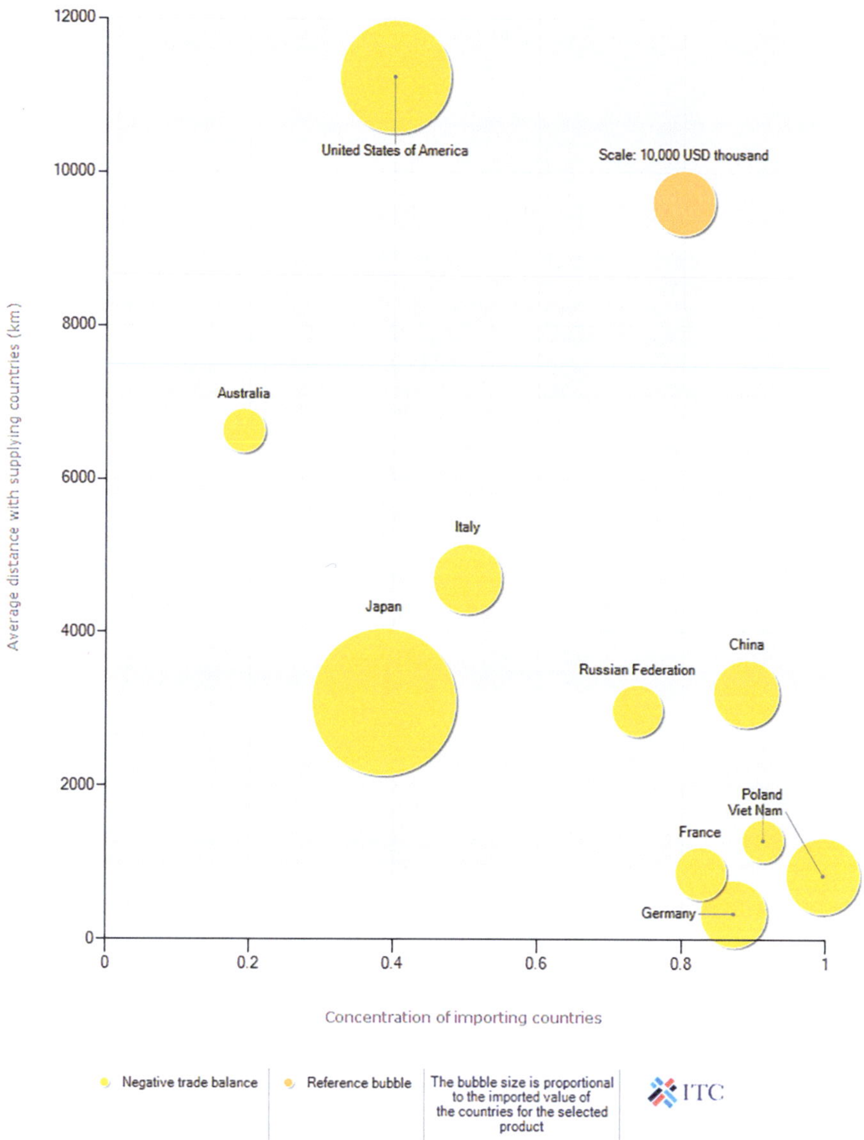

Fig. 10.3 Concentration of top cut orchid importing countries and their average distance from supplying countries in 2024

included Vietnam, New Zealand, Malaysia, China, Singapore, Belgium, and Italy, reflecting both regional specialization and diversification of supply sources.

Export performance also demonstrated contrasting dynamics in terms of trade volumes and unit values. In 2024, the Netherlands exported 4410 tons of cut orchids, achieving a high unit value of USD 15,735 per ton, while Thailand exported a much

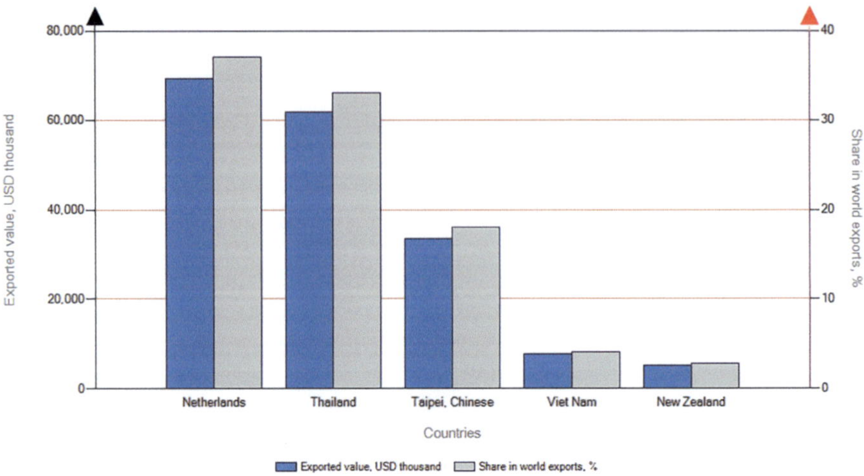

Fig. 10.4 Exported value and global share of orchids by leading countries in 2024

Fig. 10.5 Major exporters of floricultural products and their share in global export, 2024

larger volume of 19,126 tons but at a significantly lower unit value of USD 3234 per ton. By contrast, Taipei, China, exported just 2034 tons, yet commanded the highest unit value at USD 16,511 per ton. These differences highlight the varied market positioning of exporters, ranging from high-value, niche suppliers to bulk, price-competitive traders. Overall, the annual growth rate of orchid exports was 2%, suggesting relatively modest expansion compared to the broader cut flower sector. The distribution of export shares across major countries is further depicted in Fig. 10.5, offering insights into the structure and concentration of the global orchid trade.

10.7 Orchid Industry in Different Countries

10.7.1 The Netherlands Model

The Netherlands hosts one of the most advanced orchid industries with a dominant focus on *Phalaenopsis* cultivation, incorporating high technological developments and strong international relationships promoting sustainable floriculture. Orchids represent a major segment of Dutch Horticulture, contributing significantly to the national economy and shaping the international trade patterns. Historically, consignments ranging between 3500 and 4000 tonnes of orchids have been shipped annually, although volumes and values have fluctuated in response to the market dynamics. The evolution of the industry dates back to the early twentieth century, when pioneering firms such as Floricultura, established in the 1930s, initially traded orchids, tropical ornamentals, and seeds. During the 1960s and 1970s, Dutch breeders and propagators laid the foundation for industrialization through techniques such as clonal propagation and meristem culture, enabling large-scale and uniform planting material production. By the late twentieth century, greenhouse cultivation expanded rapidly, with *Phalaenopsis* leading the other genera and becoming the dominant commercial crop. Today, *Phalaenopsis* accounts for more than 80% of orchids produced in the Netherlands, supplying 90% of all *Phalaenopsis* sold across Europe. Dutch nurseries collectively deliver over 1.5 million mature orchids every week through auction and direct contracts, sustaining export values that exceed €68 million annually.

Recent adjustments in production strategy, including a 22–25% reduction in output, have stabilized supply–demand balance, leading to improved value per plant and recovery of market stability. The Dutch orchid sector is characterized by intensive greenhouse cultivation supported by automation, climate control, LED lighting and tissue culture-based propagation, which together optimize spike quality, flowering induction and post-harvest longevity. Sustainability has become the key component of Dutch floriculture, with increasing adoption of renewable energy, closed-loop irrigation, and reduced chemical inputs enhancing the industry's reputation as an environmentally responsible producer. Companies such as Tr Laak Orchids operate one of Europe's most energy-efficient greenhouses, while Levoplant has incorporated natural pest management and water cycling into its production system. Collaboration and knowledge sharing also play a vital role, with platforms such as Inspiration Days showcasing new cultivars, breeding innovations and evolving consumer preferences while strengthening networks among growers and global stakeholders. Despite its leadership, the Dutch Orchid industry continues to face structural challenges, including past episodes of overproduction and ongoing market volatility driven by seasonal demand fluctuations. The industry has demonstrated resilience by strategically reducing volumes, diversifying into niche markets, and aligning production practices with sustainability and climate-smart approaches. Continued advances in breeding, resource-efficient cultivation, and supply chain optimization will be critical for maintaining competitiveness in the global market.

10.7.2 The Case of Taiwan

Taiwan has emerged as a global leader in orchid production, breeding, and biotechnology, earning the title of the 'Orchid Kingdom'. The orchid sector is now a flagship industry within Taiwan's agricultural exports, with markets extending to over 60 countries, including the United States, Japan, Canada, and Australia. In 2024, orchid production in Taiwan was valued at NT$6.4 billion, with exports accounting for NT$6.1 billion. The Taiwan Orchid Biotechnology Park, located in Tainan, has become a recognized hub for innovation, supply, and international collaboration in the orchid industry. Taiwan's orchid industry has benefited significantly from the synergistic efforts of government, academia, and private enterprises, enabling a transition from traditional cultivation practices to a technology-driven, branded, and globally integrated production system. The incorporation of artificial intelligence (AI), smart agriculture, and precision breeding strategies reflects a revolutionary approach to sustaining competitiveness and addressing global market demands.

International recognition of Taiwan's role in the orchid sector is evident through major events such as the Taiwan International Orchid Show (TIOS), regarded as one of the world's three premier orchid exhibitions, alongside the World Orchid Show and the Tokyo Orchid Show. The triennial Asia-Pacific Orchid Conference (APOC), returning to Taiwan in 2025 after a 20-year hiatus, underscores the country's significance in global orchid diplomacy and trade. Themed 'Next 20', the 2025 APOC-TIOS combined event is being hosted at the Taiwan Orchid Biotechnology Park from March 8 to 23, featuring over 1500 orchid entries evaluated by 200 international judges, extensive landscape competitions, multidisciplinary digital exhibitions, and thematic forums.

The convergence of these events not only highlights Taiwan's scientific and technological leadership in orchid breeding and biotechnology but also reflects the cultural and economic value orchids hold for the nation.

10.7.3 China: Leading in Scale and Technological Integration

China has emerged as a global leader in orchid cultivation, particularly in the production of potted orchids like *Phalaenopsis* and *Dendrobium*. The country's orchid industry has flourished due to significant advancements in scientific research on breeding and cultivation technologies. China has made substantial progress in orchid breeding, with over 1200 applications for new orchid plant varieties and approvals for 372 new varieties as of March 2022. Additionally, more than 21 orchid species have had their high-quality genomes mapped, facilitating targeted breeding efforts. The adoption of advanced greenhouse technologies, including automated climate control systems and LED lighting, has enabled year-round production of high-quality orchids. This approach ensures a consistent supply and meets both domestic and international demand. It has implemented IPM strategies that combine biological control agents, cultural practices, and minimal chemical interventions to manage pests and diseases effectively, ensuring the production of safe and sustainable orchid products.

10.7.4 Thailand: Export-Oriented Production with Sustainable Practices

Thailand has developed a robust orchid industry focused on export markets, particularly for species like *Dendrobium* and *Vanda*. Thailand maintains its position as the world's largest orchid exporter. Thai growers utilize locally sourced substrates, such as coir and rice husk, reducing costs and promoting sustainability in orchid cultivation. The nursery growers employ a combination of biological control agents and minimal chemical treatments to manage pests, ensuring the production of high-quality, export-ready orchids. According to the Ministry of Commerce's official report on cut flower exports in 2024, Thailand exported 21,135 tons of cut flowers, reaching a total value of 2.5 billion baht (71.3 million USD), with orchids comprising the main flower export at 19,129 tons valued at 2.18 billion baht (62.1 million USD) (Thailand PRD 2024). Thai exports of decorative plants from 2023 to 2024 grew to 9.32 billion baht, with more than half coming from orchids, according to DBD director-general Oramon Sapthaweetham (Vietnam Plus 2025). Major markets for Thai decorative plants include the United States, Vietnam, Japan, and Malaysia (The Nation Thailand 2025). Thailand's competitive advantages stem from established cultivation expertise, efficient supply chain infrastructure, favourable tropical climate, and government trade promotion support through agencies like the Department of Business Development.

10.7.5 Indonesia: Community-Based Orchid Cultivation

Indonesia's orchid industry is characterized by community-based cultivation, focusing on native species and sustainable practices. In Indonesia, local communities are involved in orchid cultivation, utilizing traditional knowledge and practices to grow native species for both domestic use and export. The growers emphasized the use of organic fertilizers and natural pest control methods, aligning with global trends towards sustainable agriculture. While the industry faces challenges related to infrastructure and access to advanced technologies, there is significant potential for growth through capacity building and the adoption of modern cultivation techniques.

10.7.6 Malaysia: Specialized Positioning in High-Value Markets

Malaysia has developed a specialized approach to orchid production within its broader floriculture industry. According to official Department of Agriculture data, the floriculture industry recorded export values of RM 417.3 million in 2021 compared to RM 403 million in 2020 (The Star 2024). Historical data indicate that the export value of fresh-cut-flowers reached more than RM500 million (US$109 million) annually by 2017 (FFTC Agricultural Policy Platform 2020). According to Mohamad Nizam Malik, head of the Department of Agriculture's Floriculture Industry Unit, Malaysian orchid exports have the potential to contribute up to RM4 billion annually to the floriculture industry due to high international demand (Borneo Bulletin 2024). Malaysia's Ministry of Agriculture and Food Security has developed strategies to advance performance in the global floriculture industry, with recent export initiatives to international markets, including Dubai (The Star 2024). Malaysia leverages its biodiversity-rich natural habitats, research capabilities in orchid breeding, and strategic location for regional distribution. The focus on high-value, specialized production differentiates Malaysia from volume-based competitors.

10.7.7 Orchid Industry in India

India is recognized as one of the global centres for orchid diversity, boasting a rich orchid flora of 1256 species distributed across 155 genera, with nearly 400 species being endemic. This rich diversity is concentrated in biodiversity hotspots such as the Eastern Himalayas, North-Eastern states, and the Western Ghats. Historically prized for their ornamental appeal, medicinal value, and cultural symbolism, orchids have moved in recent decades from wild collection into organized cultivation and commercial floriculture. Over the past 10 years, the production landscape has changed significantly; the area under orchid cultivation expanded from barely 10 hectares in 2012–2013 to more than 3000 hectares by 2022–2023, while cut flower production rose from 180 metric tons to over 10,700 metric tons during the same period. Initially restricted to North-Eastern states such as Arunachal Pradesh, Sikkim, Meghalaya, and Assam, orchid cultivation has now reached Kerala, Karnataka, Tamil Nadu, Telangana, Madhya Pradesh, and Goa, with both cut flower and potted plant markets developing steadily. India's orchid industry, valued at over ₹700 crore annually, thrives in regions such as the Northeast, Kerala, and West Bengal, where high-value hybrids like *Dendrobium* and *Phalaenopsis* are cultivated (Lepcha et al. 2020). Besides aesthetics, orchids have a huge scope for their medicinal value (Kong et al. 2003; Chhetri et al. 2014; Zhang et al. 2024). India's herbal products market, valued at $64.8 billion in 2024 and projected to reach $120.2 billion by 2033 (CAGR 7.1%) (Grand View Research 2024), presents substantial opportunities for orchid commercialization. The medicinal plants sector specifically represents a market valued at Rs 4.2 billion in 2019, expected to increase to Rs 14 billion by 2026 (Ministry of AYUSH 2020). India currently imports 80% of its orchids, primarily from Thailand (80.67% of total imports) (Trade Statistics 2024), representing a significant import substitution opportunity within the country's $651 million annual herbal product exports (APEDA 2024).

Despite this impressive domestic growth, India's orchid industry is still at a nascent stage when viewed in the context of global trade. India's exports remain limited due to high production costs, lack of dedicated cold chain infrastructure, and weak integration into global supply networks. Illegal collection of wild orchids and habitat loss also remain persistent challenge, especially for orchids which are prohibited from commercialization under the Wildlife Protection Act, 1972, and CITES (Appendix II) (Tamang et al. 2023). In contrast, ex situ tissue culture, used by 70% of commercial nurseries, produces virus-free plantlets with over 90% survival rates and conserves genetic diversity through seed banking and cryopreservation (Meghalaya Basin Development Authority 2023). Nevertheless, the increasing popularity of orchids in Indian cities, growing demand from the hospitality and wedding industries, and the emerging niche of value-added products such as dried flowers, handicrafts, and ornamental arrangements suggest untapped potential. Orchid cultivation offers high returns compared to traditional crops, driven by premium pricing (₹100–₹3000 per stem) and low input requirements post-establishment (Tamang et al. 2023). With its vast genetic resources, favourable agro-climatic zones, and rising consumer base, India has the capacity to evolve from a largely domestic producer into a competitive player in the international orchid market, provided there is investment in scale-efficient production, post-harvest technologies, and export-oriented value chains. Table 10.1 summarizes key economic indicators from various studies.

10.8 Regulatory and Trade Complexities

International orchid trade is regulated under strict protocols orchids are protected by the Convention on International Trade in Endangered Species of Wild Fauna and Flora (CITES) laws and cannot be traded internationally without CITES permits (CITES Secretariat 2024). Examples of CITES-protected plants include orchids, cacti, pitcher plants, and some cycads and palms (Wildlife Protection Agency 2024), requiring extensive documentation and permits for commercial trade. Enforcement challenges persist, as much of this trade is the result of illegal harvesting, meaning that it is poorly documented and absent from official statistics, while also being of growing conservation concern (Hinsley et al. 2018). Research indicates significant non-compliance creating market instability and legal risks for legitimate commercial operations. A recent report stated that 9.9% had smuggled, 4.8% had laundered, and 10.8% had been sent orchids from online purchases without paperwork (PMC Conservation Research 2024).

Consequently, despite protection measures, 29,524 orchid species are listed under various protection schemes (IUCN 2024), yet 'orchids continue to be subject to unsustainable harvesting and undocumented trade' (Conservation International 2023), indicating the economic pressures driving both legal and illegal markets. The technical complexity of orchid cultivation requires specialized expertise, controlled environment facilities, and extended production timelines that create significant barriers to market entry. The combination of regulatory complexity, biological

Table 10.1 Orchid production in different regions of India

Region/study	Farm size	Establishment cost (₹)	Net annual income (₹)	Benefit-cost ratio (BCR)	Internal rate of return (IRR)	Payback period (years)	Reference
West Bengal	500 sq. ft	67,117	46,354	1.76	25.38%	3–5	Lepcha et al. (2020)
West Bengal	1000 sq. ft	159,746	197,760	1.76	25.38%	3–5	Lepcha et al. (2020)
Meghalaya (BioNEST Project)	12,000 *Dendrobium* plants	Not specified	80,000 (from cut flowers)	Not specified	Not specified	Not specified	Meghalaya Basin Development Authority (2023)
Andhra Pradesh (Lingamaneni case)	7 acres (1 lakh plants)	1.5 crore (total investment)	1,200,000 (monthly net profit: 100,000)	Not specified	Not specified	Not specified	Lingamaneni (2024)
Sikkim	Medium holdings (0.6–4.5 acres)	Not specified	4715–31,815 (medium income for 83% growers)	Not specified	Not specified	Not specified	Tamang et al. (2023)

sensitivity, and conservation concerns creates a challenging commercial environment where legitimate producers must navigate extensive legal requirements while competing with potentially illegal or unregulated sources. These factors collectively constrain market expansion and limit the number of viable commercial producers capable of meeting both regulatory requirements and market demands.

10.9 Conclusion

Orchid cultivation has emerged as a promising segment of global floriculture, offering enormous opportunities for developing nations to diversify their agricultural exports and strengthen rural livelihoods. The aesthetic appeal, regional diversity, and longevity of orchids, particularly as cut flowers and potted plants, have placed them as a premium commodity in national and international markets. Developing countries must meet this demand by transitioning from traditional cultivation practices to a structured, export-oriented production system (EOPS) that incorporates modern technologies, efficient production procedures, adherence to international quality standards, robust quality assurance mechanisms, sustainable practices, and supportive government policies. The architecture for such a system is based on four key pillars: access to high-quality planting material, innovative production technology, effective post-harvest management, and efficient export logistics. The integration of sustainability principles, such as reducing carbon and water footprints, managing biosecurity risks, and promoting ex situ conservation, ensures that orchid production aligns with global environmental goals. Socioeconomic inclusion, particularly the involvement of women, youth, and smallholder farmers, adds further value by fostering equitable growth and community resilience. By aligning production cycles with global demand peaks and adopting international certifications, developing nations can strengthen their foothold in the global floriculture market.

References

AboutOrchids.com (2024) Orchid temperature requirements and cultivation guidelines. Retrieved from orchid cultivation database

American Orchid Society (2024) Orchid care guidelines: environmental requirements for successful cultivation. AOS Official Publications

APEDA (2024) Agricultural and Processed Food Products Export Development Authority annual report. Ministry of Commerce and Industry, Government of India

Borneo Bulletin (2024) Orchids can generate MYR4 billion a year in export earnings, says Malaysian Department of Agriculture. Borneo Bulletin Online

Chase MW, Cameron KM, Barrett R et al (2003) DNA data and Orchidaceae systematics: a new phylogenetic classification. Natural History Publications

Chase MW, Cameron KM, Freudenstein JV, Pridgeon AM, Salazar G, van den Berg C, Schuiteman A (2015) Updated classification of orchidaceae. Bot J Linn Soc 177:151–174. https://doi.org/10.1111/boj.12234

Chhetri DR, Parajuli P, Subba GC (2014) The folklore medicinal orchids of Sikkim. Ancient Sci Life 33(4):235–242
CITES (2013) Numbers of species listed in the CITES Appendices as of October 2013. The Convention on the International Trade in Endangered Species of Flora and Fauna. http://www.cites.org/eng/disc/species.php. Accessed May 2014
CITES Secretariat (2024) Convention on International Trade in Endangered Species: plant trade regulations and permits. CITES Official Documentation
Cognitive Market Research (2024) Global orchid market analysis: growth opportunities and challenges. Market Research Report
Conservation Biology Studies (2023) Orchid conservation success rates and endangered species management. Conserv Biol J
Conservation International (2023) Global orchid trade and sustainability challenges. CI Research Publications
Darro S, Khan NS (2023) Documentation of some endangered medicinal plants growing in Indravati National Park, Bijapur district, Chhattisgarh, India. Int J Exp Res Rev 36:378–387
FFTC Agricultural Policy Platform (2020) Producer's perspective on the potential of Malaysia's floriculture industry
Ghorbani A, Gravendeel B, Naghibi F et al (2014) Wild orchid tuber collection in Iran: a wake-up call for conservation. Biodivers Conserv 23:2749–2760. https://doi.org/10.1007/s10531-014-0746-y
Govaerts R, Lughadha EN, Black N, Turner R, Paton A (2021) The WorldChecklist of Vascular Plants, a continuously updated resource for exploring global plant diversity. Sci Data 8:215
Grand View Research (2024) India herbal products market size, share & trends analysis report 2024–2033
Hinsley A, Boer HJ, Fay MF, Gale SW, Gardiner LM, Gunasekara RS, Kumar P, Masters S, Metusala D, Roberts DL, Veldman S, Wong S, Phelps J (2018) A review of the trade in orchids and its implications for conservation. Bot J Linnean Soc 186:435–455
International Trade Centre (2025) Trade statistics for international business development. https://www.intracen.org/resources/data-and-analysis/trade-statistics
IUCN (2024) Red List of Threatened Species: Orchidaceae assessment and protection status. International Union for Conservation of Nature
Kong JM, Goh NK, Chia LS, Chia TF (2003) Recent advances in traditional plant drugs and orchids. Acta Pharmacol Sin 24(1):7–21
Koopowitz H (2001) Orchids and their conservation. Timber Press, Hachette Book Group
Lal A, Pant M, Pant G, Palni LMS, Datta A, Kumar G (2023) ISSR marker assisted genetic diversity assessment in natural populations of two endemic orchids Aerides multiflora and Rhynchostylis retusa from Uttarakhand, India. South Afr J Bot 157:151–158
Lepcha N, Thapa R, Sharma G (2020) Economics of flower cultivation with special reference to orchid in hilly areas of West Bengal. Neliti Publications. https://www.neliti.com/publications/335614/economics-of-flower-cultivation-with-special-reference-to-orchid-in-hilly-areas-o
Lingamaneni R (2024) Orchid farming in Andhra Pradesh: a case study. Int J Curr Res Trends 12(3):45–52
Liu H, Luo Y, Heinen J, Bhat M, Liu ZJ (2014) Eat your orchid and have it too: a potentially new conservation formula for Chinese epiphytic medicinal orchids. Biodivers Conserv 23(5):1215–1228
Liu H, Liu Z, Jin X, Gao J, Chen Y, Liu Q, Zhang D-Y (2020) Assessing conservation efforts against threats to wild orchids in China. Biol Conserv 243:108484. https://doi.org/10.1016/j.biocon.2020.108484
Meghalaya Basin Development Authority (2023) From roots to growth: empowering women through orchid floriculture. Government of Meghalaya. https://mbda.gov.in
Ministry of AYUSH (2020) Medicinal plants sector development report. Government of India

Negi A, Chauhan K, Adhikari H, Husen A, Pant M (2024) Medicinal orchids potential, conservation, and healing traditions. In: Husen A, Pant M (eds) Exploring medicinal orchids. Taylor & Francis Group, pp 1–16. https://doi.org/10.1201/9781003436638-1

New York Botanical Garden (2024) Orchid cultivation handbook: species-specific environmental requirements. NYBG Horticultural Publications

PMC Conservation Research (2024) Orchid trade compliance and illegal harvesting patterns. PubMed Central Database

PMC Research (2024) Threats to orchid populations: habitat loss and conservation challenges. PubMed Central Database

Pridgeon AM, Cribb P, Chase MW, Rasmussen FN (2005) Genera Orchidacearum Volume 4, Epidendroideae (Part one): Epidendroideae (Part one). Oxford University Press, Oxford, p 696

Stam G (2024) Orchids emerged in the Northern Hemisphere 83 million years ago. Revista Pesquisa FAPESP. https://revistapesquisa.fapesp.br/en/orchids-emerged-in-the-northern-hemisphere-83-million-years-ago/

Swarts ND, Dixon KW (2017) Conservation status and threats to orchid species globally. Curr Biol 27(12):R571–R577

Tamang Y, Sharma S, Bhutia K (2023) A study on socio-economic profile of orchid growers in East Sikkim. SBEAR J 5(2):89–97

Thailand PRD (2024) Thai flower exports flourish worth USD71.3M in 2024. Public Relations Department, Government of Thailand

The Nation Thailand (2025) Thailand retains crown as world's No 1 orchid exporter

The Star (2024) Orchids can generate RM4bil in exports yearly, says Agriculture Dept.

Tiwari P, Sharma A, Bose SK, Park KI (2024) Advances in orchid biology: biotechnological achievements, translational success, and commercial outcomes. Horticulturae 10:152. https://doi.org/10.3390/horticulturae10020152

Trade Statistics (2024) India orchid import-export data analysis. Ministry of Commerce and Industry

Vendrame WA, Khoddamzadeh AA (2017) Orchid biotechnology. Hort Rev 44:173–228. https://doi.org/10.1002/9781119281269.ch4

Vermeulen JJ, Phelps J, Thavipoke P (2014) Notes on Bulbophyllum (Dendrobiinae; Epidendroideae; Orchidaceae): two new species and the dilemmas of species discovery via illegal trade. Phytotaxa 184(1):012–022

Vietnam Plus (2025) Thailand—world's No 1 orchid exporter. VietnamPlus News Agency

Wildlife Protection Agency (2024) CITES-protected plant species: regulatory framework for international trade. WPA Official Guidelines

Zhang D, Zhao X-W, Li Y-Y, Ke S-J, Yin W-L, Lan S, Liu Z-J (2022) Advances and prospects of orchid research and industrialization. Hortic Res 9:uhac220. https://doi.org/10.1093/hr/uhac220

Zhang Y, Li H, Wang J (2024) Bioactive compounds and anticancer activities of medicinal orchids: a comprehensive review. J Med Plants Stud 12(3):45–58

Production Strategies and Management Practices for Cymbidium Cut Flower Orchids

11

Umed Kumar Pun, Sudhir Shrestha, Nirajan Bhandari, and Januka Dahal

11.1 Introduction

Nepal is renowned for its diverse range of orchids, with over 501 species (Tiruwa et al. 2024). Their distribution extends from the plains of the Terai to the mountains, stretching from east to west across the country (Dahal and Shakya 1989). The Terai region is reported to have a higher species diversity compared to the mountainous areas, and similarly, the eastern parts of Nepal are richer in these species than the western regions (Dahal and Shakya 1989). Epiphytes are mainly found in the plains, while the limited terrestrial orchids are mostly located in the mountains. Globally, most orchids are epiphytes, primarily inhabiting tropical forests. Nevertheless, their distribution is very widespread, occurring in nearly every part of the world, from the icy regions of Alaska and the snow-covered Himalayas to the sandy deserts of Australia and Africa (Mukherjee 1983).

Orchids in Nepal have been valued for their beautiful flowers, but their use has been limited to the hilly urban areas and the rural regions of eastern Nepal. In addition to their aesthetic appeal, orchids are also used as fodder for animals, in traditional medicine, and in the manufacturing of modern health tonics (Gadal et al. 1996). The trade of native orchids for ornamental purposes, such as their beautiful

U. K. Pun (✉)
Nexus Institute of Research and Innovation (NIRI), Lalitpur, Nepal

S. Shrestha
Directorate of Agriculture, Ministry of Agriculture, Bagmati Pradesh, Nepal

N. Bhandari
Agriculture and Forestry University, College of Natural Resources Management, Pakhribas, Dhunkuta, Nepal

J. Dahal
Institute of Agriculture and Animal Science, Tribhuvan University, Gauradaha, Jhapa, Nepal

A. Husen, M. Pant (eds.), *Forest-based Orchids*,
https://doi.org/10.1007/978-981-92-0096-2_11

flowers, is not a major industry in Nepal. The most preferred orchid plants and flowers are mainly hybrids of Cymbidium, with a limited number of other genera such as Dendrobium, Phalaenopsis, Cattleya, and Cypripedium. Recently, the main business has shifted to the use of pseudobulbs and canes of *Dendrobium species* for medicinal and nutritional purposes (Pun 2014).

11.2 Commercial Cultivation of Cymbidium Orchids

Cymbidium orchids are generally evergreen epiphytes with a sympodial growth habit. There are about 70 species in this genus, along with several thousand hybrids. This genus is characterised by sword-like leaves, short and stout pseudobulbs, and fleshy roots. Typically, this genus prefers cool or intermediate temperatures, and native Cymbidiums are naturally found in the hills of Nepal. The commercial success of Cymbidium may be attributed to its beautiful and long-lasting flowers.

Commercial orchid farming for cut flowers or potted plants in Nepal began approximately 35 years ago, earlier reported to be around 25 years ago in 2016 (Pun 2016). The primary orchid genus cultivated commercially in Nepal is Cymbidium. Therefore, from now on, 'Orchid' will be referred to as Cymbidium. Over 90% of Cymbidium cultivation occurs in the Godavari region of Lalitpur district, with the remaining 10% in the hilly areas of Makwanpur district (Pun 2014). Cymbidiums are popular and among the most expensive cut flowers both domestically and internationally. This is mainly due to their exceptional flowers and long-lasting quality. Currently, more than 24 Cymbidium hybrids are grown in Nepal. The quality of Cymbidium flowers produced in Kathmandu Valley is high and suitable for export. Presently, production meets domestic demand, which is also steadily increasing. The future of Cymbidium farming in Nepal appeared quite promising, despite challenges that could hinder commercial development (Pun 2016), but the growth of this sector is not satisfactory as expected (Figs. 11.1 and 11.2).

11.3 Climate

In general, Cymbidiums are a genus of orchids that prefer a cool climate. The key environmental factors important for growing Cymbidiums are briefly outlined.

11.3.1 Temperature

Cymbidium hybrids prefer a cool climate, and the ideal temperature for most hybrid Cymbidiums is around 20 °C. Nevertheless, temperatures lower than this are better for the growth of the plants than temperatures higher than the ideal (Ichihashi et al. 2007). In current growing areas in Nepal, such as the Godavari area of Lalitpur district, the average maximum temperature is 30 °C, while the average minimum temperature is 5 °C. However, there are exceptions when the minimum temperature

Fig. 11.1 Orchid farm (Himalayan Flora Enterprises, Godavari, Lalitpur)

Fig. 11.2 Orchid farm (Himalayan Flora Enterprises, Godavari, Lalitpur)

falls below the average minimum. In Cymbidium Sweet Heart "Spring Pearl", temperatures of 15–20 °C facilitate higher stomatal conductance and CO_2 uptake (Ichihashi et al. 2007). This indicates higher photosynthesis activity at these temperatures, resulting in more photosynthates and faster plant growth. Additionally, with higher relative humidity, photosynthetic activity is greater at 20 °C than at 15 °C. Similar trends were observed in other varieties too (Ichihashi et al. 2007). This implies the importance of ideal temperature and relative humidity for best growth of *Cymbidium* plants.

11.3.2 Relative humidity

Epiphytic orchids derive water and nutrition from the atmosphere through aerial roots, aided by velamen, a spongy tissue that covers these roots. Thus, for Cymbidium, relative humidity is equally crucial for photosynthetic activity (Ichihashi et al. 2007). The higher stomatal conductance and increased CO_2 uptake at 70% humidity compared to 30% demonstrate the importance of higher relative humidity during the daytime (Ichihashi et al. 2007). It is clear that maintaining a humidity of 70% or more during the day is essential for the commercial cultivation of Cymbidium. In the Godavari area of Lalitpur, managing 70% humidity was a challenge during the winter, spring, and summer. Hence, frequent water spraying was the practice but better option needs to be explored.

11.3.3 Light Intensity

Cymbidium naturally grows on a host tree in a forest, so it does not receive full sunlight. In a greenhouse environment, a 50% shade net is used to reduce light intensity in summer and has been found effective in balancing plant growth. In cooler locations (less than 20 °C), a shade lower than 50% can be used for optimal Cymbidium growth. Light intensity is also vital because shading more than 70% negatively affects photosynthesis, growth, and flowering of Cymbidium. The ideal summer light intensity inside the growing area after shading should be around 30 Klux, which equals approximately 555 μmol m^{-2} s^{-1} of PAR.

11.4 Greenhouse Structure

Cymbidium requires a suitable environment for both vegetative growth and flowering. It is therefore cultivated inside a structure where factors such as temperature, relative humidity, and light intensity can be controlled. The greenhouse could be dome-shaped or have a side-slant design with top ventilation. It can be covered with UV plastic sheeting, and additional wiring (10′) should be installed internally to provide 50% shade during the summer months. The greenhouse should have sufficient height (15–18′) for good ventilation, an insect net on all sides with double doors for entry and exit (Pun and Shrestha 2021). It should be equipped with planting benches (3′ high and 1 m wide) and a floor that can be cleaned but also absorbs moisture when irrigated or misted. The setup should include a fogger and lighting to enable humidity and light treatments as needed. The location of the greenhouse and the maintenance of optimal environmental conditions inside it are critical to the successful cultivation of Cymbidium flowers. Every effort should be made to select a site and construct a greenhouse where the maximum average daytime temperature remains around 20 °C, with an average night temperature of about 15 °C during the growing season (Ichihashi et al. 2007). However, for better quality flowers, the maximum average daytime temperature should not exceed 30 °C, and the night temperature should be around 25 °C.

11.4.1 Planting Bench

Planting benches are structures inside the greenhouse that support the containers. These benches can be 1 m wide, with a height of 3 feet, and their length can be adjusted depending on the size of the greenhouse (Pun and Shrestha 2021). They can be constructed from a GI frame welded with GI netting or made of bamboo, including a bamboo structure and bamboo matting. Since bamboo is biodegradable, it would only last a few years.

11.4.2 Container

Cymbidium plants need various types and sizes of containers (Pun and Shrestha 2021). Different types of containers can be used for cultivating Cymbidiums, such as clay containers, plastic pots, or plastic bags. However, the size of the containers depends on the stage of the plants' growth. Deflasked Cymbidium plantlets are planted in community pots, and as the plants grow, the size of the containers is increased accordingly; individual plants (5 cm), robust-sized plants (15 cm), and flowering stage plants (30 cm) (Pun and Shrestha 2021).

(a) **Community tray (pot)**: Community trays can be used for planting small plantlets after deflasking and fungicide treatment.
(b) **Independent plant pot**: Plants from community trays (pots) can be transplanted into an individual pot (5 cm diameter), but can still be placed in a holding tray.
(c) **Good-sized plant pot**: A plant from an independent pot can be transplanted into a suitably sized plant pot (15 cm diameter) and can be positioned independently.
(d) **Flowering stage plant pot**: A plant from a good-sized pot can be transplanted into a flowering-sized pot (30 cm) and can be placed independently.

11.4.3 Media

Cymbidiums, being epiphytic, prefer media that retain moisture and release it slowly, without constantly keeping the media wet; it must be well-drained (Pun and Shrestha 2021). The various types of media used include sphagnum moss, peat moss, gravel, charcoal, brick pieces, conifer barks, and a mixture of FYM, leaf litter, and soil. These media are commonly used either as a mixture or alone for the commercial cultivation of Cymbidium. Sphagnum moss or peat moss are typically used as media for newly deflasked plantlets. Pine bark has become very popular in recent years, with good success, and is the most common medium for Cymbidium in Japan and China. However, the choice of pine bark depends on the size of the plants. The bark size should be smaller for individual plants (0.5–1 cm), increasing as the plant grows and reaches flowering size (2–3 cm). If imported, pine bark must be thoroughly washed with fresh water before use to prevent salt injury, particularly to the leaves. These media should be fumigated with hydrogen peroxide combined with nano silver (Silvox) at 5 mL/L of water prior to use. After drenching, the media should be covered to create an airtight environment for 2–3 h. Unlike other fumigants, the media can be used within hours after treatment with Silvox.

11.5 Cultivars

Cymbidium cultivars for commercial production are mainly classified into three groups. These groups are standard Cymbidium, intermediate Cymbidium, and miniature Cymbidium. Some examples of cultivars grown in Nepal under different groups are also listed.

(a) Standard Cymbidium: This group of Cymbidium has longer spikes of cut flowers (>60 cm), a greater number of florets (>12), and larger florets. Examples include Golden Girl, Blazing Red Star, Crystal White, Lovely Bonny, Monalisa, Arabian Night, etc.
(b) Intermediate Cymbidium: This group of Cymbidiums has flower spikes of intermediate length (30-60 cm), with fewer florets and smaller florets compared to standard Cymbidiums. Examples include Pink Lady, Show Girl, Cooksbridge, Amesbury, Yanki Lily, Bertha, and Age 15 etc.
(c) Miniature Cymbidium: This group of Cymbidium bears flower spikes of shorter length (<30 cm) and is more commonly used as potted plants. For example, Linen Hall, Minisari, Kenwine colour, etc.

11.6 Plant Propagation

Propagation of Cymbidiums can be achieved through tissue culture and division (Pun and Shrestha 2021). Typically, mass production using tissue culture techniques (micropropagation) of popular cultivars is undertaken by plant propagation companies. Nonetheless, division of Cymbidiums can be performed at the grower's level. Division can be carried out in two ways: splitting a larger clump that has outgrown the pot, or using back bulbs as propagation material.

(a) **Splitting of bigger bunch of plants**
This technique is used to divide a large group of healthy plants that have outgrown their pots. The division can be performed in spring but not during the flowering stage or after flowering. The overgrown plants should be watered a day in advance, and the media or container for the new cuttings should be prepared. The overgrown plants should be removed from the pot and separated using a clean secateur. The secateur should be wiped with alcohol or disinfectant after cutting a plant and before using it on another. Splitting should be done so that each section contains three generations of pseudobulbs: a back bulb or recently flowered pseudobulb, a well-developed pseudobulb, and a new sucker. The newly divided clusters should be treated with 0.5% systemic fungicide by dipping for 10 min and then allowed to dry. After drying, each cluster of pseudobulbs from these three generations should be planted in a pot, watered, and kept in a shady location for about 2 weeks. These plants can then be moved to a normal 50% shaded environment after 2 weeks, once they show no signs of stress.

(b) **Back bulb**
The use of back bulbs is not very common, but if the cultivar is popular and the plant is healthy, the back bulb (a pseudo bulb that has completed its flowering cycle) can be used for propagating new plants. Back bulbs can be removed either by twisting or cutting with a clean knife. The knife should be wiped with alcohol or disinfectant after each cut before using it on another plant. The detached back bulbs should be dipped in a 0.5% systemic fungicide solution for about 10 min, then removed and allowed to dry. Once dried, the back bulbs can be planted in a moist sand bed placed in a shady spot. In approximately 8–10 weeks, new suckers will emerge from the back bulbs, which will become papery over time. This plant can be ready for independent planting in about 6–8 months.

11.7 Market

The market centre for Cymbidium potted plants or cut flowers is primarily the Kathmandu Valley. Buyers mainly visit production centres to buy flowering potted plants directly from growers. However, cut flowers are marketed through wholesalers or directly to florists and consumers. Potted flowering plants are the main source of income for growers. The price of potted flowering plants varies depending on the cultivar, size, and number of flower stems, ranging from Rs. 2000 to Rs. 5000 or more. In contrast, the price of cut flowers depends on the cultivar, length of the stem, and number of florets, costing up to Rs. 150. The wholesale and retail prices of Cymbidium are significantly higher than those of other cut flowers

Table 11.1 Wholesale price (Nepalese Rupees) of different cut flowers during different times in Nepal

Cut flower	1998	2005	2009	2014	2015	2025
Gladiolus	6.0	5.0	5.0	8.0	9.0	-
Dutch rose	10.0	7.0	7.0	9.0	8.5	30
Carnation	8.0	7.0	7.0	8.0	8.5	20
Cymbidium	**15.0**	**31.0**	**80.0**	**100.0**	**150.0**	-

Table 11.2 Retail price (Nepalese Rupees) of different cut flowers during different times in Nepal

Cut flowers	2005	2006	2009	2014	2015	2025
Gladiolus	9.5	9.5	10	15.0	15.0	-
Dutch rose	14.0	14.0	15	20.0	15.0	70
Carnation	15.5	15.5	16.0	20.0	18.0	35
Cymbidium	**72.0**	**72.0**	**120**	**150.0**	**250.0**	-

(Tables 11.1 and 11.2), but the present trend is in selling the potted flowering plants instead of cut flowers. This is reflected in no Cymbidium cut flowers report at the wholesale and retail in 2025 because Cymbidium are mostly traded as flowering potted plants, fetching a higher price.

11.8 Cultivation Technology of Cymbidium

The information above details the prerequisites of infrastructure and other essentials for successful Cymbidium farming (Pun and Shrestha 2021). Therefore, it is essential to have these structures well-maintained, along with skills and market information, before venturing into Cymbidium cultivation.

11.8.1 Planting Materials

The selection of planting materials depends on how the grower or company wishes to start the project. It could be deflasked tissue-cultured plantlets, independent plants, suitably sized plants, or flowering-sized plants. The grower may plan a combination of these categories or opt to start with just one. The more advanced the age of the plant, the higher the cost of the planting materials, but it will also generate cash flow earlier.

11.8.2 Planting

The planting process for plants of various ages is as follows.

Tissue culture plants: Tissue-cultured plants can be used for planting from mid-spring onwards and in summer (Pun and Shrestha 2021). However,

Fig. 11.3 Newly deflasked and transplanted plantlets in sawdust medium

mid-spring is ideal because the climate is mild and the plants have a longer growing period. Assuming the greenhouse is operational, the container and media are prepared, and tissue-cultured plants can be ordered. These plantlets can be sent in a flask or after being deflasked and cleaned. In the case of flask-grown plants, remove the plantlets and wash off the agar attached to the plants with clean water, then let them dry. For deflasked and cleaned plantlets, dip them in systemic fungicide (0.5%) for about 10 min, remove, and allow to dry. Similarly, flask-grown plantlets should also be dipped in systemic fungicide (0.5%) for about 10 min. Once dried, the plantlets are ready to be transplanted into community trays (community pots) (Plate # 3). Each plantlet should be placed 2–3 mm into the media and then sprayed with a fine mist of water to thoroughly dampen the soil. The greenhouse humidity should be maintained at 70% or above.

Independent size plants: These categories of plants could be transported dry by the supplier. Upon arrival at the company premises, the consignment shall be treated with systemic fungicides (0.5%) for 10 min and then left to dry. After drying, plants will be transplanted according to their growth stage.

Planting of various stages of plants

1. Community plants
   The plants that have been recently deflasked or are deflasked and then transferred to a community tray or pot (Fig. 11.3).
2. Independent plants
   The plantlets in community pots with 3–4 well-developed leaves should be transferred to larger, separate pots in a community tray (Fig. 11.4).

Fig. 11.4 Independent plant grown in a separate pot within a community tray with slow-releasing fertiliser

Fig. 11.5 Independent plant grown in a separate and larger pot with slow-release fertiliser

3. Good size plants
   The plants with 5–6 well-developed leaves and growing pseudo bulbs should be transferred into an independent pot (Fig. 11.5).
4. Flowering stage plants
   The plant should have 6–8 well-developed leaves, a well-grown pseudo bulb, and at least one side sucker. This plant will be ready to flower in the upcoming flowering season (Fig. 11.6).

Fig. 11.6 Potted flowering plant for display and sale

11.8.3 Irrigation

Cymbidium plants need irrigation daily, especially during dry summer months. Irrigation should be carried out in the morning and early afternoon on hot summer days (Pun and Shrestha 2021). No irrigation or fogging should be done after 4 pm. Avoiding irrigation or misting after 4 pm helps prevent disease infestation. Proper irrigation not only promotes plant growth but also increases the relative humidity of the greenhouse (Pun and Shrestha 2021). A fogger can be installed in the greenhouse to maintain humidity above 70% during the day. The risk of overwatering is higher in local media where the water-holding capacity of the medium (FYM/leaf litter/soil/sand) exceeds that of newer media such as pine barks (Pun and Shrestha 2021). Conversely, underwatering may occur in media mainly composed of pine bark. Therefore, it is important to follow a strict irrigation regime tailored to the media type and stage of plant growth.

11.8.4 Nutrition

Cymbidium plants require a constant supply of nutrients, and therefore, in major Cymbidium-growing countries such as Japan or China, slow-releasing fertiliser is applied (Pun and Shrestha 2021). This slow-releasing fertiliser (Osmocote) is available in five longevity options: 3–4, 5–6, 8–9, 12–14, and 16–18 months and is convenient to use (ICL 2025). This fertiliser consists of small granules encapsulated by cellulose, allowing only a small amount of nutrients to be released each time the pot is irrigated. Osmocote is available in different combinations of NPK for the suitability of use for vegetative growth or inducing flowering. In Nepal, this fertiliser is not available; therefore, in pine bark media or local media, NPK (19:19:19) at 0.2% sprayed fortnightly showed a good response in vegetative growth (Pun and Pun 2013). An oil cake and fish meal mixture is also used in Nepal for Cymbidium (Kabiraj Rai pers communication 2nd July 2025). Similarly, NPK (30:10:10) at 0.1% sprayed fortnightly increased the length and number of leaves of Cymbidium hybrid Showgirl 'Cooksbridge' (Barman and Upadhyaya 2003a). Hastening flowering in the same hybrid was achieved using NPK (20:10:10) at 0.1%, (20: 20:10) at 0.1%, and (20:30:10) at 0.1%, all sprayed fortnightly (Barman and Upadhyaya 2003a). A mixture of oil cake and bone meal (2:1 or 1:1) provides a good source of nutrition for Cymbidium. The dosage for Cymbidium plants growing in 6 cm pots was 2 g of the mixture per pot per month, whereas for plants in 15 cm pots, it was 30 g per pot per month. Goat manure and neem cake, each diluted (1:25 manure to water), sprayed fortnightly on Cymbidium hybrid Showgirl 'Cooksbridge', significantly increased plant height and leaf length compared to the control (Barman and Upadhyaya 2003b).

11.9 Plant Management

The Cymbidium plants should be well cared for to ensure healthy growth, and no side suckers should be allowed to develop until the plant reaches the good size pot stage. The removal of these suckers (Mekaki in Japanese) helps direct all the carbon towards the plant and pseudobulb development (Pun and Shrestha 2021). Plants in the flowering stage may have three different sizes: large pseudobulb, intermediate pseudobulb, and smaller pseudobulb. This technique ensures rapid growth of the plant, shortens the time to flowering, and promotes a regular bearing habit in the potted plants. Similarly, when Cymbidiums are commercially grown in regions with hot summers and cold winters, flowering-sized plants are often moved to higher altitude areas (cooler locations during the hot summer months), a process often called 'mountain climbing', to induce flower bud initiation and prevent damage to young flower buds from heat and typhoons. Later, in early winter, when temperatures are cooler, all the plants are moved to a climate-controlled facility, which allows flowering, and they are shipped when ready for sale. This method is known as Yamagee in Japanese.

11.9.1 After Care (Weeding, Plant Placement)

Among aftercare activities, weeding may be significant in local media, but it is not often necessary in pine bark media (Pun and Shrestha 2021). However, plant placement could be an important activity because, as plants begin to grow, the leaves of two pots may become entangled. Such a situation should be avoided to ensure good aeration and prevent fungal or bacterial infections. Additionally, adequate space and proper ventilation for each plant promote effective photosynthesis and higher carbohydrate accumulation.

11.10 Plant Protection

Cymbidium plants are generally healthy, but they are also attacked by many types of diseases and pests (Pun and Shrestha 2021). Some important diseases and pests are listed here with control measures.

(a) **Leaf spot**

Symptoms: Symptoms appear as buff-coloured areas with dark, mousy grey margins. Initial sporadic spots later merge to form irregular, elongated areas.

Causal organism: *Fusarium species*

Control measures: Apply systemic fungicides at 0.1%. Use 0.2% Silvox fortnightly to manage this disease.

(b) **Plant rot/Black rot**

Symptoms: Pseudo bulbs turn yellow, followed by yellowing, wilting, and death of the affected plant. Plants affected by plant rot/black rot produce water-soaked spots on the infected parts. In the advanced stage, the spots turn black and dry, resulting in complete defoliation of the plant. This disease can become severe during the monsoon season or when humidity is high.

Causal organism: *Phytophthora species*

Control measures: Spray with Ridomil at 0.05% a.i. or Zineb at 2 g/L of water as a prophylactic measure during pre-monsoon and monsoon seasons. Remove and destroy infected plant parts.

(c) **Leaf tip burn**

Symptoms: Infected leaf tips develop spots that eventually merge, causing the entire leaf tip to die. Dead leaf tips are often covered with a mass of powdery spores.

Causal organism: *Botrytis species*

Control measures: Remove dead leaf tips. Drench the tips with fungicides such as Bavistin or Dithane M-45 at 0.1%.

(d) **Cymbidium mosaic virus**

Symptoms: Leaves show yellow or white distinct mottling.

Control measures: The best control measure for this disease is to burn all infected plants. Additionally, eradication of aphids and disinfection of cutting tools after each use are very important.

(e) **White grub**

Characteristics: Damages the roots, causes yellowing of the leaves, and wilting of the plant. The occurrence of such infestations is more commonly observed when traditional media containing farmyard manure is used.

Control: Grubs can be controlled by spraying 2 mL/L neem oil or Chloropyriphos 20 EC at 2 mL/L during the summer and monsoon periods.

(f) **Aphids**

Characteristics: Both nymphs and adults suck sap from tender shoots and flower buds. Honeydew secreted by aphids attracts ants and fungi.

Control: Aphid infestation can be prevented by prophylactic application of systemic insecticide such as Malathion before the population increases, using a lower concentration of 1 mL/L. A higher concentration of 2 mL/L may be required if the infestation is more severe.

(g) **Scale insects**

Characteristics: It appears as a small, brown, raised spot on the foliage. It feeds on the plant by sucking the sap, which causes the plant to lose vigour and growth.

Control: Remove scale insects by rubbing the scurf encrustations with a toothbrush or cotton swab dipped in 70% isopropyl alcohol or methylated spirit. If an infestation is found on roots, repotting should be carried out, and after gentle cleaning, the roots should be sprayed with insecticide. Spray malathion 2 mL, dimethoate 1.5 mL, or monocrotophos 1 mL/L of water. A mixture of 5–10 mL mineral oil plus 1 mL systemic insecticide per litre of water is also effective for pest control.

(h) **Snails/Slugs**

Characteristics: They mainly feed on tender parts such as root tips, but also damage stems, leaves, and flower spikes. The attack usually occurs at night. Presence can be identified by silvery trails created by their movement.

Control: Mechanical picking occurs at night, followed by destruction. Chemical control can also be achieved using Slugit, Methiocarb, and aluminium sulphate.

11.11 Preharvest Care

Around one third of the post-harvest longevity of orchids is reported to be affected by the methods of preharvest handling (Awasthi et al. 2024). Preharvest care starts with identifying flower buds and placing a stake near each bud. As the flower bud appears, it should be tied to the stake to ensure straight growth (Pun and Shrestha 2021). If a pot has multiple flower spikes, each spike needs its own stake. In commercial greenhouses for Cymbidium, vertical trellising is used for this purpose. The developing flower spikes and buds are closely observed to prevent pest attacks. Care is much easier inside a greenhouse with insect netting, but it is more challenging in a simple greenhouse without one.

11.12 Harvesting, Grading, and Pricing

Harvesting of flower stems is usually done in the morning using a clean secateur and disinfectant to wipe the blades while cutting spikes between plants (Pun and Shrestha 2021). For the local market, spikes are harvested when 75% of the flowers have opened. Furthermore, harvesting at this stage has shown the highest shelf life with or without postharvest treatment (Bharathi et al. 2014). After harvesting, the spikes are placed in a bucket filled with water and transported to the grading room. Grading of the spikes is based on the types of Cymbidium, such as standard, intermediate, or miniature Cymbidium, as specified in the section Cultivars.

The price of Cymbidium flowering plants is graded based on the rarity of the cultivar, plant size, number of flowering stems, and the supply and demand situation. All types of Cymbidium are sold as flowering potted plants.

11.13 Postharvest Care and Storage

Cymbidium stems are not treated locally in Nepal, but if treated with 0.2 mM silver thiosulphate (STS) for 30 min, the vase life of the cut Cymbidium could significantly increase. Silver treatment also negates the flower's response to pollinia disturbance, thereby reducing postharvest loss. It is highly recommended to treat Cymbidium cut flowers with STS for a longer shelf life. Recently, nano silver has been reported to improve the shelf life of both ethylene-sensitive cut flowers, such as carnations (Naing et al. 2017; 25 ppm), and ethylene-insensitive cut flowers like gerbera (Liu et al. 2008; 5 ppm). Therefore, nano silver could also be effective in extending the shelf life of Cymbidium. Nano silver is relatively inexpensive, easily available, and can serve a dual purpose of soil fumigation and also has the dual function of extending shelf life for both ethylene-sensitive and ethylene-insensitive cut flowers.

11.14 Packaging and Transport

The cut Cymbidium flower stems, after being graded and treated with postharvest care, are packed in a perforated plastic sleeve for each stem (Pun and Shrestha 2021). This helps cushion the florets and the cut stem, preventing injury. These sleeved stems are then placed in a cardboard box with perforated holes. Four Cymbidium stems are arranged per layer, covered with a fine paper, and another layer of four stems is added until the box is full. The box is sealed and taped to prevent movement of the stems. Currently, in Nepal, Cymbidium cut flower stems are transported in a non-refrigerated van to the wholesale centre or retail store.

11.15 Conclusion

Cymbidium is the major orchid genus, commercially cultivated for its beautiful cut flowers and potted flowering plants. Commercial cultivation of Cymbidium requires the maintenance of specific physical conditions, nutritional requirements, and an integrated pest management approach. It is also essential to identify suitable locations before embarking on a planned production of Cymbidium orchids. Besides, there should be effort to introduce better varieties and improve the production technology especially in the developing countries, having climatic conditions suitable for Cymbidium production. An integrated approach will enable strengthening the economy through floriculture.

References

Awasthi M, Chandra A, Malik S (2024) Conservation of endangered medicinal orchid species through post-harvest handling for future sustainable use. In: Husen A, Pant M (eds) Exploring medicinal orchids. Taylor & Francis Group, LLC, pp 24–36. https://doi.org/10.1201/9781003436638-1

Barman D, Upadhyaya RC (2003a) Effect of NPK on growth and development of Cymbidium hybrid. Annual Report 2002–2003 National Research Centre for Orchids, ICAR, Pakyong, Sikkim, India, p 22

Barman D, Upadhyaya RC (2003b) Effect of organic manures on growth and development of Cymbidium hybrid. Annual Report 2002–2003. National Research Centre for Orchids, ICAR, Pakyong, Sikkim, India, p 21

Bharathi TU, Barman D, Naik SK (2014) Effect of harvesting stages and chemical preservatives on postharvest life of Cymbidium hybrid 'Red Princess'. Vegetos 27(1):188–194

Dahal S, Shakya PR (1989) A glimpse of Orchid flora of Nepal. Proceedings of the National Conference of Science and Technology, RONAST, Khumaltar, Lalitpur

Gadal B, Pun UK, Maharjan PK (1996) Diversity of orchids in Makwanpur and Parsa Districts. IAAS Res Rep 1994–95:53–67

Ichihashi S, Yamauchi D, Ota K (2007) Investigation on photosynthetic aspects of Cymbidium orchid (In Japanese). Bull Aichi Univ Educ 56:29–32

ICL (2025) https://icl-growingsolutions.com/ornamental-horticulture/super-brands/osmocote/. Accessed 2 July 2025

Liu J, Zhang Z, Joyce DC, He S, Cao J, Lv P (2008) Effect of postharvest nano silver treatments on cut flowers. Acta Hortic 847:245–250

Mukherjee SK (1983) Orchids. Indian Council of Agricultural Research, New Delhi, India

Naing AH, Win NM, Hang JS, Lim KB, Kim CK (2017) Role of nano silver and the bacteria strain *Enterobacter cloacae* in increasing vase life of cut carnation 'Omea'. Front Plant Sci 8:1590

Pun UK (2014) Orchid farming potentiality and commercialization (aesthetic and medicinal orchids) in Nepal. Proceeding of National Workshop on NTFP/MAPs sector action plan

Pun UK (2016) Commercial orchid farming in Nepal: opportunities and challenges. Nepalese Floricult 20:33–35

Pun AB Pun UK (2013) Effect of media, nutrition and sucker management on growth and flowering of Cymbidium orchid. Proceeding of Orchid Workshop, Kathmandu

Pun UK, Shrestha SR (2021) Orchid Cultivation Technology, Floriculture Development Center, Directorate of Agriculture, Bagmati Pradesh, Godavari, Lalitpur, Nepal

Tiruwa BL, Neupane BD, Kadariya R, Pokharel CP, Pant B (2024) Diversity of orchids in terms of their distribution, uses and conservation in Annapurna Conservation Area of Nepal. Am J Plant Sci 15:422–440

Status and Scope of Orchid Cultivation as Pot Plants

12

L. C. De, S. S. Biswas, Suman Natta, and S. P. Das

12.1 Introduction

The ornamental plant industry is a growing sector that provides significant advantages over agricultural sector in several countries. The main attribute of floriculture sector is the production of aesthetical items (Wani et al. 2018; Ferrante and Ferrini 2023). During 2021, the floriculture at Royal Flora Holland in the Netherlands was valued at 5.6 billion euros showing an increase 4.8 billion euros in 2019 (Salachna 2022). In 2019, the flower industry in the United Sates created 2.32 million jobs for farmers, traders, and exporters. The ornamental plant trade sector contributed $17 billion which is equivalent to 10% of total horticultural crop production in the country (Wei et al. 2023).

Global floriculture exports had consistent growth over the past 5 years, with a compound annual growth rate (CAGR) of 3.9%. All categories, including cut flowers, potted plants, ground cover and balcony plants, bulbs, perennials, shrubs, and trees, have all seen an increase trend. Cut flowers are those flowers that are cut with a little stem to be arranged in arrangements, bouquets, or decorations. Cut flowers have a long shelf life and constitute a large part of the world's floriculture output (Malviya et al. 2022; Mayak et al. 1973), making them a more export-orientated commodity. In a sharp contrast, loose flowers are those flowers that are picked under the petals and used in everyday activities such as worship, weddings, and garden decoration and are mostly for sale in domestic market.

Orchids constitute one of the most popular plants in the global floriculture trade. They are popular ornamental plants because of exceptional variety of colours, forms, and flowers with long shelf life. The Orchidaceae family is divided into five subfamilies: Apostasioideae (2 genera and 16 species), Cypripedioideae (5 genera and 130 species), Vanilloideae (15 genera and 180 species), Orchidoideae

L. C. De (✉) · S. S. Biswas · S. Natta · S. P. Das
ICAR-NRC for Orchids, Pakyong, Sikkim, India

A. Husen, M. Pant (eds.), *Forest-based Orchids*,
https://doi.org/10.1007/978-981-92-0096-2_12

(208 genera and 3630 species), and Epidendroideae (more than 500 genera and 20,000 species) (Freudenstein and Chase 2015; Tsai et al. 2017; Lu et al. 2019; Song et al. 2022; Zhang et al. 2023). About 70% of orchid species grow epiphytically on trees or rocks, which makes them highly vulnerable to environmental changes (Zotz 2013). They display either monopodial or sympodial growth patterns (Teo et al. 2019).

Hybrids of orchids like Dendrobium, Vanda, Cymbidium, Phalaenopsis, etc., have a significant contribution in the world flower market as potted plants and cut flowers (Yuan et al. 2021). Besides, many species are used in traditional medicine and as food sources (Kreziou et al. 2016; Teoh 2016; Veldman et al. 2018). Owing to their high value, orchids have been illegally collected and traded for various purposes, while also facing the threats of climate change and habitat loss (Subedi et al. 2013; Hinsley et al. 2018; Williams et al. 2018; Gale et al. 2019; Besi et al. 2023).

The Convention on International Trade in Endangered Species of Wild Fauna and Flora (CITES) encompasses all orchid species to control the trade of wild orchids, accounting for over 90% of all plant species listed in its appendices (CITES 2024). More than half of the assessed orchid species are considered threatened based on the IUCN Global Red List (IUCN 2023). Additionally, around 300 orchid species are included in China's List of National Key Protected Wild Plants, making up over 25% of the country's protected plant species (Tang and Jiao 2022).

12.2 Global Scenario of Orchid Trade

In 2023, global trade of orchids reached $194M, reflecting a 2.38% decrease from 2022, when trade totalled $199M. Over the past 5 years, trade in this category has declined at an annualized rate of 4.89%. Among the 5026 products traded in 2023, orchids ranked 3907 in global trade value, accounting for <0.005% of world trade. In 2023, the leading exporters of orchids were Thailand ($69.8M), Netherlands ($59.5M), and Chinese Taipei ($35.3M). The top importers were Japan ($56.3M), United States ($24.7M), and Vietnam ($14.4M). In 2023, the countries with the highest share of Orchids in their export portfolios were Thailand (0.021%), New Zealand (0.012%), and Netherlands (0.0098%).

12.3 Dynamics of Orchid Market

12.3.1 Increasing Demand for Orchids in Interior Decoration, to Increase the Demand Globally

Orchids are prized for their beauty, elegant forms, and variety, making them a common choice for interior decor. Their use in interior design has increased steadily due to their vibrant colours, unique shapes, and long-lasting flowers. Interior designers appreciate orchids for their versatility, as they complement many styles from modern to traditional. This trend is expected to further boost the demand for orchids in the interior decoration market, especially among urban consumers seeking to improve the aesthetic of their homes and workplaces.

12.3.2 Growing Popularity of Orchid-Based Therapies and Aromatherapy to Propel Market Growth

Orchids are gaining recognition for their health benefits. The use of orchid essence and oil is increasing in the wellness sector because of their calming and revitalizing effects. Moreover, orchids are incorporated into aromatherapy for their soft, soothing aroma, believed to promote relaxation and stress relief. As consumers increasingly prioritize holistic wellness and natural remedies, the demand for orchid-based therapies and aromatic products is likely to rise, fostering further growth in the orchid market.

12.3.3 Limiting Factors for the Orchid Market

Environmental factors to limit the sales: Orchids are highly sensitive to environmental changes, which pose a significant challenge in the market. They need specific temperature, humidity, and light conditions, making them vulnerable to climate and weather variations. Climate change worsens this situation by altering their growing environment and impacting their health and development. Extreme weather events such as droughts, floods, and storms can harm habitats and disrupt their natural growth cycles. Furthermore, deforestation and habitat destruction decrease the wild orchid populations. To address these challenges, initiatives are being developed to breed more resilient orchid types and adopt sustainable cultivation methods that lessen environmental impact.

Lighting issues pose a major challenge to market expansion: Lighting issues significantly hinder the growth of the orchid market, particularly in commercial cultivation and home gardening. Orchids require very specific light conditions, which differ among species but generally fall within a narrow optimal range. These precise needs make orchids more sensitive to typical horticultural environments, posing challenges for both mass producers and individual growers. Proper lighting is vital for healthy flowering and development. Insufficient light can lead to weak, elongated stems, delayed blooming, or complete failure to flower. Conversely, excessive light can scorch leaves, cause dehydration, and induce stress-related diseases. Finding the right balance is essential, requiring thorough knowledge of each orchid species' light tolerance and how they respond to different lighting conditions, whether natural or artificial.

12.4 Scope of Potted Orchids

Potted orchids have a significant and growing presence in the global market as both ornamental plants and, increasingly, for their medicinal and other properties. The global market for orchids, including potted varieties, was valued at $751.9 million in 2023 and is projected to reach $1176.62 million by 2032, according to Zion Market Research.

Dominant in Floriculture: Orchids, including potted varieties, are a major component of the global floriculture trade, with some estimates placing their share at around 10% of the international fresh cut flower trade.

Growing Market: The market for potted orchids is expanding, driven by increasing demand for ornamental plants and the increasing global trade in orchids, including for medicinal and other uses according to ScienceDirect.com.

Leading Varieties: Phalaenopsis orchids are a particularly significant and growing segment of the potted orchid market.

Global Production: Countries like the United States, China, Germany, Japan, Netherlands, Taiwan, and Thailand are major players in the large-scale production of potted orchids according to NEHU.

Beyond Ornament: Potted orchids are also valued for their medicinal properties, being rich in polysaccharides, alkaloids, and other chemical components. They hold a special place in various cultures, particularly in East Asia, where they are symbols of refinement and grace.

Economic Importance: Potted orchids contribute significantly to the global floriculture trade, with an 8% share, and have potential for developing economies.

12.5 Production Status

In India, orchids belong to 158 genera and 1331 species, thriving up to elevations of 5000 meters (Chowdhery 2009). Terrestrial orchids are commonly found in moist, humus-rich soils under tree canopies in Northwestern India. The Western Ghats host small-flowered orchids, while epiphytic orchids are frequent in Northeast India, growing up to 2,000 meters above sea level (Singh 1990). Indian orchids with high ornamental value are used as breeding material. Native genera such as Cymbidium, Paphiopedilum, Vanda, Arachnis, and Dendrobium are cultivated extensively for cut flower production. Cymbidiums are mainly grown in the NEH Region, Sikkim, the Darjeeling hills, Arunachal Pradesh, and Assam. Tropical orchids are cultivated in Kerala, as well as in parts of Tamil Nadu, Karnataka, and Maharashtra. It is advisable to select species that flower during winter and spring to facilitate exporting flowers to temperate regions from December to May.

12.5.1 Temperate Orchids (De 2020a, b)

Cymbidium and *Paphiopedilum,* the temperate orchids, are now grown in the Sikkim, Arunachal Pradesh, Meghalaya, Himachal Pradesh, and Darjeeling Hills of India.

Cymbidium: 'Levis Duke Bella Vista', 'Madrid Forest King', 'Sparkle Late Green', 'Angelica December Gold', 'Sleeping Nymph', 'Pine Clash Moon Venus', 'Soul Hunt', 'Dr. H. C. Aurora', 'Susan Highes', 'Tia Gaig Suther Land', 'Miss Sanders', 'Amesbury', 'Kenny Wine', 'Red Star', 'Red Princess', 'Show Girl, Jungfrau 'Snow Queen', Jungfrau 'Dos Pueblos', Lilian Stewart 'Coronation', Lilian Stewart 'Party Dress', Orkney 'Pink Heather', 'Ensikhan', 'Alpha Orient', 'Winter Beach Sea Green', 'Fire Storm Ruby', 'Fire Storm Blaze', 'Angelica Adven't, Dingwall 'Lenes', Peter pan 'Green Leaves', 'Everett Stockstill Bullai', 'Samurai Hee Haw Sagun', 'Datin Doll', 'Bertha Peterson', 'Bob Marlin Lucky', 'Carypepper Peachy Keen', 'Red Beauty Evening Star', 'Rocky Creek Pebbles', 'Valley Legend Steff', 'Satin Doll', 'Pontac Flame Rocky', 'Hero Boy', 'Khan Flame OC Brown', 'Sandrige Torch'.

Paphiopedilum (Lady's Slippers Orchid): A pseudobulbless, sympodial terrestrial or lithophytic orchid is grown in cool and shady conditions. Common varieties are 'Joyce Hasegawa', 'Lynleigh Koopowitz', 'Magic Lantern', 'Harold Koopowitz', 'Magic Lantern', 'Armeni White', 'Red Glory', 'Black Pepper'.

12.5.2 Tropical Orchids (De 2020a, b)

Tropical orchids are cultivated in Kerala, Assam, Tripura, Goa, West Bengal, and some parts of Tamil Nadu, Karnataka, and Maharashtra.

Dendrobium: It consists of a large number of species (over 1500), and several hybrids have been produced and released for commercial cultivation. Those popular in Kerala are 'Sonia 17', 'Sonia 17 Mutant', 'Sonia 28 Mutant', 'Hieng Beauty', 'Dorine White', 'Ekapol Panda', 'Sakura Pink', 'Pramot Sabine', 'Emma White', 'Kasem White', 'Mme Vipar', 'Kasem Gold', 'PM 11', 'Waiphu Beauty', 'Sarifa Fatimah' and 'Jiad Gold'. In north-eastern states, popular varieties are 'Abraham', 'Madam Pompadour', 'Triple Pink', 'Thongchai Gold', 'Bangkok Blue', 'Erika', 'Lervia', 'Kating Dang', 'Emma White', and 'Dang Saard'.

Salient features and common varieties of other popular genera are as follows:

Arachnis (Scorpion orchid): It is monopodial, requires good sunlight and temperature for growth and flowering. *Renanthera* is a related genus. Common varieties are 'Maggie Oei Red Ribbon', Maggie Oei Yellow Ribbon', 'Maroon Magie', and 'Merry Maggie'.

Vanda: Its habit is that of *Arachnis*. Based on shape of leaf, terete (cylindrical leaved, pencil Vanda), semi-terete, and strap leaved vanda. *Aerides* and *Rhyncostylis* are related genera. Varieties: 'Roberts Delight', 'Dr. Anek', 'Pakchong Blue', 'Miss Joaquim', 'Fuch's Delight', 'Lumpini Red', 'Motes Indigo Blue', 'Pat Delight', 'Rasri Gold', 'Samsai Blue', 'Adisak Smile', 'Robert Sorenson', 'Kasems Delight',

'Richard Brandon', Kultana Delight', 'John Clubb', 'Josephine Van Brero', 'Norbert Alphonso', 'Wirat Uchida', 'Pures Wax', 'Berniece Miller', 'Kultana Violet', and 'Ruby Prince' are important.

Phalaenopsis (Moth orchid): Monopodial requires partially shaded conditions. *Doritis* is a related genus. Varieties: 'Strawberry', 'Manchester', 'Detroit', 'ChianXen Magpie', 'Shih Hua Long First Love', 'Taida Salu Red', 'Hsin Yang Fortune', 'Kaleidoscope', 'Rose Parade', 'Keith Shaffer', 'Temple Cloud', 'Hennessy', 'Angel Eye', 'Bilbao', 'Burgundy', 'Cover Girl', 'Day Tripper', 'Echoes', 'Friday Flirt', 'Happy Hour', 'Heart Breaker', 'Red Eye', 'Spring Glow', 'Summer Song', 'Sweet Heart', and 'Diana Pinky' are important.

Oncidium (Dancing Girl): Sympodial, with or without pseudobulb, loves partially shaded conditions. Varieties: 'Goldiana', 'Sharry Baby Sweet Fragrance', 'Taka Yellow', 'Sweet Sugar', 'Wild Cat Bob Cat', 'Dark Tower', 'Ruby Frost', and 'Gower Ramsay' are important.

Cattleya: Sympodial, loves partially shaded conditions. *Brassovola*, *Laelia*, *Sophronitis*, *Encyclia*, etc., are related genera. Varieties: 'Queen Sirikhit', 'Ahmad Sheikhi', 'Chinese Beauty Orchid Queen', 'Brazilian Star', 'Amazon Blue', 'Black Wonder', 'Little Bell', 'Suzzanne Hye', 'Purple Cascade Fragrant Beauty', 'Varut Valentine', and 'Edithe Bow Bells' are important.

Aranda (*Arachnis* x *Vanda*): 'Christine', 'Thailand Sunspot', 'Millenium Dawn', 'Broga Giant', 'Salaya Red', 'Prapin White', 'Prapin Spot', 'Lueng Cholburi', 'Ishbel Manisaki', 'Baytown', 'Chao Praya Blue', 'Chao PrayaDotCom', 'Chao Praya Beauty'. 'Madam Panne', 'Bangkok Gold', 'Chao Praya Gold', 'Chark Kuan Orange', 'Chark Kuan Pink', 'Kelvin Red', 'Kelvin Orange', 'Walter Ouame', 'Jitti Orange', 'Happy Beauty', 'Chark Kuan Red', and 'Happy Beauty.'

Vascostylis: 'Ique Pink', 'Kruewan', 'Banjong Jasmine', 'Mishima Lime', 'Vanity Fair', 'Firuz', 'Pine Rivers', 'Banjong Elite', 'Pichtawit Gold', 'Chao Praya Lime', 'Lanna Rosy', 'Jeans Delight', 'Bay Sapphire', 'Spring Hill.'

Aeridovanda: 'Bensiri', 'Noreen', 'Early Bird', 'Shiv Sidhu', 'New Dawn', 'Harrison Luke', 'Somsri Sunlight', 'Mikasa Blue', 'Mikasa Pink', 'Sanchai Spot.'

12.6 Nursery Structures for Orchid Plant Production

Orchids naturally develop under the shade of trees that shield them from direct sunlight. When cultivated under controlled conditions, they are grown in specially designed orchidaria or orchid houses that are aligned North–South. In moderate climates, tropical orchids can be grown in open fields, though some protection might be needed for large-scale cultivation. Such orchids are cultivated in fibre houses, orchid houses, simple shade net structures, or polyhouses. Temperate orchids require cool houses. The optimal environment for tropical orchids includes a 75% shade net house, with 70–80% humidity, daytime temperatures between 21 and 29 °C, and nighttime temperatures from 18–21 °C. In regions with high rainfall, the shade net house should also feature a rain shelter.

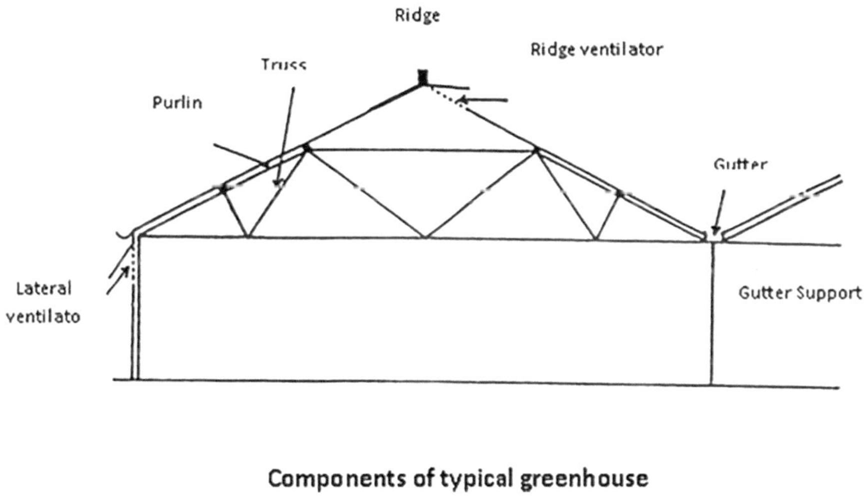

Fig. 12.1 Components of an orchid green house

Greenhouses are classified into three broad groups depending upon the cost of installation, namely

Low-cost greenhouse: Cymbidium, Paphiopedilum
Medium-cost greenhouse: Dendrobium, Cattleya, Oncidium
High-cost greenhouse: Phalaenopsis and Vanda Alliance

		Costs
1.	Less expensive greenhouse without fan and pad	Rs. 500–800/m^2
2.	Medium-cost greenhouse with pad and fan system without automation (Fig. 12.1)	Rs. 1000–1500/m^2
3.	High-cost greenhouses with fully automatic control system	Rs. 3000–5000/m^2

12.7 Types of Orchid Pots

Most orchids grow naturally in warm, humid, wooded areas like tropical rainforests, thriving in tree crotches, on decaying wood, or on shaded slopes. They absorb water from the humidity, avoiding intense sunlight that causes excessive transpiration, and heavy rains drain quickly from their roots when on trees or slopes. However, orchids bought from nurseries or garden centres are usually grown in pots, which does not reflect their wild environment. To support healthy growth, growers should choose pots that allow full potential development. Orchid pots often have drainage holes and are usually clear plastic, which has benefits such as easy inspection of roots, quick adaptation to decorative containers, and allowing roots to photosynthesize like they do in nature. Some growers use special orchid containers with slats or holes for better

Fig. 12.2 Plastic pots used for orchid plants

drainage and airflow, including wooden slatted boxes. Any pot with good drainage—ideally 4–12 holes—is suitable, with shallow pots preferred as orchids dislike excess moisture in deep pots. When using deep containers, placing lava rock or packing peanuts at the bottom can improve drainage and save on potting medium. Orchids should be repotted every 1–2 years, choosing a pot about an inch wider than the previous one. Common containers include earthenware, plastic, baskets, tree fern blocks, wooden trays, or coconut husks.

Different types of orchid pots are described below:

Plastic pots: They retain more moisture and help keep roots warmer in cooler climates. They are lightweight but can tip over easily if the plant is tall or top-heavy (Fig. 12.2).

Clay or terracotta pots: Clay pots are heavier, which reduces tipping. They also help keep orchid roots cooler in warm climates. Many clay or terracotta pots have only one drainage hole, but because the clay breathes and allows water to evaporate more quickly, they require more frequent watering (Fig. 12.3).

Wooden slatted containers or baskets: Wooden slatted containers or baskets provide excellent drainage and air flow to the roots, so they need to be watered more often. Another way orchid growers display their prized orchid plants is by mounting them on driftwood or actual trees.

Fig. 12.3 Terracotta pots for orchid plant growth

Fig. 12.4 Tree fern blocks

Terrestrial and semi-terrestrial plants such as *Paphiopedilum* and *Cymbidium* thrive better in deep pots. Basket cultivation benefits orchids like *Vanda, Rhyncostylis, and Arachnis* that have pendent flower spikes and long, dangling roots. Orchids grown in hanging pots or baskets are typically cultivated with lighter media like charcoal, coconut husk, and tree fern fibre (see Figs. 12.4 and 12.5). Clay

Fig. 12.5 Coconut husk blocks

pots are most suitable for terrestrial orchids, while plastic pots are commonly used for epiphytes. For cool-growing orchids, slabs or logs of tree fern are particularly effective.

12.8 Propagation Techniques for Orchids

12.8.1 Seed and Tissue Culture

This is best suited for commercial orchid cultivation. It is advisable to procure plants from reputable nurseries and select suitable explants (Fig. 12.6).

12.8.2 Stem Cuttings

Cuttings should be 40–50 cm in length and have at least two well-developed aerial roots. Shorter cuttings may take longer to bloom. Monopodial orchids like *Vanda, Aerides*, *Arachnis,* and *Aranda* can be propagated by top cuttings. In *Phalaenopsis*, *Phaius*, *Calanthe,* and *Thunia*, flower stalks give rise to plants.

Fig. 12.6 Tissue-cultured orchid plantlets

12.8.3 Layering

Ideal for monopodial orchids like Vanda. A 20–30 cm section below the apex is wounded and covered with moist cocopeat or coir until roots develop.

12.8.4 Division

Large clumps are divided into smaller units, each with at least 4–5 shoots. This method is effective for Dendrobiums, Cattleyas, Paphiopedilums, Oncidiums, and Zygopetalum (Fig. 12.7).

12.8.5 Offshoots (Keikis)

These small rooted plants emerge at the nodes of pseudobulbs in orchids like Dendrobiums, Epidendrums. Once they mature, they can be separated and potted (Fig. 12.8).

Fig. 12.7 Propagation of Cymbidium orchids by division method

Fig. 12.8 Dendrobium orchid keikis

12.8.6 Backbulbs

Older, dormant already flowered bulbs are severed and placed horizontally on a moist medium to stimulate root growth and ideal for sympodial orchids like Cymbidium (Fig. 12.9), Zygopetalum, Oncidium, and Cattleya.

Fig. 12.9 Cymbidium orchid backbulbs

12.9 Growing Environment

The following factors need to be controlled to ensure growth of healthy orchid plants.

Air Movement: Orchids, many of which are air plants, require consistent air circulation to maintain ideal temperature and humidity levels and to fix carbon dioxide.

Light: Orchids thrive under diffused or filtered sunlight. Too much light can burn the leaves, while too little light can cause them to yellow. Orchids have specific light intensity requirements for growth phase—5000–8000 lux and flowering phase—8000–15,000 lux. Although it varies species to species, growth habit and habitat, as the rule of the thumb, 50% shading is always advised for most of the commercial orchids.

Temperature: Orchids are divided into three temperature categories: Cool for temperate climates, intermediate for subtropical areas, and warm for tropical environments. Warm orchids, including Aerides, Vanda, Rhyncostylis, and certain Dendrobium species, thrive at daytime temperatures up to 32.2 °C and require nighttime temperatures of at least 15.5 °C. Intermediate varieties such as Cattleya, Laelia, Brassovola, Oncidium, and Miltonia prefer daytime temperatures around 26.6 °C and nights at about 12.8 °C. Cool orchids like Cymbidium, Odontoglossum, and Cypripedium favour daytime temperatures of 24 °C and must have night temperatures around 10 °C. Most orchids need cooler nights for healthy growth and to trigger blooming. Specifically, a nighttime temperature of 10–13 °C is optimal for inducing flowering in Cymbidiums.

Humidity: Optimal humidity levels range from 60 to 80%. Rainwater is ideal for orchids. Watering frequency varies depending on the planting medium and growth habit of orchids. Moss or bark requires less frequent watering than basket substrates. Monopodial orchids required frequent watering as compared to sympodial orchids. Water thoroughly in the morning and lightly in the evening to ensure dry leaves overnight.

12.10 Planting Material

Different types of growing materials are used for orchid plant production. In most cases, the substratum is selected on the basis of type of orchids being cultivated.

Terrestrial Orchids: A well-draining mix of humus, decomposed compost, leaf mould, sphagnum moss, or coir is used. The planting area should have good drainage and sunlight.

Epiphytic Orchids: Various substrates can be used, including coconut husk to retain moisture effectively, tiles/bricks for moisture retention and drainage, and charcoal to purify the growing medium and support healthy growth. A potting medium consisting of wood chips, brick pieces, and leaf mould or green moss or sphagnum moss in equal proportions is ideal for vegetative growth and flowering of epiphytic orchids. Epiphytic orchids are planted in coconut husks or hanging baskets (Figs. 12.10 and 12.11).

Fig. 12.10 Orchid plantation in coconut husk

Fig. 12.11 Planting in hanging baskets

In Cymbidium, pots should be spaced at 60 cm × 60 cm, allowing for about 3–4 pots per m^2 (gross). The total number of plants will be approximately 30,000 per hectare. For Dendrobium, the planting density should range from 100,000 to 150,000 plants per hectare.

12.11 Nutrient Management

During flower initiation and inflorescence development, plants are fed with less nitrogen, more phosphorus, and potassium. During the blooming period, a small level of nitrogen and phosphorus and high levels of potassium are maintained. In orchids, foliar feeding is found to be ideal. Frequent application of fertilizers in low concentrations is the best way to feed orchids. Nutrients may vary depending on the orchid species, potting medium, seasonal growth conditions, and growth stages.

Early growth stage: A balanced NPK fertilizer (1:1:1) at the rate of 2 g/L of water is applied once in a week.

Pre-flowering stage: A NPK fertilizer (3:1:1) at the rate of 5 g/L of water is applied once in a week.

Flowering stage: A NPK fertilizer (1:2:2) at the rate of 5 g/L of water is applied once in a week.

The chemical fertilizers are to be properly balanced with organic manures—cow dung, cows' urine, groundnut cake, fish emulsion, and neem oil cake. They are to be diluted before application. Since urine contains high level of salts, a dilution of 1:20

is necessary. For others, 1:10 dilution is ideal. One spray a week with organic manure is enough.

12.12 Plant Health Management

Since orchid plants are prone to insect infestation and other diseases, it becomes crucial to adopt suitable methods for effective management.

12.12.1 Insect-Pests Management (Rohith et al. 2024)

Slugs and snails: Slug pellets are applied around pots or on the ground.

Mites: Cause silvery discolouration. Spray miticides like Abamectin @0.06 mL, Bifenazate @0.25 ml/L of water.

Caterpillars: Damage young leaves and flowers. Use insecticides such as Spinosad @0.5–1 ml/L, Pyrethyrin1–2 ml/L of water.

Aphids: Cluster on leaves and flowers, leaving sticky residue. Rake around plants, Spray Imidacloprid @0.5–1 ml/L, neemoil@5 ml/L of water.

12.12.2 Disease Management

Brown spot (*Pseudomonas cattleyae*): Brown patches with oily spots on leaves. Use healthy plant material, reduce nitrogen levels, and spray with Streptocyclin (0.2%).

Bacterial soft rot (*Erwinia carotovora*): Foul smelling wet spots on leaves. Use healthy plant material, spray with hydrogen peroxide or streptomycin (0.1%).

Blossom blight (*Botrytis cinerea*): Small dark spots on petals in older flowers. Spray with 1 g/L or Indifil Z 2 g/L is effective.

Fusarium rot: Black triangular spots at the leaf base. Prevent overwatering, treat with Topsin or Roko.

Root rot (*Rhizoctonia* spp): Affects the roots and substrate. Improve air circulation and drainage, apply Bavistin or Benomyl drench (1 g/L).

Anthracnose (*Colletotrichum gloeosporioides*): Small oblong to circular oval sunken and reddish brown to dark brown grey-coloured spots develop on leaves. Spray with Bavistin (1 g/L) is recommended.

Viral diseases: The most common viral diseases are Cymbidium mosaic virus and Odontoglossum ring spot virus causing leaf necrosis, irregular colour pattern, mottling of leaves and stunted growth. Use healthy plant material and remove infected plants regularly.

12.13 Harvest and Post-harvest Management

Mentioned below are some standard procedures for orchid harvest and post-harvest management.

12.13.1 Harvesting and Yield

In Cymbidium, buds open at the stage when about 70% of the flowers are blooming, and this is considered the 70% bloom stage. The number of spike yields varies among different genera and varieties. A healthy, well-grown Cymbidium orchid can produce flower spikes annually, and a plant that is 10 years old may produce between 10 and 20 flower spikes (De and Rajeevan 2019). For Dendrobium orchids, spikes are harvested when about 50% of the buds are open. Typically, commercial Dendrobium varieties produce an average of 6–8 spikes. In Phalaenopsis, spikes measuring 40–60 cm long, each with at least 8–10 fully open flowers, are harvested. Usually, a plant yields an average of 6–7 stems per year. Cattleya flowers are harvested when fully open; the spike length ranges from 25 to 40 cm, depending on the species. Oncidiums are harvested when over 80% of the flowers on the spikes (which are about 60 cm long) are fully open and contain many florets. In Paphiopedilum, the inflorescences are ready for harvest after 3–4 days when the flowers have fully opened.

12.13.2 Post-harvest Management

Harvested spikes, also called stems, are immediately placed in a bucket of water, with the cut end fully submerged. They are then transported to the packing house for grading and packing. The cut ends of the spikes are wrapped with wet cotton and secured with a rubber band or inserted into a plastic tube filled with water.

The number of buds retained on each spike and grading standards vary by location. Typically, 25–50% of the total flowers buds are kept on a spike. For standard Cymbidium, 12–15 flowers are usually kept per spike, while Dendrobiums with fewer than five flowers are generally not preferred.

Packing boxes for various orchid types and grades differ in size. Use durable, appropriately sized cartons for easier handling. Approximately 50 flowers are packed in each box. During transport, store the boxes in refrigerated trucks, and chill them in refrigerated rooms. However, temperatures below 7 °C can damage the flowers through chilling injury.

Upon arrival, remove the flower spikes from the cartons and place them in water. It is recommended to cut and remove a small portion of the spike's base every 2–3 days to improve water absorption. For Cymbidium, use holding solutions like 4% sucrose plus 100 ppm Al2(SO4)3 or 2% sucrose plus 200 ppm 8-HQS (8-Hydroxy quinoline sulphate); for Dendrobium, 2% sucrose with 200 ppm 8-HQC (8-Hydroxy quinolone citrate) is used (De 2020a, b).

12.14 Requirements of Quality Cut Flowers of Orchids

Standard applied to orchid cut flower of all genera in the family Orchidaceae for commercial purposes.

12.14.1 Definitions

- An orchid cut flower refers to a fresh, complete inflorescence that includes the entire stem and flower.
- An inner box is used to contain orchid cut flowers that have already been packed in a bag or wrapped, with clear details of the orchid cut flower indicated on the box.
- A master box is utilized for transporting orchid cut flowers, with clear details of the flowers displayed on the box.
- Pest refers to living organisms such as plant pathogens, insects, animals, and weeds that are harmful and damaging to plants.

12.14.2 Description of the Product

- The orchid cut flower, according to this standard, is a type of cut flower bundled in a bag or wrap. The inner and/or master box must include detailed information such as the number of flowers, their class, and sizes.

12.14.3 Quality Requirements

- Minimum requirements: The stem must have at least 40% of its flowers open, excluding Dendrobium; there should be at least four blooming flowers. The flowers should be fresh, clean, free of insects and pests, free from defects and bruises, and without irregular shapes in stems or flowers.
- Extra Class: All parts of the orchid cut flowers should display the typical form and colour of the variety, with at least 65% of the flowers on the stem open.
- Class I: All parts of the orchid cut flowers should have the typical form and colour of the variety, with at least 55% of the flowers on the stem open.
- Class II: This class includes flowers that do not meet the criteria of higher classes but still meet the minimum requirements.
- Sizing: A code indicates the size category, based on the length of the orchid cut flower from stem base to tip (Table 12.1).

Table 12.1 Sizing of orchid cut flower

Code	Orchid cut flower length (cm)
1	Less than 35
2	35 to less than 45
3	45 to less than 55
4	55 to less than 65
5	65 to less than 75
6	75 to less than 85
7	85 to less than 95
8	95 to less than 105
9	From 105 onwards

12.14.4 Tolerances

- Extra Class 1 Quality tolerances: The tolerance must not exceed 5% based on the number of orchid cut flowers, but it must still meet Class I standards for the same. Class I Quality tolerances: The tolerance must not exceed 10% based on the number of orchid cut flowers, but it should meet Class II standards. Class II Quality tolerances: The tolerance must not exceed 10% for orchid cut flowers that do not meet the minimum requirements and must not include defective flowers.

12.14.5 Packaging

- Uniformity: Orchid cut flowers bundled in bags or wraps must be consistent in size and quality.
- Inner Box: An inner box should be made from new, clean, and high-quality materials to protect the orchids. Each inner box should hold a quantity of orchids that does not compromise their quality.
- Master Box: The master box must be new, clean, well-ventilated, free from odours and foreign substances that could harm the orchids, and durable enough for transportation.

12.14.6 Labelling

- The label or each box for orchid cut flowers must display relevant legal information clearly and visibly, including an easily seen statement that is free from errors or deception, as follows:
  Inner box information:
  - Genus, cultivar, variety, or trade name of the orchid
  - Quality class
  - Size code or specified size, as agreed with trading partners
  - Quantity of orchid flowers
  - Trademark (if applicable)

Master box information:
- Exporter or trader code
- Genus, cultivar, variety, or trade name of the orchid
- Quality class
- Country of origin
- Exporter or trader's address
- Additional details if there is no inner box

12.14.7 Official or Certification of Mark

The mark must comply with the criteria and conditions set by the Agricultural Standards Committee or certification body.

12.14.8 Sampling

The sampling shall be conducted in accordance with relevant laws, sampling standards, or the requirements of trading partners.

12.15 Plant Standards for Some Commercial Orchids

12.15.1 Cattleya

Quality Cattleya orchids are characterized by their large, showy, and often fragrant flowers, typically in vibrant colours like purple, pink, white, and yellow, with a ruffled or textured appearance. They are known for their thick, leathery leaves, and sympodial growth habit, where new growth emerges from a rhizome. Mature, well-established plants that are ready to flower are highly desirable (Fig. 12.12).

Growth habit: Cattleyas are sympodial orchids, meaning they grow from a horizontal stem (rhizome) and produce new growth (pseudobulbs) from this rhizome.

Pseudobulbs: These are thickened, bulb-like stems that store water and nutrients, allowing the plant to tolerate drought conditions. These may be cylindric, clavate, or globular.

Leaves: The leaves are typically thick and leathery, unifoliate or bifoliate, narrow oblong, and elliptic or ovate in shape.

Flowers: Cattleyas are famous for their large, showy flowers, which can be quite large relative to the plant size. They often have ruffled or textured petals and a prominent, often contrasting lip. They come in many colours, such as purple, pink, white, yellow, and combinations of these. The flowers typically last for 2–4 weeks, making them a good choice for a long-lasting display.

Fig. 12.12 Potted cattleya plant with flowering

Fig. 12.13 Potted cymbidium orchid in flowering stage

12.15.2 Cymbidium

Quality Cymbidium orchids are characterized by vibrant, long-lasting flowers, strong and healthy foliage, and a robust overall appearance (Fig. 12.13). They should exhibit well-formed, balanced flower spikes with numerous buds and blooms, typically in a spiral arrangement. The flowers should be of good size, with vibrant colours and clear markings if applicable. The foliage should be a healthy green, free from blemishes, and the pseudobulbs should be plump and firm, indicating good health.

Growth Habit: Cymbidiums are sympodial orchids, meaning they grow from a horizontal stem (rhizome) and produce new growth (pseudobulbs) from this rhizome.

Pseudobulbs: Pseudobulbs should be ovoid, conical or round, plump and firm, indicating proper nutrient and water storage. Healthy pseudobulbs are free from signs of disease or decay.

Leaves: The leaves should be long, strap-like, linear or linear-oblong, healthy, and free from blemishes or discoloration.

Flowers: Cymbidium flowers should be substantial, with well-defined petals and a prominent lip (labellum). A wide range of colours is available, from vibrant hues to subtle pastels, with clear, distinct markings where applicable. The flower spikes should be strong and erect, displaying numerous buds and blooms, preferably arranged in a spiral pattern. Standard cymbidiums have large flowers, about 10–12.5 cm wide. Miniature cymbidium blossoms are roughly half the size of the standard flowers. Cascading cymbidium orchids are eye-catching, with spikes that can grow up to 75 cm. High-quality flowers will have a long vase life, remaining vibrant and attractive for an extended period.

Overall appearance: The plant should be well-balanced, with a healthy proportion of leaves, pseudobulbs, and flower spikes. A robust plant will exhibit strong growth and a general sense of health and vitality.

12.15.3 Dendrobium

Quality Dendrobium orchids are characterized by vibrant, healthy foliage, well-developed pseudobulbs, and abundant, long-lasting blooms in various colours (Fig. 12.14). They should be free from pests and diseases and possess strong, sturdy stems. Furthermore, the flowers should be free of blemishes and have good form and symmetry.

Pseudobulbs: Woody, cylindric fleshy or clavate fleshy, well-developed and plump, indicating the plant is storing adequate water and nutrients.

Leaves: Lanceolate, elliptic or ovate in shape, healthy, vibrant green leaves, free from yellowing, spotting, or other signs of disease or stress.

Spikes: Spikes are sturdy and strong, properly trimmed, supporting the weight of the flowers without drooping or breaking. Short length with 25–35 cm, medium length with 35–50 cm, and long length with more than 50 cm.

Flowers: Flowers are well developed, abundant, intact, long-lasting flowers with vibrant colours and good form. Free from pests and diseases, with no signs of wilting.

Fig. 12.14 Potted Dendrobium plant in flowering stage

12.15.4 Oncidium

Quality Oncidium orchids are characterized by vibrant, healthy foliage, robust pseudobulbs, and an abundance of showy, well-formed flowers with intricate patterns (Fig. 12.15). These orchids are known for their diverse colour variations, including yellow, brown, red, and white, often with contrasting markings. A healthy Oncidium should also be free from pests and diseases, with firm, plump pseudobulbs and strong, healthy roots.

Plant height: 30–40 cm.

Pseudobulbs: The pseudobulbs, which store water and nutrients, should be firm and plump, indicating good hydration and overall health.

Leaves: The leaves should be a healthy green colour, free from yellowing, spotting, or other signs of disease.

Flowers: Oncidiums are known for their wide range of colours, including yellow, brown, red, white, and combinations of these, often with contrasting markings. The flowers should be numerous, well-formed, 2–3 cm in diameter and display the characteristic 'Dancing lady' shape with intricate patterns.

12.15.5 Paphiopedilum

Quality Paphiopedilum orchids, often called 'Slipper orchids', are characterized by their unique slipper-shaped labellum, vibrant and long-lasting flowers, and, in some

Fig. 12.15 Potted Oncidium orchid

species, attractively mottled leaves (Fig. 12.16). Healthy plants will have firm, green foliage (or mottled green and white), and should be free from pests and diseases.

Plant size: Paphiopedilums vary in size, with some being compact (15–30 cm) and others larger (30–60 cm).

Leaves: Leaves should be solid green or beautifully mottled, adding to the plant's ornamental value. Leaves can range from short and ovate to oblong and more linear, depending on the species.

Flowers: The most distinctive feature, the labellum is pouch-like, resembling a slipper. The dorsal sepal, which arches over the flower, is often prominent and may be streaked, spotted, or patterned. Petals can be broad or narrow and may be a different colour from the dorsal sepal and lip. Flowers come in a wide array of colours, including various shades of green, purple, pink, yellow, and brown, often with intricate patterns.

12.15.6 Phalaenopsis

Quality Phalaenopsis orchids, also known as 'Moth orchids', are characterized by vibrant, healthy blooms, robust foliage, and strong, well-developed root systems. They should have a good number of flowers and buds, with a desirable shape and colour for the variety (Fig. 12.17). The leaves should be a healthy dark green, firm,

Fig. 12.16 Potted Paphiopedilum plant in flowering stage

Fig. 12.17 Potted Phalaenopsis orchids in flowering stage

and free of blemishes. The roots should be plump, green or white, and firm, not mushy or dried out.

Leaves: Leaves should be a healthy, deep green colour, indicating good health and photosynthetic activity. The leaves should be firm and leathery, and thick. Leaves should be relatively broad and flat, with a good size for the plant. The leaves are usually arranged in two rows, overlapping each other at the base. Leaves should be free of spots, blemishes, or yellowing. Newer leaves tend to be larger than older ones, a characteristic known as 'Enlarged leaf phenomenon'.

Roots: Roots should be healthy-looking, either green (if exposed to light) or white. Roots should be firm and plump, not shrivelled or mushy. A good number of roots should be visible, indicating the plant's ability to absorb water and nutrients.

Flowers: The flower spikes (stems) should be sturdy and upright, supporting the weight of the flowers well. A quality plant will have a good number of flowers and buds, with the blooms being a desirable size and shape for the specific variety. Flowers should be vibrant and true to the variety's colour, with no fading or discoloration. The petals should be rounded and symmetrical, and the overall flower should have a pleasing, balanced appearance. The flowers are long-lasting, with some varieties remaining in bloom for up to 3 months.

Overall appearance: The plant should appear healthy and vigorous, with a balanced growth habit. There should be no signs of pests (like scale or mealybugs) or diseases (like rot). The plant should be stable in its pot and not wobbly or easily tipped over.

12.15.7 Vanda

Quality Vanda orchids are characterized by vibrant, long-lasting flowers with intricate patterns and a wide range of colours, including blue, purple, violet, yellow, and white (Fig. 12.18). They typically have large, showy blooms arranged on long, arching spikes, and some species are sweetly fragrant. Healthy Vanda orchids also exhibit robust growth, with thick, leathery leaves and strong, well-established roots.

Growth habit: Vandas are typically monopodial epiphytes, meaning they grow upwards from a single stem with roots extending outwards.

Leaves: They have thick, leathery leaves that can be strap-like (flat), channelled, or terete (pencil-like), depending on the variety.

Roots: Healthy Vandas have numerous, thick, and well-developed roots, which are important for absorbing moisture and nutrients.

Flowers: Vanda flowers are typically large and flat, 12–14 cm wide, with a short spur on the lip. They boast a wide variety of vibrant colours, including blue, purple, yellow, and white, often with contrasting markings or spots. Some Vanda species and hybrids are known for their sweet fragrance. Flowers are arranged on long, arching spikes that can bear multiple blooms. The flowers are symmetrically arranged in a cylindrical head. In good specimens, they are neither too crowded and overlapping nor too widely spaced, avoiding holes or gaps in the inflorescence.

A description and culture of orchids for pot culture is provided in Table 12.2.

Fig. 12.18 Flowering in Vanda orchid

Table 12.2 Description and varieties of other potted orchids

Genera	Description	Varieties
Aerides	*Aerides* consists of more than 60 species. These are monopodial epiphytic orchids grown on trunks or branches of trees. The stems are round, leafy, and branched with aerial roots. The strap leaves are thick, leathery, and bilobed. The inflorescence is axillary and drooping and up to 60 cm long in some species. The flowers are many and 2 cm in diameter closely set on the inflorescence and generally fragrant	'Edward', 'Brighton's Sparkle', 'Brighton Pink', 'Cagayan', 'Hollyhock', May Woo', 'Jueile', 'Pink Flush', 'Sandorata', 'Lokelani', 'Cressida'
Anoectochilus	*Anoectochilus* consists of about 25 species and commonly known as 'Jewel Orchids'. These are pseudobulbless terrestrial orchids with creeping rhizomes and petioled green, brown, copper colour veined and striated, rose, yellow, or white leaves. Flowers appear in glandular pubescent spikes	'TE', 'Alba', 'Variegated', 'Variegated Alba'

(continued)

Table 12.2 (continued)

Genera	Description	Varieties
Arachnis	*Arachnis* consists of more than 20 species and commonly known as 'Scorpion Orchid.' These are monopodial epiphytic plants, 30 cm to 4.5 m tall with many leathery leaves. Leaves are strap shaped, clasping the stem at their base. Flowers are showy, large borne on simple racemes or branched panicles, and the inflorescence is 50–65 cm long arise from the stem just above the clasping leaves. Flowers are green, yellow, brown, and purple in colour	'Maggie Oei Red Ribbon', 'Maggie Oei Yellow Ribbon', 'Maroon Maggie', 'Merry Maggie', 'Bartha Braga', 'Ishbel'
Ascocentrum	*Ascocentrum* consists of six species. These are dwarf monopodial epiphytes and characterized by large spur hangs from the tip. The plants are small, compact with small strap-shaped leaves and short stalked erect and cylindrical covered with many closely spaced flowers. Flowers are long lasting and 1–2 cm in diameter	'Sagarik Gold'
Bulbophyllum	*Bulbophyllum* consists of about 2500 species of orchids from tropical and subtropical Asia. The name indicates that the leaves are attached to the top of pseudobulbs. These orchids are sympodial epiphytes having rhizomatous stems with angled pseudobulbs. The pseudobulbs form a chain like growth on the rhizomes. Inflorescence is erect or pendent and arises from the base of the pseudobulb. The flowers are solitary or in an umbel or united on spurs. Flower size ranges from few millimeters (*Bulbophyllum minutissimum*, *B. minututum*) to 40 cm in *Bulbophyllum echinolobium*	'Ray Gabaldon', 'A dorbil Ring', 'Madeline Nelson', 'Tsiku Goldfinder', 'Fredensborg Delight', 'Swissthai Alamia', 'Jan Ragan', 'Wilmer Shear Magic', 'Jim Clarkson', 'Tsiku Taurus', 'Agathe', 'Short and Sweet', 'B.C.', 'Chua', 'Aquarius', 'Chanthaboon Glory', 'Meen Bulbul', 'Meen Candy Baby', 'Supat Sun Radiant', 'Supernova', 'Tee Dragon Fly', 'Tee Naga', 'Puch Golden Hydra', 'Tee Gigantic Condor', 'Grace Thoms'
Coelogyne	*Coelogyne* consists of 200 species of sympodial and pseudobulbous orchids. The pseudobulbs are topped by two to four leaves, slender in size and arranged along with creeping rhizomes. The leaves are coriaceous, thick and leathery with pronounced stalks. Inflorescences are erect or	'Memoria W. Micholitz', 'Linda Buckley', 'Burfordiense', 'South Carolina', 'G. Beryl Read', 'John Leathers', 'Wood Fairy', 'Carolyn Sue', 'Golden Bug', 'Spring Showers', 'Professor Georges Morel', 'Joel Strandberg'

(continued)

Table 12.2 (continued)

Genera	Description	Varieties
	pendulous. The flowers are in white or in shades of brown, yellow, and green and subtended by a papery bract	
Doritis	*Doritis* is a monotypic epiphytic orchid from Burma to Sumatra. *Doritis pulcherrima* is a small, stemless, and clump forming monopodials with oblong elliptic leathery leaves. The inflorescence is erect and arranged with 25 flowers. The flowers are 1.5–4.0 cm in diameter with pear-shaped lip. The flowers are dark lavender, rose purple, or magenta rose in colour and produced during autumn and winter season	'Anna–Larati Soekardi', 'Annie Van Tweel', 'Asahi', 'Bandung Pink', 'Bonita', 'Celebrant', 'Charm', 'Imp', 'Jim', 'Jim Chan', 'Kelsey Blush', 'Keneth Schubert', 'Musick Surprise', 'Myriam Esther', 'Profusion', 'Purple Gem', 'Purple Passion', 'Purple Sum', 'Red Elf', 'San Shia Appendo' Sandrine Dream', 'Si Kancil', Siam Treasure', 'Suka Hati', 'Summer Red', 'Sweet Gem', 'Tan Swee Eng.', 'Tarina'
Epidendrum	*Epidendrum* consists of more than 1000 species of epiphytic orchids of America. The pseudobulbs are very elongated or short with coriaceous leaves. Some leaves are thick whereas in some cases leathery with two to four leaves at the apex of each pseudobulb. In case of reed like types, the leaves are alternate. The inflorescence is terminal, simple or profusely branched, erect or pendulous. The number of flowers vary from one to many per inflorescence. The size of flower ranges from 2 to 7.5 cm in diameter and highly scented. The flowers are various in colours	Happy Valley', 'Little Valley', 'Lovely Valley', 'Spring Valley', 'Lions Mane', 'Lions Pride', 'Peace Valley', 'Lucky Valley', 'Pacific Classic', 'Pacific Mandarin', 'Pacific Parade', 'Pacific Secret', 'Pacific Volcano', 'Apricot Valley', 'Rain Valley', 'Water Valley', 'Pacific Joy', 'Pacific Globe', 'Pacific Flame', 'Pacific Heat', 'Pacific Glory', 'Walnut Valley', Fireball', 'Autumn', 'Lunar Eclipse', 'Pacific Dream', 'Pacific Lights', 'Pacific Shore', 'Pacific Delight', 'Pacific Mulberry', 'Pacific Storm', 'Candy Valley', 'Holy Valley', 'Palm Valley', 'Milky Valley', Darling Jungle', 'Pacific Girl', 'Pacific View'
Gastrochilus	*Gastrochilus* consists of 20 species of epiphytic orchids distributed in India, East Asia, and Malayasia. These are short to long stemmed epiphytic orchids with lance shaped, thick, 6–11 leathery leaves. The inflorescences are solitary or spicate with one or more flower spikes arise from the axil of the leaves. The flowers are attractive, 1 to 4 cm in diameter, and white in colour with reddish spots and blotched with yellow	

(continued)

Table 12.2 (continued)

Genera	Description	Varieties
Laelia	*Laelia* consists of 75species of epiphytic orchids from Cuba and Mexico to Argentina. These sympodial orchids possess pseudobulbs of various sizes and shapes. For example, *Laelia harpophylla* have needle-like pseudobulbs, and other groups bearing flat and round pseudobulbs are *L. rubescens*. While other groups with ovoid or egg-shaped pseudobulbs, e.g. *L anceps*. Number of leaves per pseudobulb ranges from 1 to 3. The leaves are leathery, entire, thick, and 5–30 cm in length. The inflorescence bears more than 20 flowers in some species, while some other bear solitary flowers close to the pseudobulb apex. The flowers are lavender, white, yellow, and orange in colour	'Tropical Mojo', 'Milpest', 'Rock Ruby', 'Akiras Spring', 'Bonnie Bonita', 'ViDoran', 'Newberry Glow', 'Rocket Sprite', 'Stirling Dickinson', 'Fran's Fuchsia', 'Flash', 'Gold Snood', 'William Curtis', 'Aphrodite', 'Venus'
Miltonia	*Miltonia* consists of 20 species of sympodial epiphytic orchids from Brazil. They are also called as 'Pansy Orchids' at higher elevations. These orchids are characterized by sessile, flattened pseudobulbs, and thin leaves arise from pseudobulbs. Inflorescence is single flowered or multiflorus raceme. The flowers are large, 10 cm across and yellow, reddish purple, white, pink, or brown in colour	'Puna Gold', 'Shaini's Star', 'Enzan Sunday', 'Mauri Star', 'Enzan Floss', 'Red Rock', 'Tear Drops', 'Yellow Sands', 'Black Waters', 'George W, Bush', 'Indian Hollow', 'Cast Sand', 'Emily's Delight', 'Ananda Apple', 'Whale Bay', 'Solar Storm', 'Daydream', 'Solar Storm', 'Angel Eyes', 'Top Prize Capitula Redwood', 'Bingo', 'Kismet Sena', 'Saffron Butterfly', 'Dream River', 'Moon River', 'Coconut Cream', 'Spring Jewel', 'Blackout', 'Dear Lady'
Odontoglossum	*Odontoglossum* consists of 300 species of epiphytic and lithophytic orchids from Mexico to Bolivia and Brazil. The pseudobulbs are tipped by one or two leaves. Inflorescence arises from the base of the pseudobulbs, erect or arching bears one to many large and beautiful flowers. The flower colour is variable from pure white spotted and blotched with other hues of yellow to chestnut brown	Cynthia Hill', 'Luis Salabazar', 'Mc Beans Elizabeth', 'Spotted Croc', 'Mayapan', 'Quito', 'Colocris', 'Mums Tiger', 'Ann', 'High Street', 'Uro-Day', Cormac', 'Gay Enchanment', 'Gay Starshooter'

(continued)

Table 12.2 (continued)

Genera	Description	Varieties
Renanthera	*Renanthera* consists of 12 species of epiphytic or lithophytic orchids from China, Himalayas, the Philippines, Indonesia, and New Guinea. Plants are monopodial, vinelike, pseudobulbless with ascending stems of intermediate growth. The leaves are leathery, distichous, thick, and coriaceous. A large number of aerial roots is produced from the stem of plants. The inflorescence arises from the axils of leaves (upper) and produces few to many flowers. The flowers are 9 cm across, scarlet, red orange, and crimson to yellow in colour	'SCGB Kylin', 'Datin Blanche', 'Olbery', 'Hooi Sew Yong', 'John Losgar', 'Red Leopard', 'Scarlet Belle', 'Chanachae', 'Serdang', 'Brady Crocker', '20th WOC Singapore-2011', 'Bart Motes'
Rhyncostylis	*Rhyncostylis* consists of four species of monopodial epiphytic orchids distributed in India, Burma, Thailand, Malayasia, Indonesia, Sri Lanka, East Indies, and the Philippines. These orchids are commonly known as 'Foxtail Orchids' and considered state flower of Assam and Arunachal Pradesh. These are stout, short stemmed plants with thick and leathery linear-oblong leaves closely arranged on the stem. The inflorescence is erector drooping densely covered with small colourful showy flowers. The flowers are red, magenta, blue spotting, reddish lavender, and white in colour	'Charles Marden Fitch', 'Thai Gem', 'Norma', 'Alice;s Wally', 'Bangkok Sunset', Wilton Hill', 'Jammie Harper', 'Apichart', 'Noo Noi', 'Peter Draper', 'Brighton's Albino', 'Prairie Lady'
Zygopetalum	*Zygopetalum* consists of 25 species of terrestrial, lithophytic or epiphytic orchids from Brazil, Bolivia, Peru, Paraguay, Venezuela, and New Guinea. The pseudobulbs are ovoid with distinct sheath, 5–7.5 cm tall, strong with lanceolate, distichous leaves. The inflorescence develops from the base of pseudobulbs, arching or erect in nature. The flowers are fragrant, long lasting in shades of brilliant green, blue, or purple. The flowers are 5–7.5 cm across with wavy margins. They are excellent for cut flowers and corsages	'Blue Blood', 'Blue Bear', 'Imagination', 'Tanzanite', 'Millie', 'Black Plague', 'Kiwi Black', 'Kiwi Choice', 'Pioneer', 'Tasman', 'New Era', 'Dark Star', 'Impulse', 'Intuition', 'Great Eisen', 'Blue River', 'Indigo Skies', 'Blue Banks', 'Bon Voyage', 'Centenary', 'Hot Springs', 'Blackjack', 'Night Hawk', 'Violet Moon', 'Leopard Prince', 'Blue Blood', 'Big Country', 'Hawker'

12.16 Conclusion

In international trade, orchids have gained a prominent position as potted plants and propagules due to their attractiveness, long shelf life, high productivity, appropriate blooming season, and ease of packing and transportation. The market for artificially propagated live orchid plants is mainly focused on orchids and their hybrids. Additionally, dried orchids are utilized for various purposes, including decoration in vases, baskets, and shadow boxes. Overall, orchids hold excellent potential as potted plants in both domestic and global markets and should be further explored. Successful cultivation requires identifying suitable orchid species for specific regions, managing polyhouse conditions, understanding physicochemical needs, and selecting appropriate pots and potting materials, among other important factors for growers.

References

Besi EE, Mustafa M, Yong CSY, Go R (2023) Deforestation impacts on diversity of orchids with inference on the conservation initiatives: Malaysia case study. Bot Rev 89:386–420
Chowdhery HJ (2009) Orchid diversity in northeastern states of India. J Orchid Soc India 23:19–42
CITES (2024) The CITES appendices. https://cites.org/eng/app/appendices.php
De LC (2020a) Varietal wealth in orchids. Biotica Res Today 2(4):40–46
De LC (2020b) Post-harvest management of orchid spikes and florets. Int J Curr Microbiol App Sci 9(9):2624–2638
De LC, Rajeevan PK (2019) Orchids. In: Handbook of horticulture, vol II. DKMA, ICAR, Pusa, New Delhi, pp 759–763
Ferrante A, Ferrini F (2023) Floriculture and landscapes: perspectives and challenges. Front Hortic 2. https://doi.org/10.3389/fhort.2023.1123298
Freudenstein JV, Chase MW (2015) Phylogenetic relationships in Epidendroideae (Orchidaceae), one of the great flowering plant radiations: progressive specialization and diversification. Ann Bot 115:665–681
Gale SW, Kumar P, Hinsley A, Cheuk ML, Gao J, Liu H, Liu ZL, Williams SJ (2019) Quantifying the trade in wild-collected ornamental orchids in South China: diversity, volume and value gradients underscore the primacy of supply. Biol Conserv 238:108204
Hinsley A, de Boer HJ, Fay MF, Gale SW, Gardiner LM, Gunasekara RS, Kumar P, Masters S, Metusala D, Roberts DL, Veldman S, Wong S, Phelps J (2018) A review of the trade in orchids and its implications for conservation. Bot J Linn Soc 186:435–455
IUCN (2023) The IUCN red list of threatened species, Version 2023-1. www.iucnredlist.org
Kreziou A, de Boer H, Gravendeel B (2016) Harvesting of salep orchids in north-western Greece continues to threaten natural populations. Oryx 50:393–396
Lu HC, Liu ZJ, Lan SR (2019) Genome sequencing reveals the role of MADS-box gene families in the floral morphology evolution of orchids. Hortic Plant J 5:247–254
Malviya AJ, Vala M, Mankad A (2022) Recent Floriculture in India. Int Assoc Biol Comput Digest 1(1). https://doi.org/10.56588/iabcd.v1i1.8
Mayak S, Bravdo B, Gvilli A, Halevy AH (1973) Improvement of opening of cut gladioli flowers by pretreatment with high sugar concentrations. Sci Hortic 1(4):357–365. https://doi.org/10.1016/0304-4238(73)90020-4
Rohith E, Ghazanfer G, Mogra P, Rahna Rinu KK, Ajith Kumar G (2024) Orchids: the key cultivation practices. Greenaria 2(11):31–34

Salachna P (2022) Trends in ornamental plant production. In: Horticulturae, vol 8. MDPI. https://doi.org/10.3390/horticulturae8050413
Singh F (1990) Indian orchids. Indian Hortic 35(1):14–15
Song C, Wang Y, Manzoor MA, Mao D, Wei P, Cao Y, Zhu F (2022) In-depth analysis of genomes and functional genomics of orchid using cutting-edge high-throughput sequencing. Front Plant Sci 13:1018029
Subedi A, Kunwar B, Choi Y, Dai Y, van Andel T, Chaudhary RP, de Boer HJ, Gravendeel B (2013) Collection and trade of wild-harvested orchids in Nepal. J Ethnobiol Ethnomed 9:64
Tang L, Jiao Z (2022) Analysis and comparison of national key protected wild plants list. J Fujian For Sci Technol 49:125–132
Teo ZWN, Zhou W, Shen L (2019) Dissecting the function of MADS-Box transcription factors in orchid reproductive development. Front Plant Sci 10:1474
Teoh ES (2016) Medicinal orchids of Asia. Springer International Publishing, Cham
Tsai WC, Dievart A, Hsu CC, Hsiao YY, Chiou SY, Huang H, Chen HH (2017) Post genomics era for orchid research. Bot Stud 58:61
Tulipa D, De LC, Singh DR, Promila P, Roy A (2018) Diseases in orchids. Orchid News 34:15–16
Veldman S, Kim SJ, van Andel TR, Font MB, Bone RE, Bytebier B, Chuba D, Gravendeel B, Martos F, Mpatwa G, Ngugi G, Vinya R, Wightman N, Yokoya K, de Boer HJ (2018) Trade in Zambian edible orchids—DNA barcoding reveals the use of unexpected orchid taxa for chikanda. Gen 9
Wani MA, Nazki IT, Din A, Iqbal S, Wani SA, Khan FU, Neelofar (2018) Floriculture sustainability initiative: the dawn of a new era. In: Lichtfouse E (ed) Sustainable agriculture reviews, vol 27. Springer, Cham, pp 91–127. https://doi.org/10.1007/978-3-319-75190-0_4
Wei X, Khachatryan H, Hodges A, Hall C, Palma M, Torres A, Brumfield R (2023) Exploring market choices in the US ornamental horticulture industry. Agribusiness 39(1):65 109. https://doi.org/10.1002/agr.21769
Williams SJ, Gale SW, Hinsley A, Gao JY, St John FAV (2018) Using consumer preferences to characterize the trade of wild-collected ornamental orchids in China. Conserv Lett 11
Yuan SC, Bolanos-Villegas P, Chin YT, Chen FC (2021) The breeding of Phalaenopsis hybrids. In: Chen F-C, Chin S-W (eds) The orchid genome, Compendium of plant genomes. Springer, Cham, Switzerland, pp 29–40
Zhang G, Hu Y, Huang M, Huang W, Liu D, Zhang D, Hu H, Downing JL, Liu Z, Ma H (2023) Comprehensive phylogenetic analyses of Orchidaceae using nuclear genes and evolutionary insights into epiphytism. J Integr Plant Biol 65:1204–1225
Zotz G (2013) The systematic distribution of vascular epiphytes-a critical update. Bot J Linn Soc 171:453–481

13 Nutraceutical and Cosmeceutical Value of Orchids

Kumbukani K. Nyirenda and James Nyirenda

13.1 Introduction

The Orchidaceae family comprises more than 28,000 species spanning 763 genera (Zhang et al. 2018) and is one of the most diverse groups of flowering plants in the world (Kindlmann et al. 2023). Global distribution spans all regions, except the glaciers and most arid deserts (Kindlmann et al. 2023). Since ancient times, orchids have demonstrated cultural significance in various areas, with some species associated with good health while others are used in traditional ceremonies (Cheng et al. 2019). Traditional Chinese Medicine (TCM) history for the description of orchids, their therapeutic and culinary utilizations dates back to around 2800 B.C. For ages, orchid species, especially those from the Dendrobium genus, have been incorporated into traditional remedies as a tonic for kidneys, to promote longevity, boost the immune system and treat ailments, such as hypertension, fever, and stomachaches (Cheng et al. 2019). About 40 species of the genus Dendrobium have been reported in TCM that are used for medicinal purposes; the most common ones include *Dendrobium officinale, Dendrobium candidum,* and *Dendrobium huoshanense* (Tang et al. 2017; Yatung et al. 2024). Dendrobium species are generally consumed in tea, soup, or as key constituents in herbal formulas. Additionally, several genera of orchids have been extensively researched and exploited for medicinal purposes, comprising Acampe, Aerides, Coelogyne, Crepidium,

K. K. Nyirenda (✉)
Department of Pharmacy, School of Life Sciences and Allied Health Professions, Kamuzu University of Health Sciences, Blantyre, Malawi
e-mail: knyirenda@kuhes.ac.mw

J. Nyirenda
Department of Pure and Applied Chemistry, School of Natural and Applied Sciences, University of Zambia, Lusaka, Zambia

A. Husen, M. Pant (eds.), *Forest-based Orchids*,
https://doi.org/10.1007/978-981-92-0096-2_13

Dactylorhiza, Gastrodia, Eulophia, Flickingeria, Otochilus, Pholidota, Satyrium, and Vanda (Subedi et al. 2013).

Similar to the TCM, the incorporation of orchids in the herbal medicine preparations is well established in the indigenous medicinal systems of India, mostly in the traditional Ayurvedic system. Approximately 50 species belonging to the family Orchidaceae are reportedly used in the field of medicine (Mukherjee et al. 2023). In traditional Indian folklore medicine, Ashtavarga, an essential component in various formulations, consists of four orchids: *Malaxis muscifera*, *Malaxis acuminata*, *Habenaria edgeworthii*, and *Habenaria intermedia*. These plants have shown efficacy in enhancing the immune system, cell regeneration capacity as well as improving the general health of individuals (Tang et al. 2017). In Ayurvedic and Unani medical practice, certain orchid species are valued for their traditional uses as aphrodisiacs, respiratory aids, and treatments for noncommunicable conditions such as asthma, aerosol allergies, and bronchitis. Some orchid species are used as antidotes against scorpion and red ant stings. In India, popular orchid genera like Cymbidium, Dendrobium, Phalaenopsis, and Vanda are commonly utilized for both culinary and medicinal purposes (Mukherjee et al. 2023).

Many African communities have reported diverse uses of orchids. Among the Zulus of South Africa, several species of *Eulophia* have been used to prevent miscarriages in women and as curative mixtures for infertility in men (Bhattacharyya and Van Staden 2016). Dried leaves of *Eulophia flaccida* have been used as an analgesic to relieve pain and reduce the risk of miscarriage. The pulverized leaves are applied to skin incisions. In Zulu tradition, the stems of *Ansellia gigantea* have long been used as aphrodisiacs. Meanwhile, *Eulophia aha*, or “Wild cocoa,” was brought to South Africa during the slave trade era, where it was valued for its medicinal properties in treating various health issues (Dev and Jangid 2024). In the central and eastern parts of Africa (Tanzania, Zambia, and Malawi), orchids are mostly valued for their tubers, which are used as a source of food and medicine (Dhanda et al. 2022). In Zambia, orchid tubers are cut into small pieces, ground to a homogenous powder which is mixed with ground peanut powder, salt, soda, and water and processed in a polymeric cuisine, which has a textural look like polony. The locals call it Chikanda or Chinaka in eastern Zambia and Malawi. It is consumed as a relish when fried or as a simple snack on its own. Habenaria, Satyrium, Disa, Roeperocharis, and Brachycorythis are mostly found in Zambia, where they are harvested for various uses (Hinsley et al. 2018). In Malawi, orchids are used for medicinal purposes as well as food, and 12 species have been associated with therapeutic use (Kumwenda et al. 2024). Ethnomedicinal information on two species, *Cyrtorchis arcuata* and *Eulophia cucullata,* is mostly associated with the treatment of diabetes, skin infections, and infertility, while *Eulophia cucullata* is solely used to prevent epilepsy (Hinsley et al. 2018). Some of the Malawian orchids are also used in traditional medicine for treating conditions like fontanelle, ringworms, and abdominal pains (Dev and Jangid 2024). Ethnobotanical studies of the African Traditional Medicine (ATM) system also show that some West African cultures use an amulet of leaves of *Ansellia africana* infused with a paste derived from the pseudobulbs of the same species as a short-term contraceptive.

Paradoxically, in South Africa, the same species, which is also known as the Leopard orchid, has been used to treat conditions related to the central nervous system (CNS), and possibly Alzheimer's disease (Bhattacharyya and Van Staden 2016).

Like in other traditional medicine systems, orchids have been an integral component in the herbal remedies of the indigenous communities in America. The earliest records indicate that since the fifteenth century, the native American Indians used *Vanilla planifolia* for the treatment of hysteria, fevers, impotence, rheumatism, and to increase the energy of the muscular system (Pomini et al. 2023). The native Americans used orchids primarily for medicinal purposes, and *Cypripedium calceolus*, also known as the Lady's Slipper orchid, is acclaimed for treating fevers, headaches, anxiety, and menstrual issues (Castillo-Pérez et al. 2024). According to ethnobotanical information, the roots of *Malaxis unifolia*, commonly called green adder's mouth or adder's tongue, are used as a diuretic in the treatment of hypertension (Pomini et al. 2023). Orchids appear as small herbs in eastern North America and the Caribbean. Plants from the Cypripedium genus have been used by various ethnic groups for their calming effects. These orchids were valued for their sedative and antispasmodic properties, helping to relieve insomnia and nervous tension. Popular medicinal species include *Cypripedium acaule*, *C. reginae*, *C. candidum*, and *C. parvifolium* (Prabha et al. 2024). In Mexico, orchids have been used for millennia. Uses include ceremonial rituals, as edible herbs, as medicines, and in indigenous knowledge systems, such as glue and dye making (Pomini et al. 2023). Orchids have diverse uses across cultures, reflecting the rich biodiversity and varied traditions of different ethnic groups. People in central and southern America use Malaxis orchids to treat gastrointestinal issues. *Taraxacum brachyphyllum* has been reported to induce hallucinogenic effects in some people (Castillo-Pérez et al. 2024). Moreover, the Mayan ethnic group used at least eight different orchid species for medicinal purposes, ceremonies, as a glue, and to create a mordant for dyes (Ospina 1997). The most important examples include *Vanilla* spp. for skin conditions and wounds, and *Lycaste virginalis* for spiritual and cultural practices (Pomini et al. 2023).

The traditional cultures of many nations have employed plants from the Orchidaceae family primarily for food and medicine. However, utilizing an ethnobotanical approach that involves an analysis of indigenous knowledge and bioassay guided evaluation, orchid-based products have been developed with potential applications in pharmaceuticals, cosmetics, foods, and their hybrids as illustrated in Fig. 13.1.

This chapter examines different high-value nutraceuticals and cosmeceuticals derived from orchid species. Furthermore, the diverse bioactive compounds and their corresponding pharmacological and functional properties are explored.

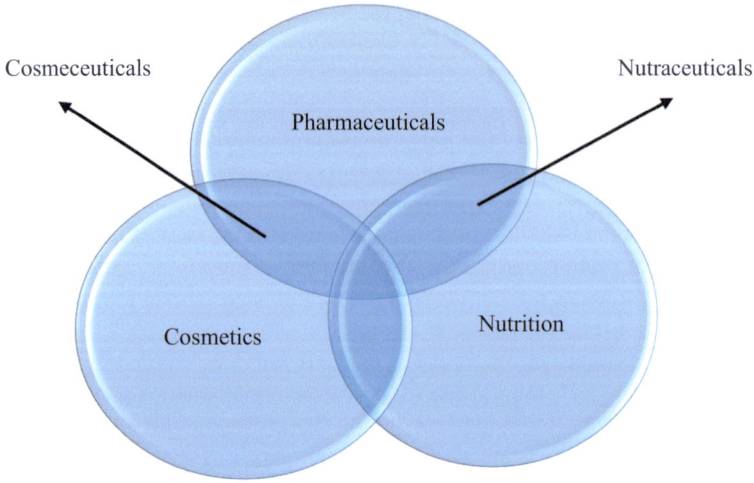

Fig. 13.1 Various applications of orchids

13.2 Nutraceutical Value

The World Health Organization (WHO) describes nutraceuticals as food or part of a food that provides medical or health benefits, including the prevention or treatment of disease, and the US Food and Drug Administration (FDA) defines a nutraceutical as any substance that is food or part of a food that has medicinal or health benefits (Thakur and Sharma 2024). Thus, both definitions highlight the potential of nutraceuticals to promote health and well-being beyond basic nutrition. Essentially, nutraceutical is a combination of nutritional and pharmaceutical and refers to products derived from natural sources that offer additional health benefits. Therefore, nutraceuticals include a variety of products, ranging from isolated nutrients and dietary supplements to genetically engineered designer foods, herbal products, and processed foods such as cereals, soups, and beverages.

The nutraceutical properties of orchids have been widely reported, revealing that edible species contain important minerals, vitamins, and phytochemicals, such as phenolics and flavonoids (Biswas et al. 2021; Kumwenda et al. 2024). Misra et al. (2023) evaluated the nutritional potential of the edible terrestrial orchid, *Eulophia nuda*. The results showed that the species contained a significant amount of ash, fiber, crude alkaloid, total phenolics, and flavonoids. The plant was also found to be rich in macro- and micro-nutrients, carbohydrates, and energy, comparable to conventional cereals and superfoods like finger millet and foxtail millet. Similarly, Kumwenda et al. (2024) studied the nutritional, phytochemical, and functional properties of four orchid species from Malawi, namely: *Disa zombica, Satyrium buchananii, Satyrium carsonii,* and *Satyrium trinerve*. The plants contained considerable and variable amounts of essential nutrients and phytochemicals important for

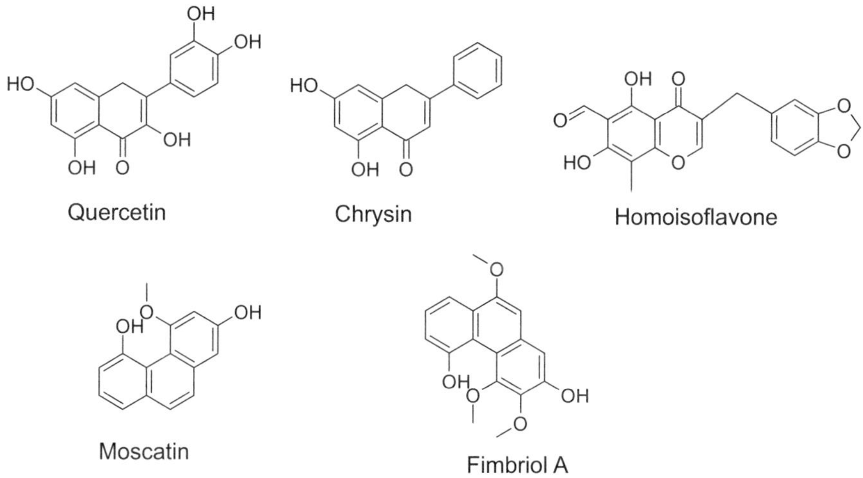

Fig. 13.2 Examples of phytochemicals isolated from *Malaxis acuminata*

human nutrition and health. In India, Chyawanprash is a popular herbal tonic, considered one of the best herbal nutrient products. The ingredient of the tonic consists of four orchid species: *Habenaria intermedia, Platanthera edgeworthii, Malaxis acuminata*), and *Dendrobium macraei* (Sharma et al. 2019). The nutraceutical formulation is known for its high Vitamin C content, numerous essential fatty acids, high bioflavonoids, carotenoids, and a large amount of bioactive phytochemicals (Aryal 2021). A phytochemical analysis of the orchid species, *Malaxis acuminata,* showed that the pseudobulbs of the plant contained alkaloids, flavonoids, glycosides, and β-sitosterol (Verma and Ranawat 2024). Figure 13.2 provides examples of compounds that have been isolated from *Malaxis acuminate.*

The most common nutraceutical orchids in China appear to comprise species from the Dendrobium genus, particularly *Dendrobium officinale* and *Dendrobium nobile*, which have been valued since 200 B.C.. Use of *Dendrobium officinale* for minor conditions such as indigestion, dehydration, and fever has sparked extensive research in finding the bioactive compounds to help elucidate the biochemical pathway and mechanism of action (Tang et al. 2017). To this effect, several studies have reported various bioactices, including polysaccharides such as O-acetylglucomannan (dendronan), phenols, and alkaloids. Some of the compounds from *D. officinale* have found commercial use such as component in nutraceutical beverages and health foods, including stir-fries, sauces, teas, juices, wine, and desserts (Cakova et al. 2017; Huang et al. 2015). According to Devaraj et al. (2019), *Dendrobium officinale* extracts have water-absorbing, thickening, and emulsifying properties when used as a food ingredient, which may be attributed to glucomannans and pectin content as well as aroma properties useful in the medicinal and food industries. A review by Kirkwood et al. (2023) reported more than 250 volatile compounds identified from *D. officinale*, some with flavor profiles and the most predominant classes being terpenoids, aldehydes, alcohols, esters,

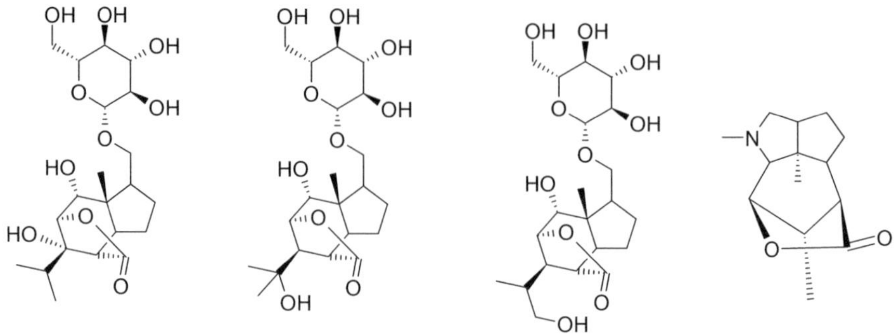

Fig. 13.3 Chemical structures of dendrobine alkaloid derivatives from *D. nobile*

ketones, and alkanes. Dong et al. (2020) studied the function-specific volatiles of *D. officinale* with high levels of vitamin E and heptacosane being detected in the roots, stems, and leaves. The authors opined that these compounds could be utilized as nutritional supplements. *Dendrobium nobile* is consumed by indigenous populations of the northeastern Himalayan states of India as well as China, where it is extensively used traditionally for treating conditions related to the kidneys, lungs, stomach, and to alleviate fever, diabetes, and thirst. Stems of the species are used as a tonic to improve digestion and promote the production of body fluids (Jain et al. 2024). Chemical analysis of *Dendrobium nobile* constituents revealed a diverse array of bioactive compounds, including alkaloids, phenols, polysaccharides, gibberellin, and flavonoids. Further, the study revealed that dendrobine-type alkaloids were the major compounds, which exhibited analgesic, anti-inflammatory, and antipyretic properties (Fan et al. 2023). Figure 13.3 shows some of the phytochemicals isolated from *D. nobile*.

The phytochemicals have been associated with various nutraceutical and health benefits, including anti-aging, immune enhancement, blood sugar regulation, anti-inflammatory, and neuroprotective effects. Although scientific studies show promise for its components, human research is insufficient to inform the safety and efficacy of *D. nobile* as a dietary supplement.

13.3 Cosmeceutical Potential

Cosmetics have been used since time immemorial, dating back to the ancient Egyptian and Sumerian civilizations (Lall et al. 2020). In Europe, during the Middle Ages, cosmetics were used for whitening the face and beautifying the cheeks. Cosmetics are products used to enhance or alter the appearance or fragrance of the human body. They are chemical compounds sourced from either natural or synthetic origins (Mita et al. 2024). Cosmeceuticals, by contrast, represent a hybrid category of cosmetics and pharmaceuticals, designed to improve both physiological health

Table 13.1 List of some orchid species used as cosmeceuticals

Botanical name	Part(s) used	Uses
Bletilla striata	Root	As a cleansing face mask and moisturizer
Cymbidium Grandiflorum	Flower and root	Skin conditioning component of cosmetic products
Dendrobium Nobile	Whole plant	Skin conditioning component of cosmetic products
Phalaenopsis Lobbii	Whole plant	Bleaching (lightens the shade of hair or skin) component of cosmetic products
Vanda coerulea Griff. ex Lindl.	Whole plant	Antioxidant component of cosmetic products

and aesthetic appearance of the skin. There are several ways of categorizing cosmeceuticals and the most common one is based on their active ingredients, their primary functions, and by product application (Carpio et al. 2021). This section discusses the nutraceutical potential of orchid species as agents for anti-aging, skin moisturizer, antioxidant, and perfumery. Globally, fundamental and applied studies have largely been conducted on the following genera: *Cymbidium*, *Dendrobium*, *Oncidium*, *Vanda*, and *Phalaenopsis*, and Table 13.1 shows some orchid species used as cosmeceuticals.

13.3.1 Anti-aging

Anti-aging products refer to a category of skincare and other goods designed to prevent, reduce, or mask the visible signs of skin aging. Two physiological mechanisms explain skin aging: chronological aging and photoaging. Chronological aging is a slow process that leads to reversible tissue degeneration and usually affects the whole body (Hadi et al. 2015). In contrast to chronological aging, photoaging, also known as premature aging, is induced by chronic exposure to ultraviolet irradiation (Carpio et al. 2021). Several studies have reported the effects of orchid extracts on the pharmacological and physiological processes associated with skin aging. Recently, Rungsang et al. (2023) examined the biological properties of extracts from *Dendrobium sulcatum*. The researchers reported that the ethanolic extract of the plant exhibited antioxidant, antiglycation, anti-inflammatory, and ant-tyrosinase properties. These results provided strong scientific evidence supporting the efficacy of the extracts as anti-inflammatory and anti-aging agents in topical cosmetic products (Rungsang et al. 2023).

Simmler et al. (2010) evaluated the skin photoprotective effect of *Vanda coerulea,* which is commonly known as the blue orchid. The stem extracts displayed significant DPPH/OH radical scavenging activity and in vitro inhibition of PGE-2 production in irradiated HaCaT cell lines. The phytochemical screening isolated five stilbenoids (Fig. 13.4), two 9,10-dihydrophenanthrenes (methoxycoelonin and coelonin), and one dibenzylgigantol. Further analysis attributed the antioxidant

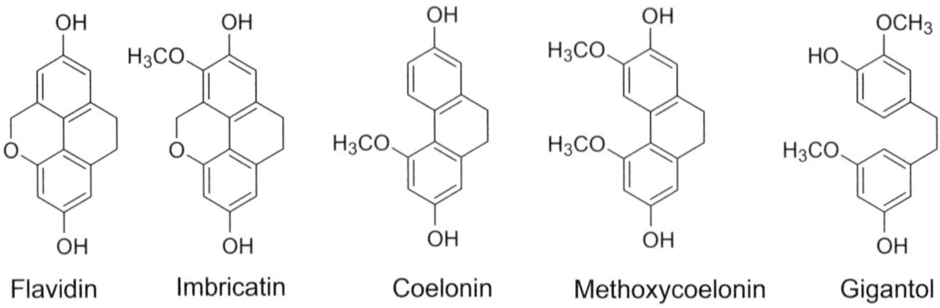

Fig. 13.4 Structures of stilbenoids isolated from *V. coerulea* stems (Simmler et al. 2010)

activity to three compounds: imbricatin, methoxycoelonin, and gigantol, thus enabling them to neutralize ROS and reduce oxidative stress (Simmler et al. 2010). Additionally, imbricatin is known to exhibit anti-inflammatory activity, which reduces UVB-induced COX-2 expression, leading to lower PGE-2 production and inhibition of collagen synthesis (Bae et al. 2010). Therefore, the findings suggest that extracts of *V. coerulea* show potential as a key ingredient for anti-aging applications.

13.3.2 Skin Moisturizing

The skin, which is the body's largest organ, has many functions, including protecting the human body against UV radiation, pollutants, and temperature and humidity fluctuations. A defect in the stratum corneum barrier, which is the superficial protective layer of the skin, results in xerosis, a medical term for dry skin (Hogan et al. 2012). Dry skin is characterized by a scaly appearance and flaking, resulting from reduced moisture retention due to atrophy of skin glands. The current cosmeceutical formulations include plant extracts believed to be beneficial in maintaining skin hydration, especially in dry skin. Hadi et al. (2015) recommended the use of elderflower, *Aloe vera*, coltsfoot, mallow, acacia, carrot, comfrey, honey, and orchid extracts for dry or sensitive skin.

Scientific studies have been conducted to evaluate the moisturizing effect of orchid species. Kim et al. (2012) investigated the hydrating effects of orchid extracts in skincare in a clinical trial. The study involved 25 participants with dry and sensitive skin who applied a moisturizer containing orchid extract twice daily for 4 weeks. The study found that participants who used orchid extract experienced a 30% increase in skin hydration and 85% reported improved skin smoothness and softness. The extract also reduced water loss, demonstrating its effectiveness as a moisturizer (Kim et al. 2012). Some researchers have opined that orchid extract, specifically *Vanda coerulea*, works by strengthening the skin's barrier function and enhancing water transport within the skin, leading to improved hydration and skin health (Hadi et al. 2015).

Table 13.2 Orchid plants with fragrant properties

Plant	Part(s) used	Product use
Dendrobium Moniliforme	Leaf/Stem, Callus Culture extract	Perfume
Red Cattleleya	Flowers	Perfume
Phalaenopsis javanica	Flowers	Scent
Ludisia discolour	Flowers and leaves	Scent candles
Dendrobium species	Flowers, leaves, Seeds, and roots	Essential oils

According to some researchers, orchid extracts moisturize the skin through two key mechanisms: strengthening the skin's barrier to reduce water loss and boosting water transport within the skin by activating proteins like aquaporin 3 (AQP3) and Lympho-epithelial Kazal-type-related inhibitor (LEKTI). Additionally, orchids contain mucilage, a natural humectant that absorbs and retains water, providing further moisturizing and softening benefits skin tissue (Kim et al. 2012).

13.3.3 Perfumery

Various orchid species contain fragrance oils that offer a luxurious, sweet, and sophisticated floral scent, which is valued for relaxation, stress relief, and mood enhancement (Patil et al. 2024). They are mainly used in perfumes, candles, soaps, massage oils, and diffusers to create an elegant and uplifting aroma. Table 13.2 shows examples of orchid plants that exhibit fragrant properties.

Despite estimates that over three-quarters of orchid species are fragrant, only a few studies have investigated the phytochemical and biological properties of their volatile oils. Baek et al. (2019) characterized the volatile floral scents and flavors from the genus Cymbidium. The floral scents varied significantly among the cultivars. However, the main components of the sepals and petals were monoterpenes (82.86–91.17%) and sesquiterpenes (6.60–7.81%). The results also showed that α-pinene, linalool, eucalyptol, and 4,8-dimethyl-1,3,7-nonatriene were the major compounds associated with the floral scent of the variety (Baek et al. 2019). Recently, de Vasconcelos et al. (2023) analyzed the chemical constituents of the volatile concentrates of eight Catasetum specimens from the Brazilian Amazon. The major volatile compounds included transgeranylgeraniol, 1,4-dimethoxybenzene, linalool, 2-phenylethyl acetate, geraniol, 7-epi-1,2-dehydrosesquicineole, 1,8-cineole, benzyl acetate, limonene, and methyl salicylate.

13.4 Orchid-Derived Valuable Products

Over the past two decades, the global market of the products from the Orchidaceae family has steadily increased, with countries like China, the Netherlands, and the USA dominating the cultivation and production (Yuan et al. 2021). According to the Cognitive Market Research, in 2024, the global orchid market size was USD 302.5

million, which was projected to expand at a compound annual growth rate (CAGR) of 4.00% from 2024 to 2031 (Yuan et al. 2021). Among the orchid products, the cosmeceutical segment reportedly experienced rapid market growth due to increasing demand for orchid extracts in skincare and beauty products. The 2022 report by the Convention on International Trade in Endangered Species of Wild Fauna and Flora (CITES) revealed the rampant international trade in food products containing orchid specimens, particularly Chikanda use in Africa and Salep use in Central Asia and Europe (Dhanda et al. 2022). Moreover, an investigation of the European market identified the presence of products derived from 39 species of the Orchidaceae family (Brinckmann 2014). The products mainly originated from India, China, and Korea and were used as cosmeceuticals, food, food supplements, and medicinal products.

13.4.1 Vanilla

The primary valuable product from orchids is vanilla, which is predominantly derived from species belonging to the genus Vanilla, such as *Vanilla planifolia*, *Vanilla pompon,* and *Vanilla tahitensis* (Sikarwar et al. 2024). Other genera, including Cymbidium, Phalaenopsis, and Dendrobium, are also commercially used. *Vanilla planifolia* is native to Mexico, Central America, Colombia, and Brazil and is one of the primary sources for vanilla flavoring. Vanilla is the second most expensive spice after saffron, used as a flavoring agent in ice creams, dairy products, biscuits, cakes, and beverages, as well as in perfumery and cosmetics (De 2023). The major chemical constituents of *V. planifolia* pods are vanillin, vanillic acid, 4-hydroxybenzaldehyde, and 4-hydroxybenzoic acid (Fig. 13.5). In addition to the major compounds, the overall vanilla flavor is influenced by smaller amounts of other chemicals in the pods of *Vanilla planifolia*, including guaiacol, 4-methylguaiacol, acetovanilone, and vanillic alcohol (Ferrara 2020).

According to the Food and Agriculture Organization (FAO), in 2023, Madagascar was the leading vanilla producer, with a yield of more than three million tons, representing 41.9% of the global output (FAO 2020). Table 13.3 lists the top producers of vanilla in the world. The global vanilla market was valued at USD 3.61 billion in 2024, and the European industry accounted for 33.6% of the revenue

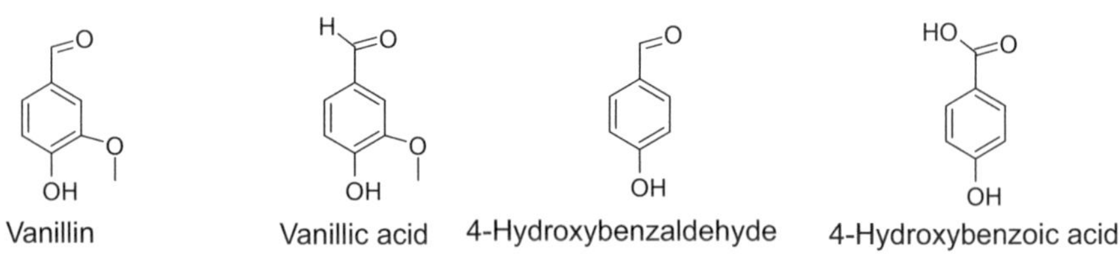

Fig. 13.5 Major chemical compounds of *Vanilla planifolia* pods

Table 13.3 Vanilla producing countries in the world (FAO 2020)

Country	Vanilla production/tons
Madagascar	3034
Indonesia	1965
Mexico	710
Papua New Guinea	491
China	432
Turkey	384
Comoros	240
Tonga	193
Uganda	187

share (Yuan et al. 2021). Projected data shows that the market value will increase to USD5.17 billion by 2030. With consumer preference shifting toward organic and natural food products, there is strong potential for orchid vanilla due to the perceived superior nutraceutical and cosmeceutical properties.

13.4.2 Salep

Salep is a mixture sold as tubers or powder used in the preparation of beverages, confectionery, and desserts, such as ice cream. The product is processed from orchid species, specifically from the Orchis genus, including *Orchis mascula* and *Orchis militaris* (Ghorbani et al. 2014). Orchid tubers contain glucomannose, which acts as a stabilizer and optimizes the melting of ice cream (Hinsley et al. 2018). The tubers for Salep originate from wild populations in mainly Turkey, Iran, and Greece and are indiscriminately collected and traded. However, international trade is restricted under the Convention on International Trade in Endangered Species of Wild Fauna and Flora (CITES) regulations (Masters et al. 2025). Nevertheless, due to rising demand for traditional, organic, and alternative foodstuffs in Greece, the price of Salep has increased to between 55 and 150 Euros per kilogram. The overall market value and production volume of Salep are not well-documented, although its presence in illicit and online trade has been reported (Masters et al. 2025; Subedi et al. 2013).

13.4.3 Chyawanprash

Chyawanprash (CP) is a traditional Indian health supplement, processed from around 50 medicinal herbs and their extracts (Sharma et al. 2019). Among the plants, four of them are orchids, viz.: *Malaxis muscifrea*, *Malaxis acuminata*, *Habenaria intermedia*, and *Habernaria edgeworthii*. Drawing on the principles of the traditional Ayurvedic system, this formulation has been commercially developed and is now available as an over-the-counter product. Chyawanprash is treasured for its multiple health benefits, including preventive, promotive, and curative aspects of health. Jayachandran Nair et al. (2017) reported the chemical constituents of CP, which included vitamin C and phenolics such as gallic acid, protocatechuic acid, catechin, vanillic acid, chlorogenic acid, syringic acid, rutin, ferulic acid, and quercitrin, which could be attributed to its therapeutic activity. Chyawanprash is a complex recipe comprising polycompounds and is manufactured by various firms. Indeed, scientific literature provides little support for the product's safety and effectiveness, which is complicated by inconsistencies in the formula.

13.4.4 Chikanda

Chikanda is a Zambian traditional meat-like snack, commonly known as "African polony," and is made from the ground tubers of wild terrestrial orchids mixed with peanut flour and food seasonings (Dhanda et al. 2022). Most African countries, including Malawi, Tanzania, and Mozambique, have food products equivalent to Chikanda (De Boer et al. 2017; Ghorbani et al. 2014). The Zambian product is mainly derived from orchid species in the genera Brachychorythis, Disa, Eulophia, Habenaria, Platycoryne, Roeperocharis, and Satyrium. A study on the distribution, diversity, and utilization patterns of terrestrial edible orchids conducted in southern Africa revealed that illegal harvesting and trading of orchid tubers are prevalent (Masters et al. 2025). There is a paucity of scientific information on the market potential and nutraceutical value of the edible orchids in the region.

13.5 Conclusion

Orchids have a long history of use in various traditional medicine systems and are being explored for their utility in food supplements, functional foods, and cosmetics. Studies have shown that species belonging to the Orchidaceae family are rich in antioxidants, vitamins, minerals, and phytochemicals. Thus, the plants possess significant nutraceutical and cosmeceutical properties, including anti-aging benefits, skin hydration, immune support, and overall wellness. Specifically, species of the genera Dendrobium and Vanilla have been extensively explored and commercialized as high-value cosmetic products. To fully exploit the potential of orchids, future research directions should include clinical validation, innovative formulations, the isolation of bioactive compounds, and the sustainable sourcing of plant materials. Indigenous knowledge and traditional claims of anti-aging, hydration, and anti-inflammatory benefits require support from rigorous clinical trials. The growing commercial value of orchids in the global market has justified the development of multifunctional products, quality control, and standardization, as well as the sustainable sourcing of plant materials.

References

Aryal (2021) Quality assessment of five different brands of Chyawanprash marketed in Nepal. J Food Qual

Bae JY, Choi JS, Kang SW, Lee YJ, Park J, Kang YH (2010) Dietary compound ellagic acid alleviates skin wrinkle and inflammation induced by UV-B irradiation. Exp Dermatol 19(8)

Baek YS, Ramya M, An HR, Park PM, Lee SY, Baek NI, Park PH (2019) Volatiles profile of the floral organs of a new hybrid cymbidium, 'sunny bell' using headspace solid-phase microextraction gas chromatography-mass spectrometry analysis. Plants 8(8)

Bhattacharyya P, Van Staden J (2016) Ansellia africana (Leopard orchid): a medicinal orchid species with untapped reserves of important biomolecules—a mini review. S Afr J Bot 106: 181–185

Biswas SS, Singh DR, De LC, Kalaivanan NS, Pal R, Janakiram T (2021) A comprehensive scenario of orchid nutrition–a review. J Plant Nutr 44(6):905–917

Brinckmann J (2014) Quick scan of Orchidaceae species in European commerce as components of cosmetic, food and medicinal products

Cakova V, Bonte F, Lobstein A (2017) Dendrobium: sources of active ingredients to treat age-related pathologies. Aging Dis 8(6):827–849

Carpio LE, Sanz Y, Gozalbes R, Barigye SJ (2021) Computational strategies for the discovery of biological functions of health foods, nutraceuticals and cosmeceuticals: a review. Mol Divers 25(3):1425–1438

Castillo-Pérez LJ, Ponce-Hernández A, Alonso-Castro AJ, Solano R, Fortanelli-Martínez J, Lagunez-Rivera L, Carranza-Álvarez C (2024) Medicinal orchids of Mexico: a review. Pharmaceuticals 17(7)

Cheng J, Dang PP, Zhao Z, Yuan LC, Zhou ZH, Wolf D, Luo YB (2019) An assessment of the Chinese medicinal dendrobium industry: supply, demand and sustainability. J Ethnopharmacol 229:81–88

De LC (2023) Pharmaceutical properties of orchids. Pharm Innov 12(6):1454–1460

De Boer HJ, Ghorbani A, Manzanilla V, Raclariu AC, Kreziou A, Ounjai S, Osathanunkul M, Gravendeel B (2017) DNA metabarcoding of orchid-derived products reveals widespread illegal orchid trade. Proc R Soc B Biol Sci 284(1863)

de Vasconcelos FM, Andrade EHA, Teixeira LOA, Figueiredo PLB, Maia JGS (2023) Volatile constituents from catasetum (Orchidaceae) species with occurrence in the Brazilian Amazon. Plants 12(4)

Dev R, Jangid AK (2024) Nature's pharmacy: orchids and their vital role in herbal medicine—a review. Int J Res Public Rev 5(2):488–493

Devaraj RD, Reddy CK, Xu B (2019) Health-promoting effects of konjac glucomannan and its practical applications: a critical review. Int J Biol Macromol 126:273–281

Dhanda S, Bullough LA, Whitehead D, Grey J, White K (2022) A review of the edible orchid trade

Dong X, Gao X, Liu L, Chen L, Liu Q, Zhang D (2020) Function-specific volatiles and volatilization characteristics of Dendrobium officinale. J King Saud Univ - Sci 32(3):2020–2028

Fan C, Sun X, Wang X, Yu H (2023) Therapeutic potential of the chemical composition of Dendrobium nobile Lindl. Front Pharmacol 14

FAO (2020) 10 biggest countries of vanilla production in the world

Ferrara L (2020) Medical reviews narrative review medicinal and pharmaceutical properties of Vanilla planifolia. Int J Med Rev 7(1):25–29

Ghorbani A, Gravendeel B, Naghibi F, de Boer H (2014) Wild orchid tuber collection in Iran: a wake-up call for conservation. Biodivers Conserv 23(11):2749–2760

Hadi H, Razali SNS, Awadh AI (2015) A comprehensive review of the cosmeceutical benefits of Vanda species (orchidaceae). Nat Prod Commun 10(8):1483–1488

Hinsley A, De Boer HJ, Fay MF, Gale SW, Gardiner LM, Gunasekara RS, Kumar P, Masters S, Metusala D, Roberts DL, Veldman S, Wong S, Phelps J (2018) A review of the trade in orchids and its implications for conservation. Bot J Linn Soc 186(4):435–455

Hogan MB, Peele K, Wilson NW (2012) Skin barrier function and its importance at the start of the atopic march. J Allergy 2012:1–7

Huang X, Nie S, Cai H, Zhang G, Cui SW, Xie M, Phillips GO (2015) Study on dendrobium officinale O-acetyl-glucomannan (Dendronan®): Part VI. Protective effects against oxidative stress in immunosuppressed mice. Food Res Int 72:168–173

Jain A, Sarsaiya S, Gong Q, Wu Q, Shi J (2024) Pharmacological and therapeutic biofunction of dendrobium nobile: a critical review. Nat Prod Commun 19(11)

Jayachandran Nair CV, Ahamad S, Khan W, Anjum V, Mathur R (2017) Development and validation of high-performance thin-layer chromatography method for simultaneous determination of polyphenolic compounds in medicinal plants. Pharm Res 9(5):S67–S73

Kim H, Lee YJ, Lee YS (2012) Hydrating effects of orchid extract in skincare: a clinical evaluation of skin moisture retention. Int J Cosmet Sci 34(4):329–355

Kindlmann P, Kull T, McCormick M (2023) The distribution and diversity of orchids. Diversity 15(7)

Kirkwood A, Fisk I, Ayed C, Xu Y, Yang N (2023) A flavour perspective of Tiepishihu (Dendrobium officinale)—an emerging food ingredient from popular traditional Chinese medicinal plants: a review. Int J Food Sci Technol 58(10):4921–4930

Kumwenda FD, Sumani J, Jere W, Nyirenda KK, Mwatseteza J (2024) Edible orchid species from Malawi. pp 1–13

Lall N, Mahomoodally MF, Esposito D, Steenkamp V, Zengin G, Steyn A, Oosthuizen CB (2020) Editorial: cosmeceuticals from medicinal plants. Front Pharmacol 11

Masters S, Bogarín D, Viruel J, van Vugt R, van Andel T, de Boer HJ, Gravendeel B (2025) Horticultural hybrid development of edible terrestrial orchids for verifiable sustainable trade. Front Sustain Food Syst 9:1–10

Misra A, Chaudhary MK, Tripathi D, Srivastava PK, Gupta V, Acharya R, Srivastava S (2023) Nutritional potential of an edible terrestrial orchid Eulophia nuda LINDL and validation of its traditional claim in arthritis. J Ethnopharmacol 306

Mita SR, Husni P, Putriana NA, Maharani R, Hendrawan RP, Dewi DA (2024) A recent update on the potential use of catechins in cosmeceuticals. Cosmetics 11(1)

Mukherjee S, Nangare N, Jagtap S (2023) Use of orchids in Ayurveda: is substitution scientific and appropriate? J Natur Remedies 23(4):1173–1190

Ospina MH (1997) Orchids and the Aztecs. An herbal documented the use of orchids in the Americas. Orchids 66:1160–1163

Patil M, Mahajan V, Chougule N (2024) A review on: orchid extract in skincare current trends and scientific insights. Int J Pharm Sci 2(11):1273–1285

Pomini AM, Sahyun SA, De Oliveira SM, De Faria RT (2023) Bioactive natural products from orchids native to the Americas—a review. An Acad Bras Cienc 95

Prabha A, Kushwaha V, Kumari A, Pathak P (2024) Cypripedium orchids and their utilization in traditional and modern medicine. In: Husen A, Pant M (eds) Exploring medicinal orchids, 1st edn. Taylor & Francis Group, LLC/CRC Press, pp 77–85. https://doi.org/10.1201/9781003436638-1

Rungsang T, Srivilai J, Rakasawapokin P, Rakasawapokin P, Mungmai L, Saesue K, Aoonboontum P, Plukham N, Siriwipanan P, Chaichanathawiki P, Khorana N, Wongwad E (2023) Assessment of antioxidant, anti-lipid peroxidation, antiglycation, anti-inflammatory and anti-tyrosinase properties of dendrobium sulcatum Lindl. Cosmetics 10(2)

Sharma R, Martins N, Kuca K, Chaudhary A, Kabra A, Rao MM, Prajapati PK (2019) Chyawanprash: a traditional Indian bioactive health supplement. Biomolecules 9(5)

Sikarwar PS, Vikram B, Sengupta JD, Singh A, Kumar N, Mishra A, Gautam DK, Kumar V (2024) Vanilla: a review of powerful herb with ayurvedic medicinal properties. Int J Environ Climate Chang 14(2):809–817

Simmler C, Antheaume C, Lobstein A (2010) Antioxidant biomarkers from vanda coerulea stems reduce irradiated HaCaT PGE-2 production as a result of COX-2 inhibition. PLoS ONE 5(10)

Subedi A, Kunwar B, Choi Y, Dai Y, van Andel T, Chaudhary RP, de Boer HJ, Gravendeel B (2013) Collection and trade of wild-harvested orchids in Nepal. J Ethnobiol Ethnomed 9(1)

Tang H, Zhao T, Sheng Y, Zheng T, Fu L, Zhang Y (2017) Dendrobium officinale Kimura et Migo: a review on its ethnopharmacology, phytochemistry, pharmacology, and industrialization. Evidence-based Complementary and Alternative Medicine 2017

Thakur S, Sharma L (2024) Nutraceuticals: new paradigm for better management of rickets. Apple Academic Press, pp 293–311

Verma G, Ranawat MS (2024) A review on phytochemistry and pharmacological activity of Ficus benjamina. Curr Funct Foods 3(3)

Yatung T, Momin KC, Bhargav V, Murthy P, Bharathi TU (2024) Medicinal and therapeutic uses of Dendrobium orchids. In: Husen A, Pant M (eds) Exploring medicinal orchids. Taylor & Francis Group, LLC, pp 86–109. https://doi.org/10.1201/9781003436638-1

Yuan SC, Lekawatana S, Amore TD, Chen FC, Chin SW, Vega DM, Wang YT (2021) The global orchid market. In: Chen FC, Chin SW (eds) The Orchid Genome. Compendium of Plant Genomes. Springer, Cham. https://doi.org/10.1007/978-3-030-66826-6_1

Zhang S, Yang Y, Li J, Qin J, Zhang W, Huang W, Hu H (2018) Physiological diversity of orchids. Plant Diversity 40(4):196–208

Value Addition of Orchid Products 14

Prasenjit Paul and Sanjiban Goswami

14.1 Introduction

Orchids, being the botanical marvels, are one of the most diverse and the largest groups of flowering plants globally, recognized for their remarkable beauty, variety, forms, and adaptability. With over 800 genera and 25,000 species, orchids represent the most highly evolved family of monocotyledons showcasing nature's complexity and elegance (Stewart and Griffiths 1995). Every continent, except Antarctica is home to them, and they thrive in a variety of environments, including semi-deserts, alpine regions, tropical rainforests, and areas close to the arctic zone. Their remarkable floral displays and adaptability to a wide range of habitats continue to captivate scientists, gardeners, and nature enthusiasts worldwide. Orchids are valued both culturally and economically, in addition to their inherent beauty. In many cultures, they have long been prized in horticulture, literature, and art as symbols of strength, richness, and refinement. The economic worth and consumer appeal of any floral product is enhanced through value addition in floriculture. Value addition is the process of enhancing a product or service's features, quality, or utility beyond its most basic form to increase its value or appeal. Processing in many forms, such as cleaning, cutting, packaging, smoking, drying, freezing, extracting, preserving, or adding new features, is one of the common value additions (De and Pathak 2015). A commodity's profitability often rises when a raw material is transformed into a distinctive product. Value addition offers the grower a high premium and produces high-quality goods for both domestic and international markets. Recently, the pattern of consumption has become more varied, shifting towards value-added goods that are developed and sold, such as essential oils, fragrances, medicinal and nutraceutical compounds, insecticidal and nematoidal compounds, natural

P. Paul (✉) · S. Goswami
Department of Botany, St. Edmund's College, Shillong, Meghalaya, India

A. Husen, M. Pant (eds.), *Forest-based Orchids*,
https://doi.org/10.1007/978-981-92-0096-2_14

dyes and pigments, and products made with ingredients like vanilla, gulkand, and rose water.

14.2 Value-Added Orchid Products

Orchids often offer humans a large number of value-added products from the plants' blossoms, stems, and roots, among others, some of which are discussed below (Fig. 14.1 and Table 14.1).

14.2.1 Cut Flowers

Several orchid species, including *Cymbidium eburneum*, *Paphiopedilum insigne*, *Paphiopedilum villosum*, *Paphiopedilum venustum*, *Paphiopedilum hirsutissimum*, *Renanthera imschootiana*, and *Vanda coerulea*, are frequently used as cut flowers (De and Pathak 2015). Along with fresh flowers and other ornamentals, flower arrangements and floral decorations using orchids are significant value additions to floristry products. Orchid flower arrangements make excellent centrepieces for many wedding ceremonies as well as decorations for an array of programs. *Cymbidium*, *Dendrobium*, *Phalaenopsis*, *Oncidium*, *Cattleya*, and other orchid hybrids with varying colours and forms are used as cut flowers, floral arrangements, and exhibits (De 2021).

Fig. 14.1 Value addition of orchids

Table 14.1 Orchids and their value-added products

Sl. No	Orchids	Value addition
1	*Cymbidium eburneum*, *Paphiopedilum insigne*, *Paphiopedilum villosum*, *Paphiopedilum venustum*, *Paphiopedilum hirsutissimum*, *Renanthera imschootiana*, *Vanda coerulea*, *Cymbidium*, *Dendrobium*, *Phalaenopsis*, *Oncidium*, *Cattleya*, etc.	Cut flowers
2	*Dendrobium*, *Cymbidium*, *Paphiopedilum*, *Cattleya*, *Pholidota*, etc.	Dry flowers
3	*Dendrobium*, *Phalaenopsis*, *Oncidium*, *Cymbidium*, *Paphiopedilum*, *Cattleya*, *Vanda*, *Epidendrum*, etc.	Potted plants
4	*Paphiopedilum*, *Dendrobium*, *Vanda*, *Cymbidium*, etc.	Orchid landscaping
5	*Rhynchostylis retusa*, *Dendrobium*	Cultural symbols
6	*Dendrobium moniliforme*, *Red Cattleya*, *Dendrobium*, etc.	Extracts and oils
7	*Dendrobium nobile*, *Cymbidium grandiflorum*, *Bletilla striata*, *Phalaenopsis lobbii*, *Vanda coerulea*, etc.	Cosmetics and skin care
8	*Dendrobium nobile*, *Dendrobium fimbriatum*, *Habernaria intermedia*, *Habernaria edgeworthii*, *Malaxis muscifera*, and *Malaxis acuminata*, etc.	Phytomedicinal products
9	*Vanilla planifolia*, *Dendrobium*, *Vanda*, *Jumellea fragrans*, etc.	Foods and beverages
10	*Dendrobium nobile*, *Dendrobium fimbriatum*, *Paphiopedilum insigne*, *Vanda*, etc.	Tissue culture plantlets

14.2.2 Dry Flowers

Another significant area is the manufacture of dried or dehydrated flowers, which has recently become more eclectic in consumption patterns (Mebakerlin and Chakravorty 2015). The best techniques for drying orchids are freeze drying or microwave drying using silica gel. Greeting cards, covers, wall hangings, calendars, artistic displays, floral balls, and other floral crafts are made with dried orchid flowers. Drying can be done with the bright flowers of orchid genera such as *Dendrobium*, *Cymbidium*, *Paphiopedilum*, *Cattleya*, and *Pholidota* (De and Pathak 2015).

14.2.3 Potted Plants

They add both aesthetics and utility to homes and workplaces. The potted plants flourish well when given the proper pot, substratum, and light, adding joyfulness and a little greenery to daily life. *Dendrobiums* are widely used as flowering potted plants worldwide due to their long vase life, year-round availability, and floriferousness, which encompasses a broad range of flower colours, sizes, and shapes (De et al. 2019). Depending on the species and hybrid, potted orchids have a shelf life of 3 weeks to 4 months, which is longer than cut flowers (Pal et al. 2019). Throughout many nations, orchids are traded frequently as gifts or commodities. Among the principal orchid genera that are sold in pots worldwide are *Phalaenopsis*, *Oncidium*, *Cymbidium*, *Paphiopedilum*, *Cattleya*, *Vanda*, and *Epidendrum* (Lopez and Runkle 2005). Thus, potted flowering orchids have become a promising crop category that offers greenhouse growers a lucrative substitute for other floriculture crops.

14.3 Orchid Landscaping

Orchid landscaping is another more recent development in orchid value addition. Orchid landscaping is the art of using orchids in outdoor garden designs to create exotic, tropical, and aesthetically pleasing settings. Orchids, particularly epiphytic types, are frequently mounted on trees, rocks, or wooden structures, in contrast to conventional bedding plants. This technique improves the attractiveness of the surrounding environment by integrating the orchids into scenes with a jungle or arboreal theme. In botanical gardens, private estates, and tropical resorts, orchid landscaping is a widespread practice. Several types of orchids, including *Paphiopedilum*, *Dendrobium*, *Vanda*, and *Cymbidium*, are frequently used in landscaping. It is more than just ornamental; it is a subtle blend of design and horticulture that values the richness and beauty of one of the most beautiful members of the plant kingdom.

14.4 Orchids as Cultural Symbols

Across many civilizations, orchids have long had cultural and symbolic value. In India, they are widely featured in literature, art, culture, and customs around the world. The foxtail orchid, *Rhynchostylis retusa*, often referred to as Kopou Phool in the local dialect, is ingrained in Assamese culture, especially during the Rongali Bihu festival (also known as Bohag Bihu). This orchid, which blooms in April to June, celebrates the Assamese New Year and the arrival of spring, and adorns the hair of young ladies. The blooms of *R. retusa* have become emblematic of Bihu, as well as joy, fertility, and represent the close relationship that exists between local cultures and nature. Some Manipur tribal groups use orchids, especially the Golden Arch Dendrobium, in their traditional clothing and headgear. In their headgear, known as Shamkeen, which is typically worn by the women as a culture, the Maring tribe of Manipur mostly uses dried orchid stems. As part of its traditional symbolism, the Tangkhul tribe of Manipur also uses dried orchid stems in their Pasi, a headdress worn by the men. In terms of culture, orchids are more than just ornamentals; they represent virtues like innocence, tenacity, and refinement, making them classic symbols of world heritage.

14.5 Orchid Extracts and Oils

Known for their aromatic and medicinal qualities, orchid oils and extracts are made from a variety of orchid species. These organic substances are frequently utilized in cosmetics, aromatherapy, and conventional medicine. Antioxidants present in orchid extracts help shield the skin from environmental stresses and are found to support a youthful, healthy appearance. Usually obtained by steam distillation or solvent extraction, orchid oils have a subtle, flowery scent and are often used in essential oil blends, perfumes, and holistic therapies. The leaf or stem extract of *Dendrobium moniliforme* is used in the preparation of perfume as compared to Red Cattleya, where the floral extract is used. Essential oils are made from the flowers, leaves, seeds, and roots of several *Dendrobium* species (Singh et al. 2016).

14.6 Orchid-Based Cosmetics and Skin Care

The orchid flowers' abundant antioxidant, moisturizing, and emollient properties have led to the rise in popularity of orchid-based skin care and cosmetics products in Europe and South Asia. Orchids' mucilaginous content makes them ideal as emollients and moisturizers due to the numerous hydrogen bonding linkages that maintain the skin's outermost layer at the appropriate water content. They are utilized in some of the after-sunscreen products and in the treatment products for sensitive skin. The skin-conditioning ingredient in some of the cosmetic products contains *Dendrobium nobile* and *Cymbidium grandiflorum*. According to the Traditional Chinese medical system, *Bletilla striata* root extracts are used as a cleansing face mask and moisturizer. The entire plant extract of *Phalaenopsis lobbii* has been used in cosmetic products as a bleaching agent, lightening the colour of skin or hair (Singh et al. 2016). *Vanda coerulea* is another significant orchid that has been used as an antioxidant source in a number of cosmetic products. In general, the elegance, effectiveness, and botanical appeal of orchid-based ingredients—which provide both noticeable results and a luxurious feel—make them valued in high-end and natural skin care lines.

14.7 Orchids-Based Phytomedicinal Products

Many orchid species, such as *Dendrobium nobile* and *Dendrobium fimbriatum*, are used in both traditional and contemporary herbal medicine to make orchid-based phytomedicinal products (Bhattacharyya et al. 2016; Paul and Kumaria 2020). Wild orchids are used as a form of folk medicine by tribal people in the northeastern hill region due to the presence of various bioactive compounds. These plants, which are abundant in bioactive substances such as alkaloids, phenols, polysaccharides, flavonoids, β-sitosterol, rhein, fimbriatone, and scopolin, have immunomodulatory, anti-inflammatory, hepatoprotective, nephroprotective, antioxidant, and antibacterial properties (Shailajan et al. 2015; Paul et al. 2017). Along with other plants,

Habernaria intermedia, *Habernaria edgeworthii*, *Malaxis muscifera*, and *Malaxis acuminata* are used in a polyherbal formulation for the preparation of Chyawanprash, which is well known for improving digestion, respiratory health, general wellness, and strengthening immunity (Singh et al. 2016).

14.8 Orchid-Infused Foods and Beverages

The delicate flavour, exotic appeal, and potential health benefits of orchid-infused foods and drinks are gaining popularity. The majority of culinary applications involve edible orchid species, such as *Vanilla planifolia*, which is the source of natural vanilla extract (De et al. 2019). Vanilla is mainly utilized as a flavouring agent in ice creams, dairy products, cakes, biscuits, and beverages. However, other decorative orchid species, like *Dendrobium* and *Vanda*, are used as garnishes or flavour enhancers in salads, desserts, cocktails, and gourmet meals. Salep, a blend of dried tubers from several orchids, is used to make dondurma, sometimes referred to as Turkish orchid ice cream (Singh et al. 2016). A popular Zambian dish called chikanda, sometimes referred to as "African polony" or "African sausage," is made from the ground tubers of terrestrial orchids combined with peanuts and spices. The orchid *Jumellea fragrans* is used to make the well-known beverage known as "Faham" or "Madagascar Tea" on the islands of Mauritius and Madagascar (De et al. 2019). Orchids are therefore a distinctive addition to innovative food and beverage trends, as evidenced by the rise in popularity of orchid-infused foods in luxury and holistic sectors, which combine traditional culinary practices with modern gastronomy.

14.9 Tissue Culture Plantlets

Tissue culture of orchids under laboratory conditions offers horticulture a substantial value-added opportunity by creating high-quality plants. In this method, disease-free orchid plants can be quickly multiplied from tiny tissue samples. Due to their outstanding quality, unique, and reliable characteristics, these year-round lab-grown orchids command high prices in the commercial market (Nandi et al. 1999). The utilization of asymbiotic seed germination for commercialization and large-scale propagation is another facet of orchid value addition (Diengdoh et al. 2017; Paul and Kumaria 2024; Wongsa et al. 2025). They are especially useful for new varieties, hybrids in export markets where plant health certifications are essential. In vitro flowering of orchids using plant tissue culture is an advanced biotechnology technique that greatly increases value addition. A number of juvenile orchids, including *Dendrobium*, have been documented to flower in vitro (Paul et al. 2017; Pujari and Babu 2022). Growers can purchase young tissue-cultured orchid plantlets, or plugs, and raise them to maturity. This lowers shipping expenses and increases production flexibility. Nurseries, florists, and commercial producers all profit from stable supply chains in the marketplace. By converting orchid production from seasonal to controlled, efficient, and year-round production, this technique significantly increases the value-added prospects in the ornamental plant sector.

14.10 Conclusion

In conclusion, orchids and their value-added products are not only exquisite botanical specimens but also representations of the intricacy and beauty of nature. Orchids are now even more significant in the wellness, cosmetics, and pharmaceutical sectors due to the rising demand for natural and organic products. Sustainable sourcing makes orchid-based goods a valuable component of high-end, environmentally friendly products, as they provide a unique combination of nature's benefits.

References

Bhattacharyya P, Kumaria S, Job N, Tandon P (2016) *En-masse* production of elite clones of *Dendrobium crepidatum*: a threatened, medicinal orchid used in Traditional Chinese Medicine (TCM). J Appl Res Med Aromat Plants 3(4):168–176

De LC (2021) Value Addition to 75 Indian Orchid Species. Biotica Res Today 3(11):982–995

De LC, Pathak P (2015) Value addition in orchids. J Orchid Soc India 29:31–37

De LC, Kumar R, Khan A, Medhi RP (2019) Social science orchid farming-a remunerative approach for farmers livelihood. Int J Sci Res 3(9):77–80

Diengdoh RV, Kumaria S, Tandon P, Das MC (2017) Asymbiotic germination and seed storage of *Paphiopedilum insigne*, an endangered lady's slipper orchid. S Afr J Bot 112:215–224

Lopez RG, Runkle S (2005) Environmental physiology of growth and flowering of orchids. HortScience 40(7):1969–1973

Mebakerlin MS, Chakravorty S (2015) Value addition in flowers. In: Value addition of horticultural crops: recent trends and future directions. Springer India, New Delhi, pp 83–99

Nandi SK, Kumar A, Palni LMS (1999) Role of plant tissue culture in biodiversity conservation and economic development. Curr Sci 77(10):1229–1231

Pal R, Meena NK, Pant RP, Dayamma M (2019) *Cymbidium*: botany, production, and uses. In: Orchids phytochemistry, biology and horticulture: fundamentals and applications. Springer International Publishing, Cham, pp 1–37

Paul P, Kumaria S (2020) Precursor-induced bioaccumulation of secondary metabolites and antioxidant activity in suspension cultures of *Dendrobium fimbriatum*, an orchid of therapeutic importance. S Afr J Bot 135:137–133

Paul P, Kumaria S (2024) Conservation of medicinal orchids through plant tissue culture. In: Husen A, Pant M (eds) Exploring medicinal orchids. Taylor & Francis Group, LLC, pp 17–23. https://doi.org/10.1201/9781003436638-1

Paul P, Joshi M, Gurjar D, Shailajan S, Kumaria S (2017) *In vitro* organogenesis and estimation of β-sitosterol in *Dendrobium fimbriatum* Hook.: an orchid of biopharmaceutical importance. S Afr J Bot 113:248–252

Pujari I, Babu VS (2022) Precocious *in vitro* flowering in threatened ornamental orchid, *Dendrobium ovatum*–decoding the causal factors. Curr Plant Biol 31:100257

Shailajan S, Kumaria S, Gurjar D, Joshi M, Paul P, Khongthaw N (2015) Variation in the marker content of five different *Dendrobium* species: comparative evaluation using validated HPTLC technique. J Appl Pharm Sci 5(10):032–038

Singh DR, Kishore R, Kumar R, Singh A, Rai D (2016) Orchid preparations. Astral International (P) Ltd., New Delhi
Stewart J, Griffiths M (1995) Manual of orchids. Timber Press
Wongsa T, Piapukiew J, Kuenkaew K, Somsanook C, Sapatee O, Linjikao J, Kunakhonnuruk B, Kongbangkerd A (2025) Asymbiotic seed germination and in vitro propagation of the Thai rare orchid species; *Eulophia bicallosa* (D Don) PF Hunt & Summerh. Plants 13(13):2212

Orchid Cultivation for Hobbyists and Beginners

15

Vasundhra and Anuprabha

15.1 Introduction

Orchids, the most intricate flowering plants, are classified within the family Orchidaceae, one of the largest groups of flowering plants and the second largest in the plant kingdom (Husen and Pant 2024). India is home to around 1256 orchid species across 155 genera (Singh et al. 2019), but globally, there are 28,484 species distributed across 850 genera (Govaerts et al. 2017). Recently, the Royal Botanic Gardens, Kew catalogued 705 genera (POWO 2023) and 29,481 validated species (WFO 2023) within the family. They may be epiphytic on tree trunks, branches, and rocks; terrestrial with a short aerial life cycle and heterotrophic, thriving on decomposing organic waste. Similar to other herbaceous plants, terrestrial varieties flourish in temperate and alpine regions, with most possessing underground bulbs, corms, or tubers. Alongside variations in their habits and habitats, as well as the form, size, orientation, and texture of the flowers, the lip/labellum exhibited variation at the species level. Their complex flower structure sets them apart from other plant groupings and encourages more complex biotic cross-pollination. Numerous orchid species, such as *Cattleya labiata* (Brazil), *C. mossiae* (Venezuela), *C. trianae* (Colombia), *Phalaenopsis amabilis* (Indonesia), and *Vanda* Miss Joaquim (Singapore), are recognized as national flowers (De and Singh 2016b).

Orchid cultivation is a fascinating and rewarding hobby that combines the beauty of nature with the joy of nurturing plants. Known for their exotic flowers and wide variety of species, orchids have long captured the admiration of gardeners and collectors worldwide. While they may seem delicate or difficult to grow, many orchids can thrive with the right basic care and environment (Friend 2004; Biswas and Pal 2021). For beginners, learning simple techniques such as proper watering, temperature, light management, and potting can make the process enjoyable and

Vasundhra · Anuprabha (✉)
Department of Botany, Goswami Ganesh Dutta Sanatan Dharma College, Chandigarh, India

A. Husen, M. Pant (eds.), *Forest-based Orchids*,
https://doi.org/10.1007/978-981-92-0096-2_15

successful. Orchids not only enhance indoor and outdoor spaces with their vibrant blooms but also bring a sense of satisfaction to growers. With patience and observation, even hobbyists can cultivate orchids that bloom year after year. Starting with hardy varieties helps build confidence, eventually encouraging exploration of more unique and rare species. This introduction to orchid cultivation will guide beginners towards a fulfilling and creative gardening journey.

Depending on the vegetative growth they exhibit, orchids have one of two types of growth habits:

15.2 Types of Orchids

15.2.1 Monopodial Orchids

Monopodial orchids possess a solitary, vertically oriented, unbranched stem that elongates with the progression of the seasons. These orchid varieties produce a succession of leaves that develop alternately along the stem, similar to *Vanda* and *Arachnis*.

15.2.2 Sympodial Orchids

When cultivated in pots, these orchids often have a horizontal rhizome (the main stem) at the base from which several growths grow upward. In several sympodial orchids, including *Cymbidiums*, *Jumeleas*, *Paphiopedilum*, and *Phragmipedium*, the new growths emerge in proximity to the base of the older growths, developing a circular arrangement around them. Seasonally, an orchid plant may simultaneously exhibit vertical and horizontal growth. These are referred to as "pseudomonopodials".

15.3 Growing Conditions

Generally, orchids are cultivated in polyhouses with sufficient cross ventilation or shade houses, which are buildings covered with shade cloth. In regions with considerable precipitation, it is recommended to cultivate in shaded polyhouses. The lateral surfaces of structures should be adorned with shade fabric. In particular, the structures are designed to shield the plants from pests, extreme light intensity, wind, and rain.

15.3.1 Greenhouse

A greenhouse is a semi-permanent building, typically enveloped in transparent or translucent material, ranging from basic handmade designs to advanced prefabricated constructions, in which the climate can be adjusted to facilitate the cultivation of plants. The materials utilized for constructing a greenhouse frame may include wood, bamboo, steel, or aluminium.

15.3.1.1 Advantages of Greenhouse

Depending on the kind of greenhouse, crop, and environmental regulating facilities, the output might be 10–12 times better than that of outside production.

1. The reliability of crops improves during greenhouse cultivation.
2. Optimally meant for vegetable and floral cultivation.
3. Continuous production of floricultural crops throughout the year.
4. Off-season cultivation of vegetables and fruit crops.
5. Continuous production of unaffected, disease-free, and genetically superior transplants is achievable.
6. Efficient use of chemicals and insecticides for pest and disease management.
7. The water requirements of crops are minimal and easily manageable.
8. Management of stock plants, cultivation of grafted plantlets, and micropropagated plantlets.
9. Acclimatization of tissue-cultured plants.
10. Primarily beneficial for overseeing and regulating the volatility of diverse ecological systems.

Modern methods like as hydroponics (soil-less growth), aeroponics, and nutrient film techniques are feasible exclusively inside the greenhouses.

Protected cultivation offers several benefits, including higher yields, higher-quality blooms, and more efficient use of land and other resources. This is particularly relevant to flower cultivators in India with limited land holdings, specifically those with holdings of less than one hectare. To get the most benefits, even with cost-effective protective structures, careful design, administration, and attention to detail are required (Fay 2018). Cultivating flowers within location-specific, appropriate structures can yield significant advantages, even for small-scale farmers (Fig. 15.1).

15.3.2 Aeration and Ventilation

Optimal air quality and enough circulation are crucial for orchid cultivation. Consistent light breezes provide a substantial source of carbon dioxide for photosynthesis. There should be a slight breeze to shift the leaves. Air circulation in the cultivation area promotes optimal growth and reduces the prevalence of illnesses and pests. Ceiling and oscillating fans are efficient in delivering mild airflow in hobby greenhouses or indoor cultivation spaces. Both can include extensive regions with a continuously varying airflow pattern without causing significant desiccation of the plants. *Paphiopedilums* require enough ventilation and higher relative humidity, especially during the summer, to promote leaf and root development and to mitigate bacterial and fungal infections. The airflow facilitates the evaporation of surplus

Fig. 15.1 Interior view of greenhouse showing orchids under cultivation

Fig. 15.2 (**a**) Epiphytic orchids in wooden baskets; (**b**) Epiphytic orchids in coconut husk basket

moisture in the compost or potting material. Evaporative air conditioners are vital for commercial greenhouses in warm areas. Orchids require ample space for air to flow freely because most cultivated orchids are epiphytic plants. Proper airflow helps maintain optimal humidity and temperature levels, supporting healthy growth. The placement may be adjusted to ensure sufficient air circulation when cultivated in a greenhouse or shade house. Hanging baskets have superior air circulation compared to pots or benches, allowing species such as *Vanda*, *Aerides*, and *Cattleya* to flourish (Fig. 15.2).

15.3.3 Light

Light is essential for the optimal growth of orchids. They appreciate diffused sunshine overall. However, leaving the leaves in direct sunlight will cause them to burn. Optimal light levels facilitate healthy growth and nutrient accumulation, hence enhancing the flowering cycle. If the leaves begin to yellow, the plant should be moved to a more shady location.

The orchid's leaf will indicate the adequacy of light, determining if it is sufficient or excessively intense for the plant.

1. The light is probably at its best if the orchid's leaves remain firm, crisp, and green.
2. Too little light is available if the foliage is dark green.
3. If the foliage has a purple hue, the light intensity is excessive for the plants. Excessive light intensity might cause the tips of the leaves to desiccate. The phenomenon is referred as "Suntan" (Husain and Eraqui 2023).

Most orchids cannot endure direct sunshine, except for 1or 2 h following dawn and 1 or 2 h following sunset. Orchids require shade during the sun's most intense hours. Adjacent trees or towering structures may offer sufficient shade. A significant decrease in light occurs from summer to winter. In winter (from mid-October to mid-February), more illumination and decreased shading should be used to counteract the natural decrease in light. While plants may acclimatize to changing environments, they will do so more effectively with reduced stress, particularly if they are gradually introduced to new circumstances, such as acclimatizing them to bright outside light over the summer. Orchids will, to a certain extent, acclimatize to their surroundings. Orchids may endure high light levels within their specific range if they get increased hydration, additional fertilization, and improved air circulation to maintain lower leaf temperatures. Increased watering frequency entails that the orchids will be irrigated about every 5 days rather than every 7 days. This does not imply that cultivators should expose their plants to excessive light. Excessive light, coupled with insufficient water and fertilizer, will induce stress in orchids. Stressed plants exhibit increased susceptibility to pests and diseases, rendering them more vulnerable to insect attacks, ultimately impairing their reproductive development. The ideal light requirement:

Optimal growth range: 5000–8000 lux
Optimal flowering range: 8000–15,000 lux

15.3.4 Temperature

Depending on the temperature they need, orchids are divided into three main categories. These are warm climate orchids, intermediate climate orchids, and cool climate orchids (Table 15.1).

Table 15.1 Classification, favourable temperature, and shade cloth requirements of some species of orchids (Lopez and Runkle 2005)

		Day time temperature (°C)		Night time temperature (°C)	
Species	Shade cloth provided (%)	Winter	Summer	Winter	Summer
Warm growing orchids					
Phalaenopsis	60	20	0	15	25
Cattleya	40				
Catasetum	50				
Calanthe	40				
Intermediate temperature growing orchids					
Oncidium	20–30	18	27	13	21
Heat tolerant *Cymbidium*	30–40				
Paphiopedilum	50				
Bulbophyllum	50				
Cool temperature growing orchids					
Cymbidium	30–40	15	25	10	18
Disa uniflora	50–70				
Dendrobium	30				
Coelogyne	50				
Zygopetalum	40				

15.3.5 Humidity and Water Management

The majority of orchids suffer from overwatering rather than under-watering. Excessive irrigation resulted in root rot and several other ailments. Most orchids like water with a pH range of 5.0–6.5 (De and Singh 2016a). Irrigation with suboptimal or higher pH values, or with excessive dissolved minerals, might impede nutrient absorption. Regular irrigation is crucial in environments of intense sunshine and higher temperatures. Plants in tiny containers desiccate more rapidly than those in larger pots. Plants in clay pots demand more watering than those in plastic pots. A hanging plant, which has superior aeration compared to one in a container, requires more regular watering. Increased hydration is essential for new potting substrates. Watering should be conducted either in the morning or in the afternoon. Monopodial orchids need greater humidity levels compared to sympodial orchids. As a general guideline, temperatures that are higher should be accompanied by increased humidity levels. The implementation of misting machines, foggers, or humidifiers will guarantee sufficient humidity levels. Retention of standing water beneath the seats may be maintained to enhance humidity. In *Cymbidium*, consistent watering throughout the year is necessary to maintain the pseudobulbs in a green and smooth condition. The water supply frequency is specified as: Summer: two to three times weekly; Autumn: once or twice weekly; Winter: once weekly; Spring: as in Autumn. The ideal relative humidity range is 50–80%, which is crucial for optimal development and flowering. In higher temperatures, spraying the plants and adjacent floors and benches sustains humidity levels. *Dendrobium* and *Phalaenopsis* orchids both thrive when humidity is maintained at about 50–70%. Placing ferns, bromeliads, or lush foliage close to the Phalaenopsis plant indoors can help raise ambient humidity. *Cattleya* orchids are generally watered about once or twice a week, with more frequent watering during the summer months. Between waterings, their potting medium should be allowed to dry out somewhat (i.e. they should be allowed to desiccate) to help prevent rot and mimic their natural cycle. It is recommended to water *Cattleya* orchids in the morning to ensure the foliage is dry by evening. They need a relative humidity of 40–60%. Generally, *Oncidium* with substantial fleshy roots or leaves demands less frequent irrigation compared to those with slender leaves or roots. Watering must be comprehensive, and the substrate should be left to dry to at least 50% of its depth before subsequent irrigation. Watering frequency may range from every 2–10 days, contingent upon weather conditions, pot dimensions and composition, orchid species, and potting media type. Plants that are not in active growth should get reduced watering. Humidity levels should be maintained between 50 and 60%. Numerous *Oncidium* orchids demand lower humidity levels compared to other orchid varieties. *Paphiopedilums* require higher relative humidity, especially during the summer, to promote leaf and root development and to prevent bacterial and fungal infections. As temperatures decline throughout the winter season, humidity must be systematically reduced. The ideal humidity for optimal development of *Paphiopedilums* is between 65% and 75%. Vandaceous plants thrive on high-quality water. The monopodial growth habit of the plants prevents them from retaining water for extended periods. It is recommended to irrigate these plants as early as possible in the morning, ensuring that the foliage dries before midday. Orchids dislike dampness and cold; thus, irrigate your plants throughout winter just on bright, clear mornings as needed. Adult plants receive water weekly in winter and daily in summer. Vandaceous orchids need humidity levels of 70–75%.

15.4 Method of Planting

Terrestrial orchids require one part rich humus with decayed leaf mould, half part dried and decomposed compost, one part sphagnum moss or coir pit mixture, and one part a combination of the two, as well as proper drainage. The medium should be placed in ceramic or similar containers with a layer of drainage material at the bottom. For healthy growth, provide full sunlight (or strong light) and stake the plants so they have support.

A variety of growth mediums are utilized for epiphytic orchids, including those mentioned below:

15.4.1 Sphagnum Moss, Coconut Husk, and Tree Bark

Sphagnum moss is great at holding water, which makes it perfect for orchids that need a consistently moist environment. It also helps the air move around the roots.

But as it wears down, it becomes more compacted and loses air flow, so it needs to be replaced often. Coconut husk and pieces of tree bark are abundantly accessible, cost-effective, and possess a significant capacity for moisture absorption. The husk and bark ratio must be modified in order to prevent the medium from being saturated.

15.4.2 Bricks, Stone Chips, Charcoal, and Tiles

Their ability to absorb and retain moisture makes them excellent for drainage. Besides its convenience and security in orchid cultivation, charcoal is effective for purifying the medium. Before incorporating the potting mixture and the plant, which requires staking, substantial pieces of potting material are positioned around the drainage hole. Growers frequently use lightweight plastic containers that help prevent algae growth and particular clay pots with drainage holes for orchid cultivation. Due to their limited vertical growth habit, Vandas are especially fond of wooden baskets. Orchids are positioned on raised platforms in commercial settings to improve drainage and air circulation by raising the pots off the ground.

Potting mixtures of some commercial orchids are listed below:

Cattleya: Cocochips + brick piece + leaf mould/leaf fern (1:1:1)
Cymbidium: Cocochips + Cocopeat + Brick pieces + Slow-release fertilizer (3:1:1: 5 g/pot)
Dendrobium: Coco peat + brick pieces + tree bark (1: 1: 1)
Oncidium: Cocochips + brick pieces + leaf moulds (1:1:1)
Phalaenopsis: Cocochips + Brickpieces + leafmoulds + green moss (1:1:1:1)
Vanda: Cocochips + brick pieces + leaf fern (1:1:1)

15.5 Nutrient Management

The kinds of nutrients, how much of them are used, and how frequently they are used vary greatly depending on the orchid's kind, potting media, season, growing conditions, development stage, etc. In the initial 2 weeks of development, a fertilizer blends with an NPK ratio of 1:1:1 is essential. Generally, mature plants require feedings in the ratios of 3:1:1 during the pre-flowering period and 1:2:2 throughout the blooming stage. Alongside bone meal, which may be used two to three times weekly in tropical climates, organic fertilizers such as mildly fermented oil cake solutions (1:10), livestock dung (1:25), and diluted cow's urine (1:25) are also highly effective. A fertilizer combination including both macronutrients and micronutrients is optimal for foliar application.

15.6 Potting

15.6.1 Potting Monopodial Orchids

Monopodial orchids grow vertically, with a single main stem that bears leaves alternately along its length. When potting such orchids, choose a plastic pot just large enough to accommodate the roots without forcing them; if using a clay pot, go one size larger to allow space yet provide the benefits of clay. Position the plant centrally, ensuring that the lowest leaf base meets the medium's level and the roots are properly placed. Secure the plant with a stake if needed for support. The potting medium should be evenly moist but not waterlogged. After potting, raise the base of the plant slightly so leaves do not touch the medium, and leave part of the upper roots exposed above the medium. Remove yellow, dried or spotted leaves, and any diseased sections. If stakes are used, they can be removed after about 2–3 months once the plant is stable and has developed sufficient roots (Nagrare and Pal 2008).

15.6.2 Potting Sympodial Orchids

These orchids possess a rhizome (primary stem) at the base (often horizontal, particularly when cultivated in pots), from which a succession of growths emerges upwards. In several sympodial orchids, including *Cymbidiums*, *Jumeles*, *Paphiopedilum*, and *Phragmipedium*, the new growths emerge in proximity to the base of the older growths, creating a circular arrangement around them. The potting technique for these young plants resembles that of monopodial orchids; however, many sympodial orchids, including *Cattleyas*, *Dendrobiums*, and *Oncidiums*, have new growths along a rhizome that typically grow in the opposite direction from the older growths (Maitra et al. 2020). The potting process for these is identical to that of monopodial orchids, except that the plant should be positioned towards one edge of the pot, allowing space on the other side for new growth.

15.7 Post-repotting Orchid Care

Repotting can be as stressful for orchids as a major procedure is for humans, necessitating careful attention for several weeks following the process. Immediately after repotting, leaves should be gently misted twice daily—once in the morning and again by midday—for 2 weeks in healthy plants, and up to 4 weeks in weak or stressed specimens. During winter, avoid irrigation after 12 noon to prevent excess moisture in cooler conditions.

Incorporating 2–3 drops of Superthrive or a rooting solution into the misting water can significantly aid recovery. If a rooting solution is unavailable, a small amount of phosphorus-rich fertilizer may be added as an alternative. Post-repotting, position the orchid in an area with reduced light exposure compared to its usual

Fig. 15.3 Potting: (**a**) Potting in earthen pots having small holes for excess water removal and aeration for epiphytic orchid roots; (**b**) Potting of terrestrial orchids in plastic pots

conditions, and then shift to normal light duration of 3–4 weeks. These modifications alleviate stress and facilitate adaptation to the environment.

During the initial week, irrigate the orchid solely to moisten the potting substrate. Avoid excessive watering, as it may result in runoff or waterlogging. After 1 week, restore the standard watering regimen, ensuring the substrate is moist but not overly saturated. Fertilizers should be avoided for the first three to four waterings; instead, use a rooting solution to support root establishment. Staking, if necessary, may be employed for a duration of 2–3 months and subsequently removed once the plant has developed sufficient roots to sustain itself independently (Fig. 15.3).

15.8 Pest Management in Orchids

Prevention is the most effective form of defence; thus, plants should be inspected regularly for pests and diseases. Scale insects are little creatures (1–2 mm) that excrete a waxy covering for protection and resemble scales. The hard scale resembles miniature turtles. Soft scales can be located in desiccated sheaths near the base of pseudobulbs of *Cattleya* or analogous plants. Hard scale is typically located behind the leaves of *Phalaenopsis* or *Cattleya*, sometimes concealed within the pot. The visible insects should be eliminated followed by the use of a pesticide.

Mealybugs are white, like cotton, and may be around 1/4 inch in size. They may be located on or beneath the leaves, on flower stalks, on buds, behind blooms, or within the container. Visible insects should be removed, followed by the application of a pesticide. Aphids are highly persistent, reproduce on a 3-day cycle, are little sap-sucking insects, exhibit a variety of colours, and possess the ability to fly. They are located on new growth, fresh leaves, flower stalks, and flower buds. In bright, sunny weather, the plant should be carried outdoors and rinsed with a garden hose to dislodge them from the plants, and then may be treated with a pesticide. Fungus gnats resemble little black flies. They will continue to conceal themselves in the container, drawn to the moist potting medium and decomposing plant matter (dead roots and leaves). They may affect the roots, particularly those of *Cymbidiums*. The pot may be treated by submerging it in an insecticide solution. Spider mites are small and cannot be observed individually without magnification. They inhabit the undersides of leaves, in little silvery depressions, where they extract the plant sap. They can be controlled by regulating humidity and using an insecticide or miticide. The pot may be submerged in an insecticide/miticide solution. Thrips are little, elongated insects characterized by their fringed wings. They possess a length of 1 mm or less. They are present in malformed or speckled blooms. They are challenging to eliminate due to their tendency to reside in flower buds and behind sheaths, where they are shielded from pesticide applications. Deformed or discoloured flowers must be removed and disposed of.

For the aforementioned sucking insects, pesticides such as acephate, dimethoate, and imidacloprid may be utilized at a concentration of 1 g/L solution. Slugs are famous for their nocturnal activity and can be found in any place. Sluggo, available in either liquid or granular form, can be utilized for their control. Alternatively, they may be physically extracted and removed.

15.8.1 Insects Control with an Insecticide

An extensive range of insecticides is available on the market, including Chlorpyrifos-methyl, Imidacloprid, Acetamiprid, Dinotefuran, Thiamethoxam, Malathion, Pirimicarb, Carbosulfan, Lambda-cyhalothrin, Esfenvalerate, Pymetrozine, and Diafenthiuron.

When using commercially available pesticides, it is essential to follow the instructions on the label strictly and implement all necessary measures to prevent poisoning oneself and others in the vicinity. One should refrain from exceeding the prescribed dosages of the insecticide. A safe and efficient pesticide may be produced at home. A soap solution is created by combining one teaspoon of a gentle liquid dishwashing detergent with 1 L of lukewarm water. Numerous pesticides eliminate just adult insects, sometimes failing to affect the eggs or larvae (immature insects). Insects can acquire resistance to pesticides. This indicates that certain insects are resistant to the pesticide and will continue to reproduce. Applying the same pesticide to such insects will not eliminate them. To prevent resistance, one may alternate insecticides: the initial treatment with one insecticide, the subsequent application with a different one, and the third application with either the first insecticide or a third option. Rotation is unnecessary with the homemade insecticidal soap solution, as it operates by smothering the insects. The soap solution must be applied frequently to control insects emerging from eggs effectively. If the infestation is minimal, an insecticide solution will be applied meticulously to the new growth, leaves (on both surfaces), flower stems, and the undersides of buds and blooms. In cases of extensive infestation, the plant should be submerged in an insecticide solution for 15 min. For the therapy to be efficacious, plants should be treated (either by spraying or immersion) a minimum of three times, with an interval of 1 week (or every 3–4 days for aphids). Multiple applications are necessary, since the pesticide will eliminate the adults, but the eggs will hatch after a few days, restarting the cycle unless treatment is reapplied to eradicate them as well.

15.9 Common Orchids Cultural Challenges and Their Possible Solutions

15.9.1 Blooming and Flower Spikes

Extreme temperatures, excessive or insufficient light, dry air, poor watering, a lack of micronutrients, or a weak root system can all cause buds to turn yellow and fall. Temperature and light must be regulated to optimal levels, and appropriate fertilizer management practices, as previously outlined, have to be followed. If the root system becomes compromised owing to deteriorating potting materials, the plant should be repotted promptly. Flowers may fail to fully open due to genetic factors, insufficient temperature, inadequate humidity, or damage caused by thrips. The humidity and temperature require adjustment. If thrips cause the damage, provide a pesticide as previously specified. The blooms are tiny; their colours are less vibrant than earlier, likely resulting from inadequate light and/or excessive temperatures. The illumination and temperature must be altered. Flowers wither rapidly, a phenomenon potentially attributable to excessive or insufficient warmth, direct solar exposure, low humidity, vitamin deficiencies, inadequate irrigation, or compromised root system health. Regulating temperature and light to optimal levels, together with following the previously outlined fertilizer management practices, may effectively address the condition. If the root system becomes compromised owing to deteriorating potting materials, the plant should be repotted immediately.

Insufficient flower production may result from weak plants, inadequate light, or phosphorus shortage. Modifying the illumination, and according to the appropriate fertilizer application protocol, would rectify the issue. A virus may be the cause of brown striations or mosaic patterns on flowers. The inadequate presentation of flowers diminishes the appealing appeal of orchid blooms; as buds begin to develop on the flower spikes, one should refrain from altering the orientation of the flower spike to achieve optimal floral appearance (Fig. 15.4).

15.9.2 Emerging Growths

Excessive nitrogen may result in soft, quick development; thus, the application rate of nitrogen should be modified and decreased. New growths are tiny and stunted, lacking the plumpness of prior ones, and failing to develop straight. This indicates that the plant is under stress, possibly due to a compromised root system, inadequate light, extreme temperatures, nitrogen deficiency, or a combination thereof. Light intensity, temperature parameters, and fertilizer application rates must be assessed.

Fig. 15.4 Flowering spike in *Cymbidium* and *Phalaenopsis* orchids

Repotting should be conducted if necessary. Adequate shade and humidity must be provided to the distressed orchids. Absence or restricted new growth may be due to deficiencies in nitrogen and/or phosphorus, or from injury or decay of growth, particularly if the plant was divided and retained just a single growth. Nitrogen and phosphorus fertilizers should be applied.

15.9.3 Leaves

The leaves are dark green and appear quite healthy; yet, the plant fails to bloom. Insufficient light likely necessitates a reduction in shadow and an increase in light levels. The leaves lack lustre and ultimately wither. The plant is insufficiently absorbing water. The root system requires inspection. If the roots are many, robust, strong, and white, the plant is experiencing insufficient watering. Should the root system be compromised, the plant must be repotted promptly. Chlorosis, indicated by leaf yellowing, may arise from excessive light and/or deficiencies in nitrogen and/or sulphur; shade should be provided, and nutrients must be supplied. Transparent or aqueous lesions on foliage typically arise from fungal or bacterial infections. The plant requires repotting, may be administered fungicide (Ridox, Phentom, Compass, etc.), and should be maintained in a somewhat dry condition for several weeks. The discoloured region on the upper surface of curled leaves, exposed to light, is likely a result of sunburn or excessive illumination. Adequate shade must be ensured.

15.9.4 Roots

Brown or black roots are signs of rotted or damaged roots (fungus that causes root rot). Damaged and decayed roots must be pruned. The plant may be repotted and maintained in a slightly drier condition for several weeks. Fungicide may be used if there is a significant amount of decayed roots. Chewed or absent tips result from pest activity (millipedes, sowbugs, snails, or slugs); a pesticide should be employed to manage these insects. Dead root tips may result from salt accumulation because of excessively hard water, over-fertilization, or insufficient leaching of the medium. Consistent leaching of the substrate, use of high-quality water, and adherence to appropriate fertilizer dosages help rectify this issue. Chlorine deficit may result in deformed roots. Stunted roots may result from an imbalance of micronutrients. Effective fertilizer management may resolve these issues.

15.10 Conclusion

Orchids are renowned for their natural complex floral structures and ecological flexibility, illustrating the interaction of evolution, diversity, and beauty in the plant kingdom. Effective orchid cultivation requires a keen understanding of light, humidity, temperature, airflow, nutrients, and suitable potting media. Cultivation under controlled environments, such as greenhouses and shade houses, has revolutionized orchid growth by enabling precise microclimate regulation, extending flowering periods, and minimizing physiological stress. Beginners should start with hardy varieties such as Phalaenopsis or Dendrobium, which are easier to manage and highly rewarding in bloom. With patience, observation, and consistent attention, orchid cultivation can become not only a fulfilling hobby but also an inspiring journey into the beauty and complexity of plant life.

References

Biswas SS, Pal R (2021) A practical guide for successful orchid cultivation. Indian Hortic 66(2)

De LC, Singh DR (2016a) Post-harvest management and value addition in orchids. Int J Biol Sci 3(1):14–35

De LC, Singh DR (2016b) Year-round production of orchids, 1st edn. Pointer Publishers

Fay MF (2018) Orchid conservation: how can we meet the challenges in the twenty-first century? Bot Stud 59(16). https://doi.org/10.1186/s40529-018-0232-z

Friend RG (2004) Growing orchids in your garden. Timber Press

Govaerts R, Bernet P, Kratochvil K, Gerlach G, Carr G, Alrich P, Pridgeon AM, Pfahi J, Campacci MA, Holland BD, Tiggers H, Shaw J, Cribb P, George A, Creuz K, Wood JJ (2017) World checklist of Orchidaceae. Royal Botanic Gardens, Kew

Husain MF, Eraqui S (2023) Orchids: a wonderful ornamental plant. Asian J Res Crop Sci 8(4): 167–172. https://doi.org/10.9734/ajrcs/2023/v8i4177

Husen A, Pant M (2024) Exploring medicinal orchids. Taylor & Francis Group, LLC. https://doi.org/10.1201/b23375

Lopez RG, Runkle ES (2005) Environmental physiology of growth and flowering of orchids. HortScience 40(7):1969–1973. https://doi.org/10.21273/HORTSCI.40.7.1969

Maitra S, Sairam M, Shankar T, Gaikwad DJ (2020) Growing of dendrobium orchids in greenhouse. In: Protected cultivation and smart agriculture. New Delhi Publishers

Nagrare VS, Pal R (2008) Cultivating potted orchids fetches more. Indian Hortic 53(2):24

Plants of the World Online (POWO) (2023) Plants of the world online. Royal Botanic Gardens, Kew. http://www.plantsoftheworldonline.org

Singh SK, Agrawal DK, Jalal JS, Dash SS, Mao AA, Singh P (2019) Orchids of India: a pictorial guide. Botanical Survey of India

World Flora Online (WFO) (2023) World Flora Online. http://www.worldfloraonline.org